INTERNATIONAL UNION OF PURE AND APPLIED CHEMISTRY

ANALYTICAL CHEMISTRY DIVISION
COMMISSION ON SOLUBILITY DATA

SOLUBILITY DATA SERIES

Volume 15

ALCOHOLS WITH WATER

SOLUBILITY DATA SERIES

Volume 1 H. L. Clever, *Helium and Neon*

Volume 2 H. L. Clever, *Krypton, Xenon and Radon*

Volume 3 M. Salomon, *Silver Azide, Cyanide, Cyanamides, Cyanate, Selenocyanate and Thiocyanate*

Volume 4 H. L. Clever, *Argon*

Volume 5/6 C. L. Young, *Hydrogen and Deuterium*

Volume 7 R. Battino, *Oxygen and Ozone*

Volume 8 C. L. Young, *Oxides of Nitrogen*

Volume 9 W. Hayduk, *Ethane*

Volume 10 R. Battino, *Nitrogen and Air*

Volume 11 B. Scrosati and C. A. Vincent, *Alkali Metal, Alkaline Earth Metal and Ammonium Halides. Amide Solvents*

Volume 12 C. L. Young, *Sulfur Dioxide, Chlorine, Fluorine and Chlorine Oxides*

Volume 13 S. Siekierski, T. Mioduski and M. Salomon, *Scandium, Yttrium, Lanthanum and Lanthanide Nitrates*

Volume 14 H. Miyamoto, M. Salomon and H. L. Clever, *Alkaline Earth Metal Halates*

Volume 15 A. F. M. Barton, *Alcohols with Water*

Volume 18 O. Popovych, *Tetraphenylborates*

Selected Volumes in Preparation

W. Gerrard and H. L. Clever, *Hydrogen Halides in Non-Aqueous Solvents*

H. Miyamoto, *Copper and Silver Halates*

W. Hayduk, *Propane, Butane and 2-Methylpropane*

E. M. Woolley, *Aqueous Silver Chloride Systems*

A. L. Horvath and F. W. Getzen, *Halobenzenes and Halophenols with Water*

S. Siekierski and J. K. Navratil, *Actinide Nitrates*

Z. Galus and C. Guminski, *Metals in Mercury*

J. W. Lorimer, *Beryllium, Strontium, Barium and Radium Sulfates*

E. Tomlinson, *Beta-Lactams*

C. L. Young and H. L. Clever, *Carbon Dioxide*

H. L. Clever and C. L. Young, *Methane*

SOLUBILITY DATA SERIES

Editor-in-Chief
A. S. KERTES

Volume 15

ALCOHOLS WITH WATER

Volume Editor

A. F. M. BARTON

School of Mathematical and Physical Sciences
Murdoch University, Western Australia 6150

Contributors

M. C. HAULAIT-PIRSON
University of Leuven
Belgium

G. T. HEFTER
Murdoch University
Australia

G. JANCSO
Central Institute for Physics
Budapest, Hungary

A. MACZYNSKI
Institute of Physical Chemistry
Warsaw, Poland

Z. MACZYNSKA
Institute of Physical Chemistry
Warsaw, Poland

A. SZAFRANSKI
Institute for Industrial Chemistry
Warsaw, Poland

N. TSUCHIDA
National Institute of Research
in Inorganic Materials
Ibaraki, Japan

S. C. VALVANI
Upjohn Co.
Kalamazoo, MI, USA

S. H. YALKOWSKY
Arizona University College
of Pharmacy
Tucson, AZ, USA

FU JUFU
Beijing Institute of Chemical Technology
China

PERGAMON PRESS

OXFORD · NEW YORK · TORONTO · SYDNEY · PARIS · FRANKFURT

U.K.	Pergamon Press Ltd., Headington Hill Hall, Oxford OX3 0BW, England
U.S.A.	Pergamon Press Inc., Maxwell House, Fairview Park, Elmsford, New York 10523, U.S.A.
CANADA	Pergamon Press Canada Ltd., Suite 104, 150 Consumers Road, Willowdale, Ontario M2J 1P9, Canada
AUSTRALIA	Pergamon Press (Aust.) Pty. Ltd., P.O. Box 544, Potts Point, N.S.W. 2011, Australia
FRANCE	Pergamon Press SARL, 24 rue des Ecoles, 75240 Paris, Cedex 05, France
FEDERAL REPUBLIC OF GERMANY	Pergamon Press GmbH, Hammerweg 6, D-6242 Kronberg-Taunus, Federal Republic of Germany

First edition 1984

Library of Congress Cataloging in Publication Data

Alcohols with water.
(Solubility data series ; v. 15)
Includes bibliographical references and indexes.
1. Alcohols — Solubility — Tables. I. Barton, Allan F. M.
II. Haulait-Pirson, M.-C. III. Series.
QD305.A4A4 1984 547'.031045422 84-6505

British Library Cataloguing in Publication Data

Barton, A. F. M.
Alcohols with water. — (Solubility data series; v. 15)
1. Alcohols — Solubility — Tables
I. Title II. Series
547'.03104542'0212 QD305.A4
ISBN 0–08–025276–1

Printed in Great Britain by A. Wheaton & Co. Ltd., Exeter

CONTENTS

SOLUBILITY DATA SERIES

Editor-in-Chief

A. S. KERTES
The Hebrew University, Jerusalem, Israel

INTERNATIONAL UNION OF PURE AND APPLIED CHEMISTRY

IUPAC Secretariat: Bank Court Chambers, 2-3 Pound Way,
Cowley Centre, Oxford OX4 3YF, UK

FOREWORD

If the knowledge is
undigested or simply wrong,
more is not better

How to communicate and disseminate numerical data effectively in chemical science and technology has been a problem of serious and growing concern to IUPAC, the International Union of Pure and Applied Chemistry, for the last two decades. The steadily expanding volume of numerical information, the formulation of new interdisciplinary areas in which chemistry is a partner, and the links between these and existing traditional subdisciplines in chemistry, along with an increasing number of users, have been considered as urgent aspects of the information problem in general, and of the numerical data problem in particular.

Among the several numerical data projects initiated and operated by various IUPAC commissions, the *Solubility Data Project* is probably one of the most ambitious ones. It is concerned with preparing a comprehensive critical compilation of data on solubilities in all physical systems, of gases, liquids and solids. Both the basic and applied branches of almost all scientific disciplines require a knowledge of solubilities as a function of solvent, temperature and pressure. Solubility data are basic to the fundamental understanding of processes relevant to agronomy, biology, chemistry, geology and oceanography, medicine and pharmacology, and metallurgy and materials science. Knowledge of solubility is very frequently of great importance to such diverse practical applications as drug dosage and drug solubility in biological fluids, anesthesiology, corrosion by dissolution of metals, properties of glasses, ceramics, concretes and coatings, phase relations in the formation of minerals and alloys, the deposits of minerals and radioactive fission products from ocean waters, the composition of ground waters, and the requirements of oxygen and other gases in life support systems.

The widespread relevance of solubility data to many branches and disciplines of science, medicine, technology and engineering, and the difficulty of recovering solubility data from the literature, lead to the proliferation of published data in an ever increasing number of scientific and technical primary sources. The sheer volume of data has overcome the capacity of the classical secondary and tertiary services to respond effectively.

While the proportion of secondary services of the review article type is generally increasing due to the rapid growth of all forms of primary literature, the review articles become more limited in scope, more specialized. The disturbing phenomenon is that in some disciplines, certainly in chemistry, authors are reluctant to treat even those limited-in-scope reviews exhaustively. There is a trend to preselect the literature, sometimes under the pretext of reducing it to manageable size. The crucial problem with such preselection - as far as numerical data are concerned - is that there is no indication as to whether the material was excluded by design or by a less than thorough literature search. We are equally concerned that most current secondary sources, critical in character as they may be, give scant attention to numerical data.

On the other hand, tertiary sources - handbooks, reference books and other tabulated and graphical compilations - as they exist today are comprehensive but, as a rule, uncritical. They usually attempt to cover whole disciplines, and thus obviously are superficial in treatment. Since they command a wide market, we believe that their service to the advancement of science is at least questionable. Additionally, the change which is taking place in the generation of new and diversified numerical data, and the rate at which this is done, is not reflected in an increased third-level service. The emergence of new tertiary literature sources does not parallel the shift that has occurred in the primary literature.

With the status of current secondary and tertiary services being as briefly stated above, the innovative approach of the *Solubility Data Project* is that its compilation and critical evaluation work involve consolidation and reprocessing services when both activities are based on intellectual and scholarly reworking of information from primary sources. It comprises compact compilation, rationalization and simplification, and the fitting of isolated numerical data into a critically evaluated general framework.

The *Solubility Data Project* has developed a mechanism which involves a number of innovations in exploiting the literature fully, and which contains new elements of a more imaginative approach for transfer of reliable information from primary to secondary/tertiary sources. *The fundamental trend of the Solubility Data Project is toward integration of secondary and tertiary services with the objective of producing in-depth critical analysis and evaluation which are characteristic to secondary services, in a scope as broad as conventional tertiary services.*

Fundamental to the philosophy of the project is the recognition that the basic element of strength is the active participation of career scientists in it. Consolidating primary data, producing a truly critically-evaluated set of numerical data, and synthesizing data in a meaningful relationship are demands considered worthy of the efforts of top scientists. Career scientists, who themselves contribute to science by their involvement in active scientific research, are the backbone of the project. The scholarly work is commissioned to recognized authorities, involving a process of careful selection in the best tradition of IUPAC. This selection in turn is the key to the quality of the output. These top experts are expected to view their specific topics dispassionately, paying equal attention to their own contributions and to those of their peers. They digest literature data into a coherent story by weeding out what is wrong from what is believed to be right. To fulfill this task, the evaluator must cover <u>all</u> relevant open literature. No reference is excluded by design and every effort is made to detect every bit of relevant primary source. Poor quality or wrong data are mentioned and explicitly disqualified as such. In fact, it is only when the reliable data are presented alongside the unreliable data that proper justice can be done. The user is bound to have incomparably more confidence in a succinct evaluative commentary and a comprehensive review with a complete bibliography to both good and poor data.

It is the standard practice that the treatment of any given solute-solvent system consists of two essential parts: I. Critical Evaluation and Recommended Values, and II. Compiled Data Sheets.

The Critical Evaluation part gives the following information:

(i) a verbal text of evaluation which discusses the numerical solubility information appearing in the primary sources located in the literature. The evaluation text concerns primarily the quality of data after consideration of the purity of the materials and their characterization, the experimental method employed and the uncertainties in control of physical parameters, the reproducibility of the data, the agreement of the worker's results on accepted test systems with standard values, and finally, the fitting of data, with suitable statistical tests, to mathematical functions;

(ii) a set of recommended numerical data. Whenever possible, the set of recommended data includes weighted average and standard deviations, and a set of smoothing equations derived from the experimental data endorsed by the evaluator;

(iii) a graphical plot of recommended data.

The Compilation part consists of data sheets of the best experimental data in the primary literature. Generally speaking, such independent data sheets are given only to the best and endorsed data covering the known range of experimental parameters. Data sheets based on primary sources where the data are of a lower precision are given only when no better data are available. Experimental data with a precision poorer than considered acceptable are reproduced in the form of data sheets when they are the only known data for a particular system. Such data are considered to be still suitable for some applications, and their presence in the compilation should alert researchers to areas that need more work.

The typical data sheet carries the following information:

(i) components - definition of the system - their names, formulas and Chemical Abstracts registry numbers;
(ii) reference to the primary source where the numerical information is reported. In cases when the primary source is a less common periodical or a report document, published though of limited availability, abstract references are also given;
(iii) experimental variables;
(iv) identification of the compiler;
(v) experimental values as they appear in the primary source. Whenever available, the data may be given both in tabular and graphical form. If auxiliary information is available, the experimental data are converted also to SI units by the compiler.

Under the general heading of Auxiliary Information, the essential experimental details are summarized:

(vi) experimental method used for the generation of data;
(vii) type of apparatus and procedure employed;
(viii) source and purity of materials;
(ix) estimated error;
(x) references relevant to the generation of experimental data as cited in the primary source.

This new approach to numerical data presentation, developed during our four years of existence, has been strongly influenced by the diversity of background of those whom we are supposed to serve. We thus deemed it right to preface the evaluation/compilation sheets in each volume with a detailed discussion of the principles of the accurate determination of relevant solubility data and related thermodynamic information.

Finally, the role of education is more than corollary to the efforts we are seeking. The scientific standards advocated here are necessary to strengthen science and technology, and should be regarded as a major effort in the training and formation of the next generation of scientists and engineers. Specifically, we believe that there is going to be an impact of our project on scientific-communication practices. The quality of consolidation adopted by this program offers down-to-earth guidelines, concrete examples which are bound to make primary publication services more responsive than ever before to the needs of users. The self-regulatory message to scientists of 15 years ago to refrain from unnecessary publication has not achieved much. The literature is still, in 1984, cluttered with poor-quality articles. The Weinberg report (in 'Reader in Science Information', ed. J. Sherrod and A. Hodina, Microcard Editions Books, Indian Head, Inc., 1973, p. 292) states that 'admonition to authors to restrain themselves from premature, unnecessary publication can have little effect unless the climate of the entire technical and scholarly community encourages restraint...' We think that projects of this kind translate the climate into operational terms by exerting pressure on authors to avoid submitting low-grade material. The type of our output, we hope, will encourage attention to quality as authors will increasingly realize that their work will not be suited for permanent retrievability unless it meets the standards adopted in this project. It should help to dispel confusion in the minds of many authors of what represents a permanently useful bit of information of an archival value, and what does not.

If we succeed in that aim, even partially, we have then done our share in protecting the scientific community from unwanted and irrelevant, wrong numerical information.

A. S. Kertes

PREFACE

This volume is concerned with binary systems containing only water and a monohydroxy alcohol. Occasionally multicomponent systems are mentioned (for example Ringer solution, a salt solution isotonic with blood plasma) but these systems are not treated exhaustively or critically. Because the critical evaluations have been prepared by different authors, style and content vary considerably, and the editor has made no attempt to modify them for uniformity.

Amongst binary water-organic systems, 1-butanol exhibits a particularly high value for the amount of water which can be accommodated in the organic-rich phase at equilibrium. Given the number of carbon atoms in 1-butanol, this alcohol and water have a surprisingly high mutual solubility. Indeed, when 1-butanol and water are saturated with one another, "they are about as much alike as two separate phases can be" (ref 1). As the carbon chain length is shortened to 1-propanol there is complete miscibility, except between $-11^{\circ}C$ and $-2^{\circ}C$ and no reports have been found indicating gaps for ethanol and methanol with water at any temperature or pressure. Lengthening the alcohol carbon chain sharply decreases the mutual solubility of alcohol and water. The 1-butanol/water system shows low sensitivity to changes in solute character for partition equilibria, but the sensitivity to the solute nature increases with increasing chain length until it reaches a maximum at 1-octanol, and then levels out. Thus 1-octanol is often chosen as a reference point in studies of partition of solutes between water and organic liquids.

It may not be appreciated that although the proportion of alcohol in the aqueous phase falls off rapidly as the alcohol chain length increases, the proportion of water in the alcohol-rich phase remains relatively high, even for long-chain alcohols. For example, an alcohol/water mole ratio of 3 is observed for 1-dodecanol, compared with 1 for 1-butanol (ref 2).

The water-solubility of homologous series of alcohols (as well as of other classes of organic liquid) is observed to decrease in geometric progression as the carbon number increases in arithmetic progression, with the ratio of water solubility of one member to that of the next being of the order of four (ref 3,4). Both experimentally (ref 5-8) and theoretically (ref 9) there is evidence for an approximately linear relationship between the logarithm of alcohol solubility in the aqueous phase and the carbon number of the alcohol (Figure 1).

For the lower alcohols, which have a substantial solubility in water and which have been studied in some detail, these correlations are probably of limited practical value. However, in the case of higher alcohols having lower solubilities, requiring novel and more difficult methods of determination and having been less studied, it is possible that the precision of prediction by this type of equation exceeds that currently available for experimental results.

Thus for the straight-chain alcohols with n carbon atoms, the following expressions have been obtained:

$$25^{\circ}C: \quad \log (c/\text{mol L}^{-1}) = -0.58n + 2.3 \quad (n = 4 \text{ to } 16, \text{ ref } 8) \qquad 1$$

$$25^{\circ}C: \quad \log (c/\text{mol L}^{-1}) = -0.57n + 2.14 \quad (n = 6 \text{ to } 16, \text{ ref } 7) \qquad 2$$

$$25^{\circ}C: \quad \ln (c/\text{mol L}^{-1}) = -1.39n + 5.53 \quad (n = 4 \text{ to } 10, \text{ ref } 6) \qquad 3$$

or

$$\log (c/\text{g(1)/100g sln}) = -0.60n + 2.40 \quad (n = 4 \text{ to } 10)$$

$$20^{\circ}C \quad \log (c/\text{g(1)/100g sln}) = -0.56n + 3.2 \quad (n = 4 \text{ to } 10, \text{ ref } 5) \qquad 4$$

Maczynski (ref 14) has combined the data of ref 6 and ref 7 in a second order polynomial correlation:

$$\log (c/\text{mol L}^{-1}) = 2.722 - 0.6988n + 0.006418n^2 \quad (n = 4 \text{ to } 16) \qquad 5$$

(continued next page)

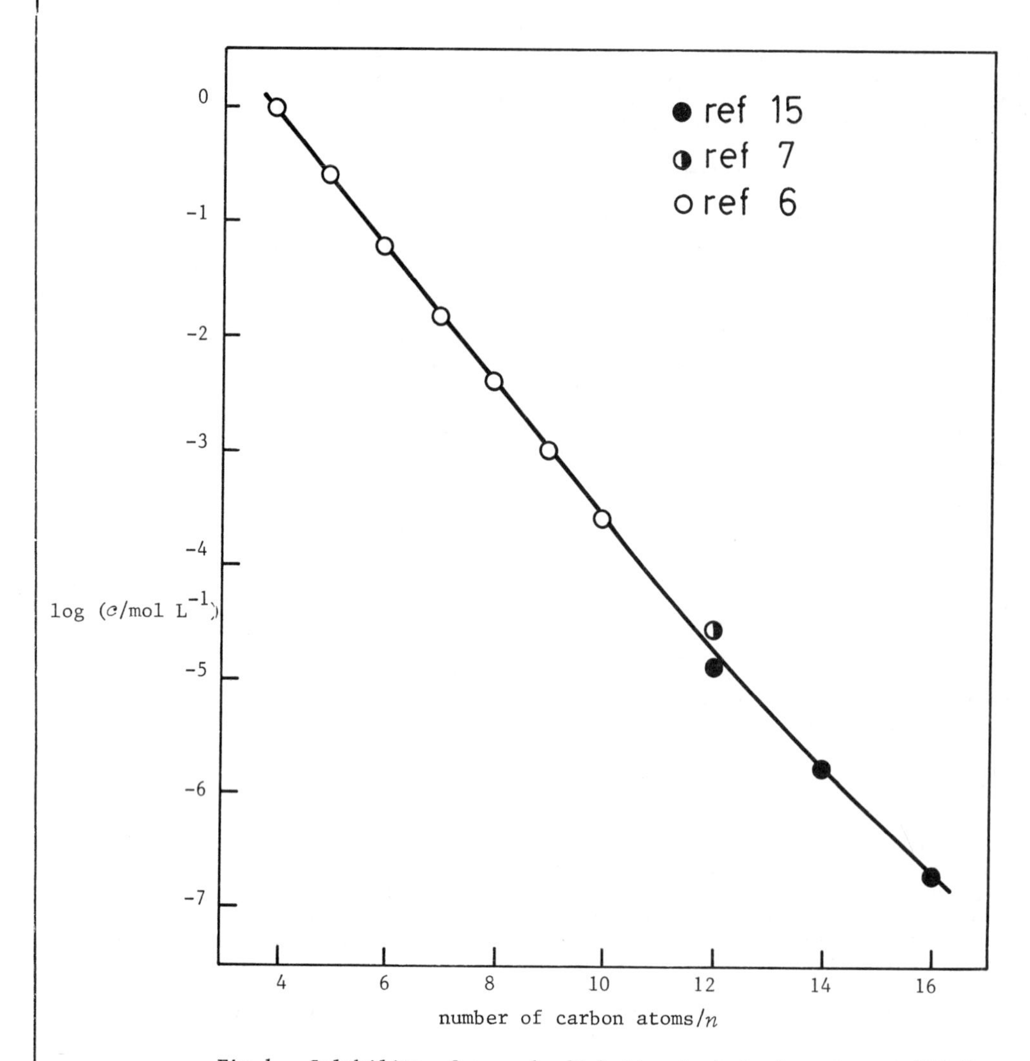

Fig 1. Solubility of normal aliphatic alcohols in water at 298 K as a function of chain length n

Related studies have dealt with the linear free energy relationships involving partition coefficients: a linear relationship exists between the logarithm of the aqueous solubility of organic liquids and their octanol-water partition coefficient (ref 10).

Similar expressions have been found useful for secondary and tertiary alcohols as well. Ratouis and Dodé (ref 11) considered alcohols with the general formula

$$R - (CH_2)_p - \underset{\underset{OH}{|}}{\overset{\overset{R'}{|}}{C}} - R''$$

where R = alkyl group

R' = H or alkyl group, with carbon number less than or equal to that of R

R'' = H or CH_3

(continued next page)

and found for pentanols and higher alcohols

$$\log \frac{c_p/(\text{g(1)}/100\text{g sln})}{c_o/(\text{g(1)}/100\text{g sln})} = -0.531p$$

or

$$\frac{c_o/(\text{g(1)}/100\text{g sln})}{c_p/(\text{g(1)}/100\text{g sln})} = 3.4^p$$

where c_o is the solubility of the alcohol with $p = 0$. Also, there was found a general relationship between the solubilities of

$$H-\underset{\underset{OH}{|}}{\overset{\overset{R}{|}}{C}}-R' \quad \text{and} \quad CH_3-\underset{\underset{OH}{|}}{\overset{\overset{R}{|}}{C}}-R'$$

such that the ratio of the solubilities was 1.65 for $R' = H$, 1.49 for $R' = CH_3$ and 1.30 for $R' = C_2H_5$.

Alcohols, like alkylamines and alkanoic acids, commonly exhibit a temperature at which a minimum solubility in the aqueous phase is observed. Although the curves for alcohols are rather broad, the minimum of each may be regarded as a characteristic point with respect to that solute in water (ref 12). Temperature-solubility plots for water in the alcohol-rich phase are also curved, the solubility decrease with decreasing temperature becoming less steep, but a minimum is not usually reached before phase separation (ref 13). However, for 1-alkanols with carbon chain length 10 atoms and greater the direction of the temperature dependence seems to be reversed (ref 2).

In general, it should be said that the solubility data in alcohol-water systems are frustrating to the compiler, the lower alcohols having a large amount of imprecise and conflicting data, and the higher alcohols having insufficient information to provide recommended values. Of the more common alcohols, the 2-butanol system in particular requires further critical study.

This volume is the result of a careful search of the literature with the aim of finding and including all published information for the systems indicated in the title. Undoubtedly some published measurements will have been overlooked, and the editor will be pleased to have these brought to his attention.

The editor would like to express his appreciation to all compilers and evaluators, but in particular to Andrzej Maczynski who has been with the project from the beginning, and to Glenn Hefter who brought much needed encouragement recently. Assistance with translations from Lu Zhen-Ya, E. Marian and Naoyuki Tsuchida is acknowledged, and the editor wishes to thank Fu Jufu, M.-C. Haulait-Pirson, G.T. Hefter, J.W. Lorimer, D.G. Shaw, and C.L. Young for their careful reviewing of the volume.

The typing of repeated drafts by Eileen Rowley and Lorelei Nottage is greatly appreciated.

A. F. M. Barton

(continued next page)

References

1. Leo, A.; Hansch, C.; Elkins, D. *Chem. Rev.* 1971, *71*, 525.

2. Tokunaga, S.; Manabe, M.; Koda, M. *Niihama Kogyo Koto Senmon Gakko Kiyo, Rikogaku Hen (Memoirs of the Niihama Technical College, Science and Engineering)*, 1980, *16*, 96.

3. Führer, H. *Ber. Dtsch. Chem. Ges.* 1924, *57B*, 510.

4. Ferguson, J. *Proc. Roy. Soc. (London) Ser. B* 1939, *127*, 387.

5. Addison, C.C.; Hutchinson, S.K. *J. Chem. Soc.* 1949, 3387.

6. Kinoshita, K.; Ishikawa, H.; Shinoda, K. *Bull. Chem. Soc. Jpn.* 1958, *31*, 1081.

7. Robb, I.D. *Aust. J. Chem.* 1966, *19*, 2281.

8. Bell, G.H. *Chem. Phys. Lipids* 1973, *10*, 1.

9. Huyskens, P.; Mullens, J.; Gomez, A.; Tack, J. *Bull. Soc. Chim. Belg.* 1975, *84*, 253.

10. Hansch, C.; Quinlan, J.E.; Lawrence, G.L. *J. Org. Chem.* 1968, *33*, 347.

11. Ratouis, M.; Dodé, M. *Bull. Soc. Chim. Fr.* 1965, 3318.

12. Nishino, N.; Nakamura, M. *Bull. Chem. Soc. Jpn.* 1978, *51*, 1617.

13. Nishino, N.; Nakamura, M. *Bull. Chem. Soc. Jpn.* 1981, *54*, 545.

14. Maczynski, A.; personal communication.

15. Krause, F.P.; Lange, W. *J. Phys. Chem.* 1965, *69*, 3171.

LIQUID-LIQUID SOLUBILITY: INTRODUCTORY INFORMATION

Allan F.M. Barton

The Solubility Data Series is made up of volumes of comprehensive and critically evaluated solubility data on chemical systems in clearly defined areas. Data of suitable precision are presented on data sheets in a uniform format, preceded for each system by a critical evaluation if more than one set of data is available. In those systems where data from different sources agree sufficiently, recommended values are proposed. In other cases, values may be described as "tentative" or "rejected".

This volume is primarily concerned with liquid-liquid systems, but related gas-liquid and solid-liquid systems are included when it is logical and convenient to do so. Solubilities at elevated and low temperatures and at elevated pressures may be included, as it is considered inappropriate to establish artificial limits on the data presented if they are considered relevant or useful.

For some systems the two components are miscible in all proportions at certain temperatures or pressures, and data on miscibility gap regions and upper and lower critical solution temperatures are included where appropriate and if available.

TERMINOLOGY

In this volume a mixture (1,2) or a solution (1,2) refers to a single liquid phase containing components 1 and 2. In a mixture no distinction is made between solvent and solute.

The *solubility* of a substance is the relative proportion of 1 in a mixture which is saturated with respect to component 1 at a specified temperature and pressure. (The term "saturated" implies the existence of equilibrium with respect to the processes of mass transfer between phases).

QUANTITIES USED AS MEASURES OF SOLUBILITY

Mole fraction of component 1, x_1 or $x(1)$:

$$x_1 = n_1/\sum_i n_i = \frac{m_1/M_1}{\sum_i (m_i/M_i)}$$

where n_i is the amount of substance (number of moles) of component i, m_i is the mass of substance i, and M_i is its molar mass.

Mole per cent of component 1 is $100x_1$

Mass fraction of component 1, w_1

$$w_1 = m_1/\sum_i m_i$$

where m_i is the mass of component i.

(continued next page)

Mass per cent of component 1 is $100w_1$, and may be described as g(1)/100g sln which makes it clear that it is mass percent of solute relative to solution and not solvent. The equivalent terms "weight fraction" and "weight per cent" are not used. The mole fraction solubility is related to the mass fraction solubility in a binary system by

$$x_1 = \frac{w_1/M_1}{w_1/M_1 + (1 - w_1)/M_2}$$

Amount of substance concentration of component 1 in a solution of volume V,

$$c_1 = n_1/V$$

is expressed in units of mol L^{-1}. The terms "molarity" and "molar" and the unit symbol M are not used.

Mass ratio is occasionally used in a two-component solution in the form g(1)/g(2), mg(1)/g(2), etc. The term "parts per million" (ppm) is not used, but may be expressed as mg(1)/kg sln or $g(1)/10^6$g sln, etc.

Molality of component 1 in component 2 is often used in solid-liquid systems defined as $m_1 = n_1/n_2M_2$, but is not used in liquid-liquid systems where the distinction between "solute" 1 and "solvent" 2 is inappropriate. The term molality alone is inadequate, and the units (mol kg^{-1}, mmol kg^{-1}) must be stated.

Mole fractions and mass fractions are appropriate to either the "mixture" or the "solution" point of view; the other quantities are appropriate to the solution point of view only.

ORDERING OF SYSTEMS

It is necessary to establish a method of ordering chemical compounds, to be used for the lists of saturating components which define each chemical system. This order is also used for ordering systems within volumes.

The systems are ordered first on the basis of empirical formula according to the Hill system (ref 2). The organic compounds within each Hill formula are ordered as follows:

(i) by degree of unsaturation, then

(ii) by order of increasing chain length in the parent hydrocarbon, then

(iii) by order of increasing chain length of hydrocarbon branches, then

(iv) numerically by position of unsaturation, then

(v) numerically by position by substitution, then

(vi) alphabetically by IUPAC name.

For example,

C_5H_8
- cyclopentene
- 2-methyl-1,3-butadiene
- 1,4-pentadiene
- 1-pentyne

C_5H_{10}
- cyclopentane
- 3-methyl-1-butene
- 2-methyl-2-butene
- 1-pentene
- 2-pentene

C_5H_{12}
- 2,2-dimethylpropane
- 2-methylbutane
- pentane

(continued next page)

$C_5H_{12}O$	2,2-dimethyl-1-propanol 2-methyl-1-butanol 2-methyl-2-butanol 3-methyl-1-butanol 3-methyl-2-butanol 1-pentanol 2-pentanol 3-pentanol
$C_6H_{12}O$	cyclohexanol 4-methyl-1-penten-3-ol 1-hexen-3-ol 4-hexen-3-ol

Deuterated (^2H) compounds immediately follow the corresponding ^1H compounds.

GUIDE TO THE COMPILATIONS AND EVALUATIONS

The format used for the compilations and evaluations has been discussed in the Foreword. Additional information on the individual sections of each sheet is now provided in the following.

"*Components*". Each component is listed by its IUPAC name (ref 1), chemical formula according to the Hill system, and Chemical Abstracts Registry Number. Also included are the "Chemical Abstracts" name if this differs from the IUPAC name, and trivial name or names if appropriate. IUPAC and common names are cross-referenced to "Chemical Abstracts" names in the System Index.

"*Original Measurements*". References are expressed in "Chemical Abstracts" style, journal names being abbreviated, and if necessary transliterated, in the forms given by the "Chemical Abstracts Service Source Index" (CASSI). In the case of multiple entries (for example, translations) an asterisk indicates the publication used for the data compilation.

"*Variables*". Ranges of variations of temperature, pressure, etc. are indicated here.

"*Prepared by*". The compiler is named here.

"*Experimental Values*". Components are described as (1) and (2), as defined in "Components". The experimental data are presented in the units used in the original paper. Thus the temperature is expressed $t/^{o}C$ or $t/^{o}F$ as in the original, and conversion to T/K is made only in the critical evaluation. However, the author's units are expressed according to IUPAC recommendations (ref 3,4) as far as possible.

In addition, compiler-calculated values of mole fractions and/or mass per cent are included if the original data do not use these units. 1975 or 1977 atomic weights (ref 5) are used in such calculations. If densities are reported in the original paper, conversions from concentrations to mole fractions are included in the compilation sheets, but otherwise this is done in the evaluation, with the values and sources of the densities being quoted and referenced.

Details of smoothing equations (with limits) are included if they are present in the original publication and if the temperature or pressure ranges are wide enough to justify this procedure.

Errors in calculations, fitting equations, etc. are noted, and where possible corrected. Material inserted by the compiler is identified by the word "compiler" in parentheses.

The precision of the original data is preserved when derived quantities are calculated, if necessary by the inclusion of one additional significant figure. In some cases graphs have been included, either to illustrate presented data more clearly, or if this is the only information in the original. Full grids are not usually inserted as it is not intended that users should read data from the graphs.

(continued next page)

"Method". An outline of the method is presented, reference being made to sources of further detail of these are cited in the original paper. "Chemical Abstracts" abbreviations are often used in this text.

"Source and Purity of Materials". For each component, referred to as (1) and (2), the following information (in this order and in abbreviated form) is provided if it is available in the original paper:

source and specification

method of preparation

properties

degree of purity

"Estimated Error". If this information was omitted by the authors, and if the necessary data are available in the paper, the compilers have attempted to estimate errors (identifiedby "compiler" in parentheses) from the internal consistency, the type of apparatus, and other relevant information. Methods used by the compilers for reporting estimated errors are based on the papers by Ku and Eisenhart (ref 6).

"References". These are the references (usually cited in the original paper) which the compiler considers particularly useful in discussing the method and material.

"Evaluator". The information provided here is the name of the evaluator, the evaluator's affiliation, and the date of the evaluation.

"Critical Evaluation". The evaluator aims, to the best of his or her ability, to check that the compiled data are correct, to assess their reliability and quality, to estimate errors where necessary, and to recommend numerical values. The summary and critical review of all the data supplied by the compiler include the following information:

(a) *Critical text.* The evaluator produces a text evaluating *all* the published data for the particular system being discussed, reviewing their merits or shortcomings. Only published data (including theses and reports) are considered, and even some of the published data may only be referred to in this text if it is considered that inclusion of a data compilation sheet is unjustified.

(b) *Fitting equations.* If the use of a smoothing equation is justifiable, the evaluator may provide an equation representing the solubility as a function of the variables reported in the compilation sheets, stating the limits within which it should be used.

(c) *Graphical summary.* This may be provided in addition to tables and/or fitting equations.

(d) *Recommended values.* Data are *recommended* if the results of at least two independent experimental groups are available and are in good agreement, and if the evaluator has no doubt as to the adequacy and reliability of the experimental and computational procedures used.

Data are reported as *tentative* if only one set of measurements is available, or if the evaluator is uncertain of the reliability of some aspect of the experimental or computational method but judges that it should cause only minor error, or if the evaluator considers some aspect of the computational or experimental method undesirable but believes the data to have some value in those instances when an approximate value of the solubility is needed.

Data determined by an inadequate method or under ill-defined conditions are *rejected*, the reference being included in the evaluation together with a reason for its rejection by the evaluator.

(e) *References.* All pertinent references are listed here, including all those publications appearing in the accompanying compilation sheets and also those which have been rejected and not compiled.

(continued next page)

(f) *Units*. The final recommended values are reported in SI units (ref 3). It should be noted that in most cases the rounded absolute temperature values (e.g. 298 K) actually refer to 298.15 K, etc, although very few solubilities are known with such precision that the differences are significant.

Continuation Sheets. These are used for both compilations and evaluations, and include sections listing the "Components" and also the "Original Measurement" or "Evaluator" as well as the word "continued". Compilation continuation sheets may include a section headed "Comments and/or Additional Data".

REFERENCES

1. Rigaudy, J.; Klesney, S.P. *Nomenclature of Organic Chemistry* (IUPAC), ("The Blue Book"), Pergamon, Oxford, 1979.

2. Hill, E.A. *J. Am. Chem. Soc.* 1900, *22*, 478.

3. Whiffen, D.H., ed. *Manual of Symbols and Terminology for Physicochemical Quantities and Units* (IUPAC), ("The Green Book") Pergamon, Oxford, 1979; *Pure Appl. Chem.* 1979, *51*, 1.

4. McGlashan, M.L. *Physicochemical Quantities and Units*, 2nd ed. Royal Institute of Chemistry, London, 1971.

5. IUPAC Commission on Atomic Weights, *Pure Appl. Chem.* 1976, *47*, 75; 1979, *51*, 405.

6. Ku, H.H., and Eisenhart, C., in Ku, H.H., ed. *Precision Measurement and Calibration*, NBS Special Publication 300, Vol. 1, Washington D.C., 1969.

COMPONENTS:

(1) 1-Propanol (*n-propyl alcohol*); C_3H_8O; [71-23-8]

(2) Water; H_2O; [7732-18-5]

ORIGINAL MEASUREMENTS:

Rosso, J-C.; Carbonnel, L.
C.R. Hebd. Seances Acad. Sci., 1969, *269*, 1432-5

VARIABLES:

Temperature: -150 to 0°C

PREPARED BY:

A.F.M. Barton

EXPERIMENTAL VALUES:

The region of liquid-liquid immiscibility extends between compositions 26.5 and 73.2 g(1)/100g sln from -10.5°C to an upper critical miscibility temperature of -1.7°C at 50 g(1)/100g sln. The corresponding mole fraction compositions (x_1) calculated by the compiler are 0.108 , 0.450 and 0.231 respectively.

A full phase diagram is presented, revealing a solid pentahydrate between -52 and -134°C:

	t/°C	g(1)/100g sln	x_1(compiler)
Eutectic	-134	95.1	0.853
Peritectic	-52.0	85.7	0.643
Monotectic	-10.5	26.0	0.0953
t_c	-1.7	50.0	0.231

AUXILIARY INFORMATION

METHOD/APPARATUS/PROCEDURE:

Thermal analysis of crystallised mixtures from liquid nitrogen temperatures.

SOURCE AND PURITY OF MATERIALS:

(1) Merck;
f.pt. -127°C.

(2) not specified

ESTIMATED ERROR:

Not specified

REFERENCES:

COMPONENTS:	ORIGINAL MEASUREMENTS:
(1) 2-Propanol *(isopropanol)*; C_3H_8O; [67-63-0] (2) Water; H_2O; [7732-18-5]	Rosso, J.-C.; Carbonnel, L. *C.R. Hebd. Seances Acad. Sci.* 1969, *268*, 1012-5.
VARIABLES:	PREPARED BY:
Temperature: (-115) to 0°C	A.F.M. Barton

EXPERIMENTAL VALUES:

The region of liquid-liquid immiscibility extends between compositions 40.80 and 60.0 g(1)/100g sln at -20°C to an upper critical miscibility temperature of -12°C at 51 g(1)/100g sln. The corresponding mole fraction (x_1) compositions calculated by the compiler are 0.1713, 0.310 and 0.238 respectively.

A full phase diagram is presented, revealing a solid trihydrate between -37 and -100°C and an entirely metastable pentahydrate down to -50°C.

	t/°C	g(1)/100g sln	x_1(compiler)
Eutectic	-100	96.30	0.886
Metastable eutectic	-108	95.2	0.856
Peritectic	-37	82.0	0.577
Metastable peritectic	-50	86.7	0.662
Monotectic	-20	40.7	0.171
t_c	-12	51.0	0.238

AUXILIARY INFORMATION

METHOD/APPARATUS/PROCEDURE:	SOURCE AND PURITY OF MATERIALS:
Thermal analysis of crystallised mixtures allowed to warm up from liquid nitrogen temperatures.	(1) Merck, containing 1% water allowed for in composition; f.p. -91.5°C (2) not stated
	ESTIMATED ERROR:
	Not specified
	REFERENCES:

COMPONENTS:	EVALUATOR:
(1) 2-Methyl-1-propanol (*isobutanol*); $C_4H_{10}O$; [78-83-1] (2) Water; H_2O; [7732-18-5]	G.T. Hefter, School of Mathematical and Physical Sciences, Murdoch University, Perth, Western Australia. November 1982.

CRITICAL EVALUATION:

Solubilities in the system comprising 2-methyl-1-propanol (1) and water (2) have been reported in the following publications:

Reference	T/K	Solubility	Method
Alexejew (ref 1)	374-405	mutual	synthetic
Michels (ref 2)	263-406	mutual	synthetic
Brun (ref 3)	273	mutual	synthetic
Janecke (ref 4)	312-406	mutual	synthetic
Jasper (ref 5)	363	mutual	refractometric
Alberty and Washburn (ref 6)	298	mutual	refractometric
Booth and Everson (ref 7)	298	(1) in (2)	titration
Donahue and Bartell (ref 8)	298	mutual	analytical
Morachevskii *et al.* (ref 9)	293	mutual	titration
Ratouis and Dodé (ref 10)	298, 303	(1) in (2)	analytical
Mozzhukhin *et. al.* (ref 11)	293, 303	mutual	titration
Mullens (ref 12)	298	mutual	interferometric
De Santis *et al.* (ref 13)	298	mutual	analytical
Moriyoshi *et al.* (ref 14)	303-407	mutual	refractometric
Lyzlova (ref 15)	293, 363	mutual	refractometric
Fu *et al.* (ref 16)	363	mutual	titration
Lutugina and Reshchetova (ref 21)	363	mutual	analytical

Apart from four further publications (ref 17-20) which did not contain sufficient information to justify their inclusion, all original data are compiled in the data sheets immediately following this Critical Evaluation.

In the Critical Evaluation the data of Booth and Everson (7), in volume fractions, and of Mullens (ref 12), in w/v fractions have been excluded from consideration.

The data of Brun (ref 3), Jasper *et al.* (ref 5), Donahue and Bartell (ref 8), de Santis *et al.* (ref 13), Lutugina and Reshchetova (ref 21), and Moriyoshi *et al.* (ref 14) (alcohol-rich phase only) are in marked disagreement with all other studies and are rejected.

The following individual points in otherwise satisfactory studies are also in marked disagreement with other studies and have been rejected: in the water-rich phase, 303, 308 and 313 K (ref 14); in the alcohol-rich phase 398 and 403 K (ref 4) and 298 K (ref 9). All other reported data are included in the table below. Values obtained by the evaluator by graphical interpolation or extrapolation from the data sheets are indicated by an asterisk (*).

"Best" values have been obtained by simple averaging. The uncertainty limits (σ_n) attached to these values do not have statistical significance and should be regarded only as a convenient representation of the spread of values rather than as error limits.

(continued next page)

COMPONENTS:	EVALUATOR:
(1) 2-Methyl-1-propanol (*isobutanol*); $C_4H_{10}O$; [78-83-1] (2) Water; H_2O; [7732-18-5]	G.T. Hefter, School of Mathematical and Physical Sciences, Murdoch University, Perth, Western Australia. November 1982.

CRITICAL EVALUATION (continued)

The letter *(R)* designates "recommended" data. Data are "recommended" if two or more apparently reliable studies are in reasonable (± 5% relative) agreement. All other data are regarded as tentative only.

Tentative and recommended (*R*) values for the solubility of 2-methyl-1-propanol (1) in water (2)

T/K	Solubility, g(1)/100g sln Reported values	"Best" value ($\pm\sigma_n$)
263	13.45* (ref 2)	13.5
268	12.4* (ref 2)	12.4
273	11.5* (ref 2)	11.5
278	10.7* (ref 2)	10.7
283	10.0* (ref 2)	10.0
288	9.3* (ref 2)	9.3
293	8.6*(ref 2), 8.3 (ref 9), 8.5 (ref 11), 8.9 (ref 15)	8.5 ± 0.1 (*R*)
298	8.15* (ref 2), 8.02 (ref 6), 8.14 (ref 10)	8.1 ± 0.1 (*R*)
303	7.8* (ref 2), 7.80 (ref 10), 7.5 (ref 11)	7.7 ± 0.1 (*R*)
308	7.5* (ref 2)	7.5
318	7.0* (ref 14)	7.0
323	6.5* (ref 14)	6.5
328	6.6* (ref 14)	6.6
333	6.8* (ref 14)	6.8
338	7.0* (ref 14)	7.0
343	7.2* (ref 14)	7.2
348	7.2* (ref 14)	7.2
353	7.5* (ref 2), 7.4* (ref 14)	7.5 ± 0.1 (*R*)
358	7.9* (ref 2), 7.6* (ref 14)	7.8 ± 0.2 (*R*)
363	8.2*(ref 2),7.9*(ref 4),7.9*(ref 14),8.1(ref 15),7.43(ref 16)	7.9 ± 0.3 (*R*)
368	8.8*(ref 2), 8.7*(ref 4), 8.8* (ref 14)	8.8 ± 0.1 (*R*)
373	9.5*(ref 2), 9.4* (ref 4), 9.7* (ref 14)	9.5 ± 0.1 (*R*)
378	10.4*(ref 2, 10.2* (ref 4), 10.7* (ref 14)	10.4 ± 0.2 (*R*)
383	11.4*(ref 1), 11.3*(ref 2),11.0*(ref 4), 11.5*(ref 14)	11.3 ± 0.2 (*R*)
388	12.2*(ref 1), 12.5*(ref 2), 11.9*(ref 4), 13.2*(ref 14)	12.5 ± 0.5 (*R*)
393	13.6*(ref 1), 14.2*(ref 2), 13.3*(ref 4), 15.0*(ref 14)	14.0 ± 0.6 (*R*)
398	17.0*(ref 1), 16.4*(ref 2), 15.3*(ref 4), 18.7*(ref 14)	17 ± 1.
403	24.8*(ref 1), 20.1*(ref 2), 19.9*(ref 4), 23.5*(ref 14)	22 ± 2.

(continued next page)

COMPONENTS:

(1) 2-Methyl-1-propanol (*isobutanol)*; $C_4H_{10}O$; [78-83-1]

(2) Water; H_2O; [7732-18-5]

EVALUATOR:

G.T. Hefter, School of Mathematical and Physical Sciences, Murdoch University, Perth, Western Australia.

November 1982.

CRITICAL EVALUATION (continued)

Tentative and recommended (*R*) values for the solubility of water (2) in 2-methyl-1-propanol (1)

T/K	Solubility, g(2)/100g sln Reported values	"Best" values ($\pm\sigma_n$)
263	13.7* (ref 2)	13.7
268	14.3* (ref 2)	14.3
273	14.7* (ref 2)	14.7
278	15.0* (ref 2)	15.0
283	15.5* (ref 2)	15.5
288	15.9* (ref 2)	15.9
293	16.2*(ref 2), 16.6*(ref 6), 16.4*(ref 11), 16.6 (ref 15)	16.4 ± 0.2 (*R*)
298	16.7*(ref 2), 17.01 (ref 6)	16.7 ± 0.1 (*R*)
303	17.2* (ref 2), 17.5 (ref 11)	17.2
308	17.7* (ref 2)	17.7
313	18.3* (ref 2)	18.3
318	19.3* (ref 2)	19.3
323	19.4* (ref 2)	19.4
328	19.8* (ref 2)	19.8
333	20.6* (ref 2)	20.6
338	21.4* (ref 2)	21.4
343	22.3* (ref 2)	22.3
348	23.3* (ref 2)	23.3
353	24.4* (ref 2)	24.4
358	25.5* (ref 2)	25.5
363	26.6* (ref 2), 26.8 (ref 15), 26 (ref 16)	26.5 ± 0.3 (*R*)
368	27.8* (ref 2)	27.9
373	29.4* (ref 1), 29.4* (ref 2)	29.4 (*R*)
378	31.2* (ref 1), 31.1 (ref 2)	31.1 ± 0.1 (*R*)
383	34* (ref 1), 33.1* (ref 2)	33.5 ± 0.4
388	36* (ref 1), 35.2* (ref 2)	35.6 ± 0.4
393	39* (ref 1), 39.4* (ref 2)	39.2 ± 0.2
398	43.4* (ref 1), 42.7* (ref 2)	43.0 ± 0.4 (*R*)
403	52* (ref 1), 49 (ref 2)	50 ± 2 (*R*)

The "best" values from the above tables are plotted in Figure 1. As will be obvious from the tables these values are largely the interpolated values of Michels (ref 2).

Moriyoshi *et al.* (ref 14) have also determined the mutual solubility of (1) and (2) at pressures of 200-2500 atm. (20-250 MPa).

The upper critical solution temperature at 1 atm has been reported as 396.5 K (3) and, probably more reliably, as 407.25 K (ref 14). The corresponding critical compositions are 37.5 (ref 3) and 36.3 (ref 14) g(1)/100g sln.

(continued next page)

COMPONENTS:	EVALUATOR:
(1) 2-Methyl-1-propanol (*isobutanol*); $C_4H_{10}O$; [78-83-1] (2) Water; H_2O; [7732-18-5]	G.T. Hefter, School of Mathematical and Physical Sciences, Murdoch University, Perth, Western Australia. November 1982.

CRITICAL EVALUATION (continued)

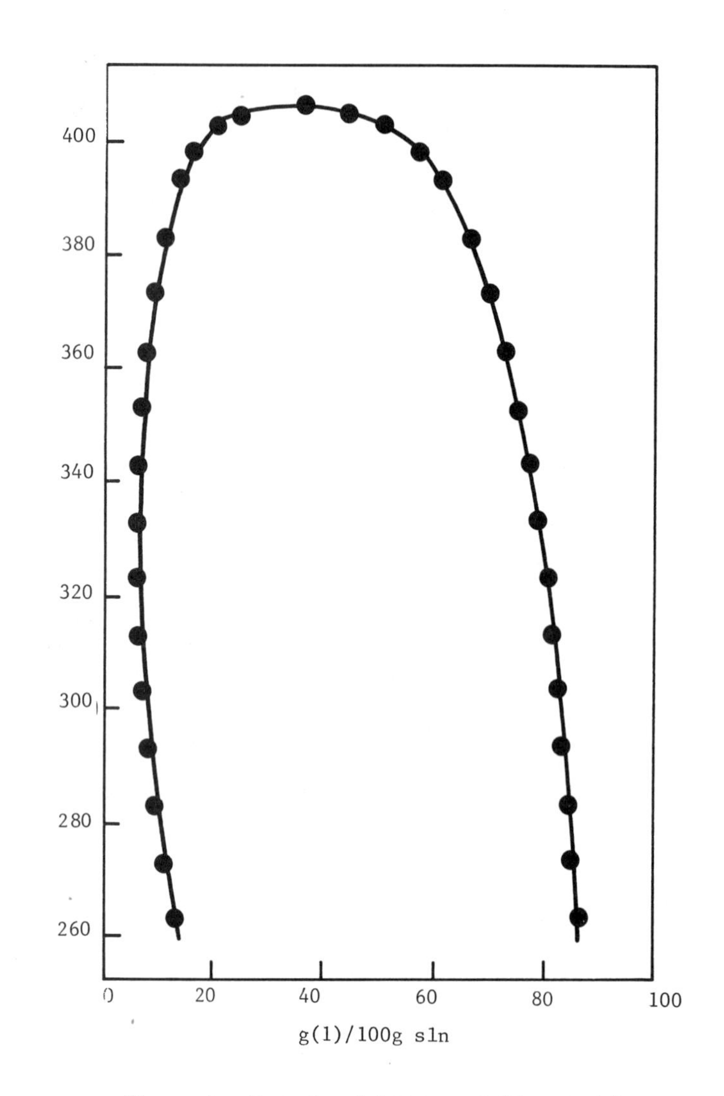

Figure 1: Mutual solubility of (1) and (2)

(continued next page)

COMPONENTS:

(1) 2-Methyl-1-propanol (*isobutanol*); $C_4H_{10}O$; [78-83-1]

(2) Water; H_2O; [7732-18-5]

EVALUATOR:

G.T. Hefter, School of Mathematical and Physical Sciences, Murdoch University, Perth, Western Australia.
November 1982.

CRITICAL EVALUATION (continued)

References

1. Alexejew, W. *Ann. Phys. Chem.* 1886, *28*, 305.
2. Michels, A. *Arch. Neerl. Sci. Exactes Nat. Ser. 3A*, 1923, *6*, 127.
3. Brun, P. *C.R. Hebd. Seances Acad. Sci.* 1925, *180*, 1745, 1926 , *183*, 207.
4. Jänecke, E. *Z. Phys. Chem.* 1933, *164*, 401.
5. Jasper, J.J.; Campbell, C.J.; Marshall, D.E. *J. Chem. Educ.* 1941, *18*, 540.
6. Alberty, R.A.; Washburn, E.R. *J. Phys. Chem.* 1945, *49*, 4.
7. Booth, H.S.; Everson, H.E. *Ind. Eng. Chem.* 1948, *40*, 1491.
8. Donahue, D.J.; Bartell, F.E. *J. Phys. Chem.* 1952, *56*, 480.
9. Morachevskii, A.G.; Smirnova, N.A.; Lyslova, R.V. *Zhur. Prikl. Khim. (Leningrad)* 1965, *38*, 1262.
10. Ratouis, M.; Dodé, M. *Bull. Chim. Fr.* 1965, 3318.
11. Mozzhukhin, A.S.; Serafimov, L.A.; Mitropolskaja, W.A.; Rudakovskaja, T.S. *Khim. Tekhnol. Topl. Masel* 1966, *11(4)*, 11.
12. Mullens, J. *Alcoholassociaten*, Doctoraatsproefschrift, Leuven, 1971; Huyskens, P.; Mullens, J.; Gomez, A.; Tack, J. *Bull. Soc. Chim. Belg.* 1975, *84*, 253.
13. De Santis, R.; Morrelli, L.; Muscetta, P.N. *Chem. Eng. J.* 1976, *11*, 207.
14. Moriyoshi, T.; Aoki, Y.; Kamiyama, H. *J. Chem. Thermodyn.* 1977, *9*, 495.
15. Lyzlova, R.V. *Zh. Prikl. Khim. (Leningrad)* 1979, *52*, 545; *J. App. Chem. USSR* 1979, *52*, 509.
16. Fu, C.F.; King, C.L.; Chang, Y.F.; Xeu, C.X. *Hua Kung Hsueh Pao* 1980 (3), 281.
17. Wratschko, F. *Pharm. Presse* 1929, *34*, 143.
18. Zhuravleva, I.K.; Zhuravlev, E.F.; Peksheva, N.P. *Zh. Fiz. Khim.* 1970, *44*, 1515; *Russ. J. Phys. Chem.* 1970, *44*, 846.
19. Fuhner, H. *Ber. Dtsch. Chem. Ges.* 1924, *57*, 510.
20. Timmermans, J. *J. Chim. Phys. Physicochim Biol.* 1923, *20*, 491.
21. Lutugina, N.V.; Reshchetova, L.I. *Vestnik Leningr. Univ.* 1972, *16*, 75.

COMPONENTS:

(1) 2-Methyl-1-propanol *(isobutanol)*; $C_4H_{10}O$; [78-83-1]

(2) Water; H_2O; [7732-18-5]

ORIGINAL MEASUREMENTS:

Alexejew, W.

Ann. Phys. Chem. 1886, *28*, 305-38.

VARIABLES:

Temperature: 101-132 °C

PREPARED BY:

A. Maczynski; Z. Maczynska; A. Szafranski.

EXPERIMENTAL VALUES:

Mutual solubility of 2-methyl-1-propanol (1) and water (2)

$t/^{\circ}C$	g(1)/100g sln		x_1 (compiler)	
	(2)-rich phase	(1)-rich phase	(2)-rich phase	(1)-rich phase
101.0	-	70.3	-	0.365
103.5	-	69.3	-	0.354
113	11.8	-	0.0315	-
123	15.3	-	0.0420	-
125	-	56.6	-	0.240
126.5	-	56.8	-	0.242
127	19.4	-	0.0552	-
129	22.39	-	0.0655	-
131.5	32.23	41.58	0.1036	0.1474

AUXILIARY INFORMATION

METHOD/APPARATUS/PROCEDURE:

The synthetic and/or analytical method was used, the latter only when the solubility diminished with temperature.

Into a tared glass tube (1) was introduced and weighed, then (2) was added through a capillary funnel. The tube was sealed, reweighed, fastened to the bulb of a mercury thermometer and repeatedly heated and cooled in a water (or glycerol) bath until the mixture became respectively homogeneous and turbid.

SOURCE AND PURITY OF MATERIALS:

(1) not specified.

(2) not specified.

ESTIMATED ERROR:

Not specified.

REFERENCES:

COMPONENTS:

(1) 2-Methyl-1-propanol (*isobutanol*); $C_4H_{10}O$; [78-83-1]

(2) Water; H_2O; [7732-18-5]

ORIGINAL MEASUREMENTS:

Michels, A.

Arch. Neerl. Sci. Exactes Nat., *Ser. 3A*, 1923, *6*, 127-46.

VARIABLES:

Temperature: (-11) to 133 °C

PREPARED BY:

A. Maczynski; G.T. Hefter

EXPERIMENTAL VALUES:

Solubility of 2-methyl-1-propanol (1) in the water-rich phase.

$t/^{\circ}C$	g(1)/100g sln	x_1 (compiler)	$t/^{\circ}C$	g(1)/100g sln	x_1 (compiler)
-10.6	13.61	0.03687	94.1	8.82	0.0230
-8.6	13.12	0.03539	99.1	9.21	0.0241
-3.6	12.18	0.03260	105.3	10.42	0.02748
-2.9	12.00	0.03207	117.1	13.11	0.03536
-2.4	11.92	0.03184	121.9	15.03	0.04121
7.0	10.42	0.02748	126.2	17.00	0.04741
14.9	9.21	0.0241	128.2	19.41	0.05529
18.95	8.82	0.0230	131.8	22.93	0.06742
21.6	8.43	0.0219	131.9	21.89	0.06375
27.4	8.00	0.0207	132.5	29.84	0.09366
36.3	7.48	0.0193	132.6	32.12	0.10313
80.2	7.48	0.0193	132.9	33.88	0.11072
87.0	8.00	0.0207	132.6	35.09	0.11610
92.6	8.43	0.0219	132.7	37.14	0.12554

(continued next page)

AUXILIARY INFORMATION

METHOD/APPARATUS/PROCEDURE:

The synthetic method was used.

A sealed glass tube, described in the paper, was used. No further details were given about the technique used.

SOURCE AND PURITY OF MATERIALS:

(1) source not specified; distilled; b.p. range 107.46-107.48 °C, d^{0} 0.8197 g cm^{-3}.

(2) not specified.

ESTIMATED ERROR:

Not specified.

REFERENCES:

COMPONENTS:	ORIGINAL MEASUREMENTS:
(1) 2-Methyl-1-propanol *(isobutanol)*; C_4H_{10}; 78-83-1	Michels, A.
(2) Water; H_2O; [7732-18-5]	*Arch. Neer. Sci. Exactes Natur., Ser. 3A,* 1923, *6*, 127-46.

EXPERIMENTAL VALUES: (continued)

Solubility of 2-methyl-1-propanol (1) in the alcohol-rich phase

$t/^{\circ}C$	g(1)/100g sln	x_1 (compiler)	$t/^{\circ}C$	g(1)/100g sln	x_1 (compiler)
-14.7	86.71	0.6132	67.3	77.70	0.4585
-10.8	86.37	0.6062	76.7	75.79	0.4320
-7.9	86.03	0.5994	85.0	75.20	0.4242
-5.9	85.79	0.5946	86.9	74.17	0.4110
-2.9	85.50	0.5889	87.8	73.91	0.4077
1.0	85.28	0.5853	94.6	72.19	0.3868
7.1	84.91	0.5776	94.9	72.11	0.3858
9.4	84.57	0.5711	99.8	70.76	0.3703
12.9	84.22	0.5646	107.0	68.00	0.3405
18.2	83.91	0.5589	113.9	65.39	0.3146
23.1	83.63	0.5538	114.5	65.21	0.3129
23.9	83.44	0.5504	115.1	64.87	0.3091
26.7	83.13	0.5449	117.2	63.58	0.3044
28.5	82.91	0.5410	119.3	62.03	0.2842
29.5	82.74	0.5381	120.7	61.98	0.2837
29.9	82.59	0.5355	121.5	60.46	0.2709
33.1	82.57	0.5351	129.3	56.30	0.2384
35.7	82.32	0.5308	128.8	53.87	0.2210
36.6	82.27	0.5300	129.1	53.02	0.2152
38.7	81.72	0.5207	131.2[a]	49.99	0.1954
39.0	81.88	0.5250	131.8	48.26	0.1848
40.7	81.82	0.5225	131.7	46.80	0.1761
41.0	81.55	0.5178	132.4	46.71	0.1756
45.6	80.66	0.5033	132.0	44.94	0.1655
53.1	80.54	0.5014	132.6	43.08	0.1553
63.8	78.82	0.4749	132.8	40.06	0.1397
68.3	77.79	0.4598	132.8	38.13	0.1302

[a] Recorded as 139.2 in original, i.e. greater than U.C.S.T. Value estimated by compilers by graphical interpolation.

COMPONENTS:

(1) 2-Methyl-1-propanol (*isobutanol*) $C_4H_{10}O$; [78-83-1]

(2) Water; H_2O; [7732-18-5]

ORIGINAL MEASUREMENTS:

Brun, P.

C.R. Hebd. Seances Acad. Sci. 1925, *180*, 1745-7; 1926, *183*, 207-10

VARIABLES:

One temperature: 0°C

PREPARED BY:

A.F.M. Barton

EXPERIMENTAL VALUES:

The mass percentage of isobutanol in the water-rich phase at 0°C was reported as 12.0 g(1)/100g sln; the corresponding mole fraction solubility recalculated by the compiler is x_1 = 0.032.

The mass percentage of isobutanol in the alcohol-rich phase at 0°C was reported as 82.0 g(1)/100g sln; the corresponding mole fraction solubility recalculated by the compiler is x_1 = 0.526.

Graphical results were reported in the 1925 paper for 40, 80, and 120°C. The critical temperature was 123.5°C, at a critical composition of 37.5 g(1)/100g sln (compiler: x_1 = 0.127).

AUXILIARY INFORMATION

METHOD/APPARATUS/PROCEDURE:

The synthetic method was used, in which the turbidity temperature is determined for mixtures of known composition.

SOURCE AND PURITY OF MATERIALS:

Not specified.

ESTIMATED ERROR:

Not given.

REFERENCES:

COMPONENTS:

(1) 2-Methyl-1-propanol *(isobutanol)*; $C_4H_{10}O$; [78-83-1]

(2) Water; H_2O; [7732-18-5]

ORIGINAL MEASUREMENTS:

Jänecke, E.

Z. Phys. Chem. 1933, *164*, 401-16.

VARIABLES:

Temperature: 38-133°C

PREPARED BY:

A. Maczynski; Z. Maczynska; and A. Szafranski

EXPERIMENTAL VALUES:

$t/^\circ C$	g(1)/100g sln		x_1 (compiler)	
	(2)-rich phase	(1)-rich phase	(2)-rich phase	(1)-rich phase
38.9 , 38.4	-	81.7	-	0.520
90.4 , 90.4	8.0	-	0.0207	-
108.0 , 107.7	10.6	-	0.0280	-
126.4 , 126.5	-	57.8	-	0.250
126.7 , 126.7	16.3	-	0.0452	-
127.7 , 127.6	17.5	-	0.0490	-
130.6 , 130.3	20.7	-	0.0602	-
132.8 , 132.4	-	49.1	-	0.190
133.0 , 133.0	26.7	-	0.0813	-

AUXILIARY INFORMATION

METHOD/APPARATUS/PROCEDURE:

The synthetic method was used.

Weighed amounts of (1) and (2) were sealed in a glass tube and heated and cooled to observe the clear and turbid points.

SOURCE AND PURITY OF MATERIALS:

(1) not specified.

(2) not specified.

ESTIMATED ERROR:

Not specified.

REFERENCES:

COMPONENTS:

(1) 2-Methyl-1-propanol *(isobutanol)*; $C_4H_{10}O$; [78-83-1]

(2) Water; H_2O; [7732-18-5]

ORIGINAL MEASUREMENTS:

Jasper, J.J.; Campbell, C.J.; Marshall, D.E. *J. Chem. Educ.* 1941, *18*, 540-2.

VARIABLES:

One temperature: 90°C

PREPARED BY:

A.F.M. Barton

EXPERIMENTAL VALUES:

The mole fractions of 2-methyl-1-propanol (1) at equilibrium at the 1 atm boiling point (89.65°C) of the binary mixture with water (2) were x_1 = 0.010 in the water-rich phase and x_1 = 0.755 in the alcohol-rich phase.

The corresponding mass percentage solubilities, calculated by the compiler, are 4.0 g(1)/100g sln and 92.7 g(1)/100g sln, respectively.

(Actual temperatures and pressures observed were 89.6°C at 758.9 mm Hg and 89.4°C at 754.2 mm Hg, respectively; 1 atm = 101.325 kPa; 1 mm Hg(0°C) = 133.322 Pa).

AUXILIARY INFORMATION

METHOD/APPARATUS/PROCEDURE:

The vapor-liquid temperature-composition diagram was determined by sampling distillate and residue. Samples from distillations with both excess (2) and excess (1) were analyzed by refractive index. Ethanol was added to the samples to maintain homogeneity during analysis, and compositions were determined from a calibration curve. The results reported were obtained while two layers remained in the distillation flask.

SOURCE AND PURITY OF MATERIALS:

Not stated

ESTIMATED ERROR:

Not stated

REFERENCES:

COMPONENTS:

(1) 2-Methyl-1-propanol (*isobutanol*), $C_4H_{10}O$; [78-83-1]

(2) Water; H_2O; [7732-18-5]

ORIGINAL MEASUREMENTS:

Alberty, R.A.; Washburn, E.R.

J. Phys. Chem. 1945, *49*, 4-8.

VARIABLES:

Temperature: 23.3 - 26.5°C

PREPARED BY:

A.F.M. Barton

EXPERIMENTAL VALUES:

$t/°C$	g(1)/100g sln	x_1 (compiler)
	Alcohol-rich phase	
23.33	83.12	0.545
25.00	82.99 [a]	0.543
26.14	82.93	0.542
26.51	82.85	0.540
	Water-rich phase	
25.0	8.02	0.0208

[a] By interpolation.

AUXILIARY INFORMATION

METHOD/APPARATUS/PROCEDURE:

For (1)-rich studies, homogeneous solutions of approximately equlibrium composition were sealed in glass tubes and cloud points (± 0.07°C) observed on cooling while they were rocked in a thermostat bath. For (2)-rich studies where cloud points were indistinct, the refractive indexes of several nearly saturated solutions were measured at 25.00°C with an immersion refractometer. The composition-refractive index curve was then extrapolated to the refractive index of a saturated solution.

SOURCE AND PURITY OF MATERIALS:

(1) Eastman Kodak;
refluxed with lime (24 h), fractionally distilled, then refluxed with Ca (4 h), refractionated;
n_D^{20} 1.39615, d_4^0 0.8172, d_4^{25} 0.79811

(2) not specified

ESTIMATED ERROR:

Not specified

REFERENCES:

COMPONENTS:

(1) 2-Methyl-1-propanol (*isobutanol*); $C_4H_{10}O$; [78-83-1]

(2) Water; H_2O; [7732-18-5]

ORIGINAL MEASUREMENTS:

Booth, H.S.; Everson, H.E.
Ind. Eng. Chem. 1948, *40*, 1491-3.

VARIABLES:

One temperature: 25°C
Sodium xylene sulfonate

PREPARED BY:

S.H. Yalkowsky; S.C. Valvani; A.F.M. Barton

EXPERIMENTAL VALUES:

It was reported that the solubility of 2-methyl-1-propanol in water was 11.1 mL(1)/100mL(2) at 25°C.

The corresponding figure in 40% sodium xylene sulfonate solution as solvent was > 400 mL(1)/100 mL solvent.

AUXILIARY INFORMATION

METHOD/APPARATUS/PROCEDURE:

A known volume of solvent (usually 50 mL) in a tightly stoppered calibrated Babcock tube was thermostatted. Successive measured quantities of solute were added and equilibrated until a slight excess of solute remained. The solution was centrifuged, returned to the thermostat bath for 10 mins, and the volume of excess solute measured directly. This was a modification of the method described in ref 1.

SOURCE AND PURITY OF MATERIALS:

(1) not specified ("C.P. or highest grade commercial").

(2) distilled

ESTIMATED ERROR:

Solubility within 0.1 mL/100 mL.

REFERENCES:

1. Hanslick, R.S. Dissertation, Columbia University, 1935.

COMPONENTS:	ORIGINAL MEASUREMENTS:
(1) 2-Methyl-1-propanol (*isobutanol*); $C_4H_{10}O$ [78-83-1] (2) Water; H_2O; [7732-18-5]	Donahue, D.J.; Bartell, F.E. *J. Phys. Chem.* 1952, *56*, 480-484.
VARIABLES:	PREPARED BY:
One temperature: 25°C	A.F.M. Barton

EXPERIMENTAL VALUES:

It was reported that at 25°C the mole fraction of water in the alcohol-rich phase (density 0.8328 g cm^{-3}) was x_2 = 0.449 and the mole fraction of 2-methyl-1-propanol in the water-rich phase (density 0.9849 g cm^{-3}) was x_1 = 0.0195.

The corresponding mass percentage values calculated by the compiler are 82.1 g(1)/100 sln and 7.56 g(1)/100g sln.

AUXILIARY INFORMATION

METHOD/APPARATUS/PROCEDURE:	SOURCE AND PURITY OF MATERIALS:
Samples of (1) with (2) were placed in glass stoppered flasks and were shaken intermittently for at least three days in a water-bath at 25 ± 0.1°C. The organic phase was analyzed for water by the Karl Fischer method and the aqueous phase was analyzed interferometrically. The main purpose of the study was interfacial tension determination.	(1) stated to be "best reagent grades"; "purified by fractional distillation, by treatment with silica gel, and by other appropriate treatments" (2) "purified".
	ESTIMATED ERROR:
	Not stated
	REFERENCES:

COMPONENTS:

(1) 2-methyl-1-propanol *(isobutanol)*; $C_4H_{10}O$; [78-83-1]

(2) Water; H_2O; [7732-18-5]

ORIGINAL MEASUREMENTS:

Morachevskii, A.G.; Smirnova, N.A.; Lyzlova, R.V.

Zh. Prikl. Khim. (Leningrad) 1965, *38*, 1262-7.

VARIABLES:

One temperature: 20 °C

PREPARED BY:

A. Maczynski

EXPERIMENTAL VALUES:

The solubility of 2-methyl-1-propanol in water at 20 °C was reported to be x_1 = 0.023.

The corresponding mass percentage calculated by the compiler is 8.3 g(1)/100g sln.

The solubility of water in 2-methyl-1-propanol at 20 °C was reported to be x_2 = 0.45.

The corresponding mass percentage calculated by the compiler is 17 g(2)/100g sln.

AUXILIARY INFORMATION

METHOD/APPARATUS/PROCEDURE:

The titration method was used.

No details were reported in the paper.

SOURCE AND PURITY OF MATERIALS:

(1) CP reagent, source not specified; distilled; b.p. 107.8 °C, n_D^{20} 1.3958, d_4^{20} 0.8020.

(2) not specified.

ESTIMATED ERROR:

Not specified.

REFERENCES:

COMPONENTS:

(1) 2-Methyl-1-propanol (*isobutanol*); $C_4H_{10}O$; [78-83-1]

(2) Water; H_2O; [7732-18-5]

ORIGINAL MEASUREMENTS:

Ratouis, M.; Dodé M.

Bull. Soc. Chim. Fr. 1965, 3318-22.

VARIABLES:

Temperature: 25-30°C

Ringer solution also studied

PREPARED BY:

S.C. Valvani; S.H. Yalkowsky; A.F.M. Barton

EXPERIMENTAL VALUES:

Solubility of 2-methyl-1-propanol (1) in water (2)

$t/^{\circ}C$	g(1)/100g sln	x_1 (compiler)
25	8.14	0.0211
30	7.80	0.0202

Solubility of 2-methyl-1-propanol in Ringer solution

$t/^{\circ}C$	g(1)/100g sln
25	7.65
30	7.40

AUXILIARY INFORMATION

METHOD/APPARATUS/PROCEDURE:

In a round bottom flask, 50 mL of water and sufficient quantity of alcohol was introduced until two separate layers were formed. The flask assembly was equilibrated by agitation for at least 3 h in a constant temp. bath. Equilibrium solubility was attained by first supersaturating at a slightly lower temperature (solubility of alcohols in water is inversely proportional to temperature) and then equilibrating at the desired temperature. The aqueous layer was separated after an overnight storage in a bath. The alcohol content was determined by reacting the aqueous solution with potassium dichromate and titrating the excess dichromate with ferrous sulfate solution in the presence of phosphoric acid and diphenylamine barium sulfonate as an indicator.

SOURCE AND PURITY OF MATERIALS:

(1) Prolabo, Paris; redistilled b.p. 107.8-107,9/754.2 mm Hg; n_D^{25} = 1.39391

(2) twice distilled from silica apparatus or ion exchanged with Sagei A20.

ESTIMATED ERROR:

Solubility: relative error of 2 determinations less than 1%.

Temperature: ±0.05°C.

REFERENCES:

COMPONENTS:

(1) 2-Methyl-1-propanol *(isobutanol)*; $C_4H_{10}O$; [78-83-1]

(2) Water; H_2O; [7732-18-5]

ORIGINAL MEASUREMENTS:

Mozzhukhin, A.S.; Serafimov, L.A.; Mitropolskaja, W.A.; Rudakovskaja, T.S.

Khim. Tekhnol. Topl. Masel, 1966, 11(4), 11-15.

VARIABLES:

Temperature: 20 and 30°C

PREPARED BY:

A. Maczynski

EXPERIMENTAL VALUES:

Mutual solubility of 2-methyl-1-propanol (1) and water (2)

t/°C	g(1)/100g sln		x_1 (compiler)	
	(2)-rich phase	(1)-rich phase	(2)-rich phase	(1)-rich phase
20	8.5	83.6	0.022	0.553
30	7.5	82.5	0.019	0.534

AUXILIARY INFORMATION

METHOD/APPARATUS/PROCEDURE:

The titration method was used.

No details were reported in the paper.

SOURCE AND PURITY OF MATERIALS:

(1) source not specified; dried and distilled; b.p. 107.4°C, n_D^{20} 1.3358, 0.17% water.

(2) not specified.

ESTIMATED ERROR:

Temperature: ± 0.1°C.

REFERENCES:

COMPONENTS:

(1) 2-methyl-1-propanol *(isobutanol)*; $C_4H_{10}O$ [78-83-1]

(2) water; H_2O; [7732-18-5]

ORIGINAL MEASUREMENTS:

Mullens, J.

Alcoholassociaten; Doctoraatsproefschrift, Leuven, 1971.

Huyskens, P.; Mullens, J., Gomez, A.,Tack,J. *Bull. Soc. Chim. Belg.* 1975, *84*, 253-62.

VARIABLES:

One temperature: 25°C

PREPARED BY:

M.C. Haulait-Pirson; A.F.M. Barton.

EXPERIMENTAL VALUES:

At 25°C solubility of 2-methyl-1-propanol(1) in the water-rich phase was reported as 1.029 mol(1)/L sln, and the solubility of water(2) in the alcohol-rich phase was reported as 9.438 mol(2)/L sln.

The corresponding values on a weight/volume basis are 76.27 g(1)/L sln, and 170.1 g(2)/L sln (compiler).

AUXILIARY INFORMATION

METHOD/APPARATUS/PROCEDURE:

The partition of the two components was made using a cell described in ref 1. The Rayleigh Interference Refractometer M154 was used for the determination of the concentrations.

Standard solutions covering the whole range of concentration investigated were used for the calibration.

SOURCE AND PURITY OF MATERIALS:

(1) Merck product (p.a.)

(2) distilled.

ESTIMATED ERROR:

Soly ± 0.001 mol(1)/L sln.

REFERENCES:

1. Meeussen, E.; Huyskens, P. *J. Chim. Phys.* 1966, *63*, 845

COMPONENTS:

(1) 2-Methyl-1-propanol *(isobutanol)*; $C_4H_{10}O$; [78-83-1]

(2) Water; H_2O; [7732-18-5]

ORIGINAL MEASUREMENTS:

De Santis, R.; Marrelli, L.; Muscetta, P.N.

Chem. Eng. J., 1976, *11*, 207-14.

VARIABLES:

One temperature: 25°C

PREPARED BY:

A. Maczynski

EXPERIMENTAL VALUES:

The solubility of 2-methyl-propanol in the water-rich phase at 25°C was reported to be 9.4 g(1)/100g sln.

The corresponding mole fraction, x_1, calculated by the compiler is 0.025.

The solubility of water in the alcohol-rich phase at 25°C was reported to be 17.3 g(2)/100g sln.

The corresponding mole fraction, x_2, calculated by the compiler is 0.463.

AUXILIARY INFORMATION

METHOD/APPARATUS/PROCEDURE:

The determinations were carried out using separator funnel with a thermostatic jacket. The extracter was loaded with (1) and (2) and after an extended period of mixing and quantitiative gravity separation, samples were withdrawn from the aqueous phase. The concentration of (1) in (2) was determined by colorimetric analysis (double-beam Lange colorimeter) of the cerium complex. The concentration of (2) in (1) was derived from a material balance based upon starting quantities and compositions. Each of the determinations was carried out several times.

SOURCE AND PURITY OF MATERIALS:

(1) Merck, analytical purity; fractionated before use.

(2) doubly distilled

ESTIMATED ERROR:

Temp. ± 0.1°C.

REFERENCES:

COMPONENTS:

(1) 2-Methyl-1-propanol (*isobutanol*); $C_4H_{10}O$; [78-83-1]

(2) Water; H_2O; [7732-18-5]

ORIGINAL MEASUREMENTS:

Moriyoshi, T.; Aoki, Y.; Kamiyama, H.

J. Chem. Thermodyn. 1977, *9*, 495-502.

VARIABLES:

Temperature: 303-407 K
Pressure: 1-2500 atm (0.1-250 MPa)

PREPARED BY:

A.F.M.Barton; G.T. Hefter

EXPERIMENTAL VALUES:

Mutual solubility of 2-methyl-1-proponol (1) and water (2)

T/K	p/atm	g(1)/100g sln		x_1 (compiler)	
		(2)-rich phase	(1)-rich phase	(2)-rich phase	(1)-rich phase
302.95	1	8.4	81.8	0.0218	0.522
302.95	500	9.5	80.6	0.0249	0.503
302.95	952	10.3	79.4	0.0272	0.484
302.95	1524	10.8	77.3	0.0286	0.453
302.95	1633	11.1	76.8	0.0295	0.446
302.95	1837	11.4	75.9	0.0303	0.434
302.95	1973	11.7	75.0	0.0312	0.422
302.95	2245	12.4	73.9	0.0333	0.408
302.95	2449	12.7	73.5	0.0342	0.403
312.85	1	7.6	82.0	0.0196	0.526
312.85	503	8.7	80.6	0.0226	0.503
312.85	680	9.0	80.0	0.0235	0.493
312.85	1000	9.4	79.4	0.0246	0.484
312.85	1571	10.3	77.6	0.0272	0.457
312.85	1728	10.6	77.0	0.0280	0.449

(continued next page)

AUXILIARY INFORMATION

METHOD/APPARATUS/PROCEDURE:

The method, described in ref 1, was that used for 2-butanol/water studies. Both components were placed in a cut-off glass syringe of about 20 cm^3 capacity used as a sample vessel, which was placed in a stainless steel pressure vessel mechanically shaken in an oil thermostat bath (± 0.05K). After stirring for at least 6h and then standing for another 6h at the desired temperature and pressure a sample of the upper layer was withdrawn. Subsequently the pressure vessel was moved, the contents allowed to settle, and the lower layer sampled.
The analysis of samples was made by refractive index, methanol being added by weight to produce homogeneity.

SOURCE AND PURITY OF MATERIALS:

(1) "best grade reagent";
dried by refluxing over freshly ignited calcium oxide, distilled twice;
n^{25} 1.3939, d^{25} 0.7983 g cm^{-3}

(2) deionized, distilled from alkaline $KMNO_4$ and then redistilled;
n^{25} 1.3327

ESTIMATED ERROR:

temp. ± 0.05K
solubility ± 0.21 g(1)/100g sln
(type of error not specified)

REFERENCES:

1. Moriyoshi, T.; Kaneshina, S.; Aihara, K.; Yabumoto, K. *J. Chem. Thermodyn.* 1975, *7*, 537.

COMPONENTS:

(1) 2-Methyl-1-propanol *(isobutanol)*; $C_4H_{10}O$; [78-83-1]

(2) Water; H_2O; [7732-18-5]

ORIGINAL MEASUREMENTS:

Moriyoshi, T.; Aoki, Y.; Kamiyama, H.

J. Chem. Thermodyn. 1977, *9*, 495-502.

EXPERIMENTAL VALUES (continued)

T/K	p/atm	g(1)/100g sln		x_1 (compiler)	
		(2)-rich phase	(1)-rich phase	(2)-rich phase	(1)-rich phase
312.85	2041	11.6	75.9	0.0309	0.434
312.85	2200	11.7	75.0	0.0312	0.422
312.85	2449	12.0	74.5	0.0321	0.415
322.75	1	6.5	81.5	0.0166	0.517
322.75	500	7.9	80.3	0.0204	0.498
322.75	952	8.0	78.5	0.0207	0.470
322.75	1361	8.8	77.6	0.0229	0.457
322.75	1565	9.0	77.0	0.0235	0.449
322.75	1840	10.1	75.6	0.0266	0.430
322.75	2041	10.4	74.2	0.0275	0.412
322.75	2163	10.6	73.3	0.0280	0.400
322.75	2381	11.1	71.1	0.0295	0.374
332.65	1	6.8	80.9	0.0174	0.507
332.65	500	8.2	79.4	0.0213	0.484
332.65	973	9.0	77.9	0.0235	0.462
332.65	1367	9.8	76.5	0.0257	0.442
332.65	1483	9.9	75.6	0.0260	0.430
332.65	2000	10.9	73.6	0.0289	0.404
332.65	2200	12.2	72.0	0.0327	0.385
332.65	2381	13.8	70.3	0.0374	0.365
352.35	1	7.4	77.0	0.0191	0.449
352.35	500	8.8	75.3	0.0229	0.426
352.35	1000	10.1	73.6	0.0266	0.404
352.35	1245	10.4	72.8	0.0275	0.394
352.35	1432	11.6	71.7	0.0309	0.381
352.35	1905	13.0	69.2	0.0350	0.353
352.35	2200	13.7	66.6	0.0371	0.327
352.35	2354	14.5	64.1	0.0396	0.303
352.35	2530	17.4	55.6	0.0487	0.233
357.35	1500	12.0	69.5	0.0321	0.357
357.35	1610	12.5	69.0	0.0336	0.351
357.35	1836	13.5	67.4	0.0364	0.334
357.35	1973	14.0	66.9	0.0380	0.329
357.35	2231	14.7	63.6	0.0402	0.298
357.35	2313	15.3	61.8	0.0420	0.282
357.35	2400	16.2	59.4	0.0449	0.262
357.35	2450	16.7	57.5	0.0465	0.248
362.25	1	7.7	74.7	0.0199	0.418
367.15	2000	16.2	59.6	0.0449	0.264

(continued next page)

COMPONENTS

(1) 2-Methyl-1-propanol *(isobutanol)*; $C_4H_{10}O$; [78-83-1]

(2) Water; H_2O; [7732-18-5]

ORIGINAL MEASUREMENTS

Moriyoshi, T.; Aoki, Y.; Kamiyama, H.

J. Chem. Thermodyn. 1977, *9*, 495-502.

EXPERIMENTAL VALUES (continued)

T/K	p/atm	g(1)/100g sln		x_1 (compiler)	
		(2)-rich phase	(1)-rich phase	(2)-rich phase	(1)-rich phase
367.15	2007	16.3	59.2	0.0452	0.261
367.15	2200	21.3	46.0	0.0617	0.172
367.15	2320	26.8	39.5	0.0817	0.1370
370.15	1560	13.2	65.8	0.0356	0.319
370.15	1578	13.4	64.6	0.0361	0.307
370.15	1810	15.2	59.9	0.0471	0.267
370.15	1986	20.2	54.0	0.0579	0.222
370.15	2109	26.8	45.9	0.0817	0.171
372.15	1	9.9	71.7	0.0260	0.381
372.15	667	10.8	67.7	0.0286	0.338
372.15	1000	11.4	66.4	0.0303	0.325
372.15	1500	14.2	62.8	0.0386	0.291
372.15	1769	18.9	56.8	0.0536	0.242
372.15	1878	22.9	52.2	0.0673	0.210
372.15	1980	30.8	43.4	0.0976	0.156
372.15	2000	39.5	-	0.1370	-
377.15	1330	15.8	61.1	0.0436	0.276
377.15	1568	23.2	53.6	0.0684	0.218
377.15	1710	34.0	44.0	0.1113	0.160
382.05	1	11.1	67.7	0.0295	0.338
382.05	500	11.6	64.3	0.0309	0.305
382.05	677	11.7	64.1	0.0312	0.303
382.05	1219	16.8	61.3	0.0468	0.277
382.05	1374	22.5	55.2	0.0659	0.231
382.05	1487	32.5	45.9	0.1048	0.171
386.95	262	12.7	-	0.0342	-
386.95	500	13.8	61.3	0.0374	0.277
386.95	1000	19.4	54.7	0.0552	0.227
386.95	1080	22.2	51.5	0.0649	0.205
386.95	1184	29.9	43.4	0.0915	0.156
391.85	1	14.5	62.3	0.0396	0.287
391.85	340	16.0	59.9	0.0443	0.267
391.85	500	16.8	58.2	0.0468	0.253
391.85	670	19.0	56.6	0.0539	0.239
391.85	870	22.9	51.3	0.0673	0.204
391.85	982	33.5	42.2	0.1091	0.151
396.85	1	17.5	58.7	0.0490	0.257
396.85	500	20.6	54.5	0.0593	0.226
396.85	680	23.6	50.2	0.0699	0.197
396.85	719	27.7	45.5	0.0852	0.169

(continued next page)

COMPONENTS:	ORIGINAL MEASUREMENTS:
(1) 2-Methyl-1-propanol *(isobutanol)*; $C_4H_{10}O$; [78-83-1]	Moriyoshi, T.; Aoki, Y.; Kamiyama, H.
(2) Water; H_2O; [7732-18-5]	*J. Chem. Thermodyn.* 1977, *9*, 495-502.

EXPERIMENTAL VALUES (continued)

		g(1)/100g sln		x_1(compiler)	
T/K	p/atm	(2)-rich phase	(1)-rich phase	(2)-rich phase	(1)-rich phase
396.85	816	30.6	40.7	0.0968	0.143
398.75	1	19.9	56.3	0.0570	0.238
398.75	350	23.2	51.8	0.0684	0.207
398.75	500	32.9	42.4	0.1065	0.152
401.75	1	21.8	52.4	0.0635	0.211
401.75	201	25.9	48.9	0.0783	0.189
401.75	298	29.7	44.4	0.0932	0.162
403.75	1	24.3	48.9	0.0724	0.189
405.75	1	28.6	44.4	0.0888	0.162
407.25	1	35.4	38.1	0.118	0.130

Properties of the critical solutions

p_c/atm	T_c(UCST)/K	x_{1c}
1	407.25	0.123
300	402.75	0.124
346	401.75	0.127
560	398.75	0.128
700	398.15	0.128
829	396.85	0.124
993	391.85	0.128
1000	392.25	0.128
1208	386.95	0.127
1500	382.75	0.132
1512	382.05	0.131
1726	377.15	0.131
2000	372.35	0.131
2004	372.15	0.130
2156	370.15	0.129
2394	367.15	
2400	366.15	0.132

$(dT_c/dp) = -(0.017 \pm 0.002)$ K atm^{-1}

$(dx_c/dT) = -(0.0002 \pm 0.00001)$ K^{-1}

1 atm = 101.325 kPa

COMPONENTS:

(1) 2-Methyl-1-propanol (*isobutanol*); $C_4H_{10}O$; [78-83-1]

(2) Water; H_2O; [7732-18-5]

ORIGINAL MEASUREMENTS:

Lyzlova, R.V.

Zh. Prikl. Khim. (Leningrad) 1979, *52*, 545-50; *J. App. Chem. USSR* 1979, *52*, 509-14

VARIABLES:

Temperature: 20 and 90°C

PREPARED BY:

A.F.M. Barton

EXPERIMENTAL VALUES:

Mutual solubility of 2-methyl-1-propanol (1) and water (2)

Temperature	x_1		g(1)/100g sln (compiler)	
$t/°C$	Alcohol-rich phase	Water-rich phase	Alcohol-rich phase	Water-rich phase
20	0.55	0.023	83.4	8.9
90	0.399	0.021	73.2	8.1

AUXILIARY INFORMATION

METHOD/APPARATUS/PROCEDURE:

The analytical method was used, samples being withdrawn from coexisting liquid phases at equilibrium for analysis by refractometry. An IRF-23 refractometer measured the refracture index correct to 0.00002. The study was concerned with phase equilibria in the ternary system 1-butanol/2-methyl-1-propanol/water, and only a few experiments dealt with mutual solubilities in the binary systems.

SOURCE AND PURITY OF MATERIALS:

(1) CP grade dried over freshly ignited K_2CO_3; distilled twice in 1.5m glass packed fractionating column; n_D^{20} = 1.39574; d_4^{20} = 0.8020 g cm^{-3}; b.p 107.8°C

(2) not specified

ESTIMATED ERROR:

Not specified for binary systems; error below ± 1.5% for alcohol ratios in ternary systems.

REFERENCES:

COMPONENTS:

(1) 2-Methyl-1-propanol (*isobutanol*); $C_4H_{10}O$; [78-83-1]

(2) Water; H_2O; [7732-18-5]

ORIGINAL MEASUREMENTS:

Fu, C.F.; King, C.L.; Chang, Y.F.; Xeu, C.X.

Hua Kung Hsueh Pao 1980 (3), 281-92.

VARIABLES:

One temperature: 89.8°C

PREPARED BY:

C.F. Fu.

EXPERIMENTAL VALUES:

The proportion of 2-methyl-1-propanol(1) in the water-rich phase at equilibrium at 89.8°C was reported to be 7.43 g(1)/100g sln. The corresponding mole fraction solubility, x_1, is 0.0191.

The proportion of water(2) in the alcohol-rich phase at equilibrium at 89.8°C was reported to be 25.96 g(2)/100g sln. The corresponding mole fraction solubility, x_2, is 0.5905.

AUXILIARY INFORMATION

METHOD/APPARATUS/PROCEDURE:

The turbidimetric method was used. Homogeneous solutions were prepared and boiled at 760 mm Hg in a specially designed flask attached to a condenser of negligible hold-up compared with the volume of liquid solution. The solution was stirred by a magnetic stirrer and titrated with (1) or (2). The end point of titration was judged both by cloudiness and constancy of boiling temperature.

SOURCE AND PURITY OF MATERIALS: (Some information not in the published paper has been supplied by the compiler).

(1) Riedel-DE HAEN AG reagent for chromatography; used as received; b.p. 107.8°C (760 mm Hg) n_D^{15} 1.3976, d_4^{15} 0.8055.

(2) distilled.

ESTIMATED ERROR: (supplied by compiler)

Temperature: ± 0.02°C

Solubility: 0.4%

REFERENCES:

COMPONENTS:

(1) 2-Methyl-1-propanol (*isobutanol*)
$C_4H_{10}O$; [78-83-1]

(2) Water; H_2O; [7732-18-5]

ORIGINAL MEASUREMENTS:

Lutugina, N.V., Reshchetova, L.I.
Vestnik Leningr. Univ. 1972, *16*, 75-81

VARIABLES:

One temperature: 89.9°C

PREPARED BY:

C.F. Fu and G.T. Hefter

EXPERIMENTAL VALUES:

The solubility of 2-methyl-1-propanol (1) in the water-rich phase at 89.9°C was reported to be 0.059 mole fraction.

The solubility of 2-methyl-1-proponal in the alcohol-rich phase at 89.9°C was reported to be 0.677 mole fraction.

The corresponding mass solubilities calculated by the compilers are 20.3 g(1)/100g sln and 89.6 g(1)/100g sln.

AUXILIARY INFORMATION

METHOD/APPARATUS/PROCEDURE:

Vapour-liquid-liquid equilibrium still described in ref 1 was used for liquid-liquid equilibrium determination at boiling point. Immediately after cutting off heating and stopping of violent boiling, samples were taken from both layers for analysis. Analytical method was described in detail.

SOURCE AND PURITY OF MATERIALS:

(1) Source not specified;
distilled over a column of 20 theoretical plates (according to C_6H_6 - CCl_4);
b.p. 108-0°C (760 mm Hg), n_D^{20} 1.3960
d_4^{20} 0.8020

(2) distilled

ESTIMATED ERROR:

Not specified

REFERENCES:

1. Morachevski, A.G., Smirnova, N.A.
Zh. Prinkl. Khim. 1963, *36*, 2391.

COMPONENTS:

(1) 2-Methyl-2-propanol (*tert-butyl alcohol, tert-butanol)*; $C_4H_{10}O$; [75-65-0]

(2) Water; H_2O; [7732-18-5]

EVALUATOR:

A.F.M. Barton, School of Mathematical and Physical Sciences, Murdoch University, Perth, Western Australia

July 1983.

CRITICAL EVALUATION:

In the two studies (ref 1,2) reported, no region of liquid-liquid immiscibility has been observed at atmospheric pressure over the temperature range 193 K to 298 K, and at 298 K in the pressure range 0.1 - 142 MPa.

References:

1. Rosso, J.-C.; Carbonnel, L. *C.R. Hebd. Seances Acad. Sci., Ser. C.* 1968, *267*, 4.

2. Nakagawa, M.; Inubushi, H.; Moriyoshi, T. *J. Chem. Thermodyn.* 1981, *13*, 171.

COMPONENTS:

(1) 2-Methyl-2-propanol; (*tert-butanol, tert-butyl alcohol*); $C_4H_{10}O$; [75-65-0]

(2) Water; H_2O; [7732-18-5]

ORIGINAL MEASUREMENTS:

Rosso, J.-C.; Carbonnel, L.
C.R. Hebd. Seances Acad. Sci., 1968, *267*, 4-6.

VARIABLES:

Temperature: (-80) - 25 °C

PREPARED BY:

A.F.M. Barton

EXPERIMENTAL VALUES:

No region of liquid-liquid immiscibility was observed. A full phase diagram is presented, revealing a solid dihydrate between -8.2 and -6 °C, and a dihydrate from -6 to 0.7 °C.

	t/°C	g(1)/100g sln	x_1
Eutectic	- 8.2	16.60	0.0462
Metastable eutectic	- 9.6	18	0.051
Peritectic	- 6.0	21.20	0.0613
Congruent melting of dihydrate	0.7	67.36	0.334
Eutectic	- 3.3	89.10	0.665
Melting point	25.5	100.0	1.000

AUXILIARY INFORMATION

METHOD/APPARATUS/PROCEDURE:

Thermal analysis of crystallized mixtures allowed to warm up from -80 °C.

SOURCE AND PURITY OF MATERIALS:

(1) Prolabo R.P.;
three successive fractional crystallizations;
m.p. 25.5 ± 0.2 °C

ESTIMATED ERROR:

REFERENCES:

COMPONENTS:

(1) 2-Methyl-2-propanol (*tert-butyl alcohol, tert-butanol*); $C_4H_{10}O$; [75-65-0]

(2) Water; H_2O; [7732-18-5]

ORIGINAL MEASUREMENTS:

Nakagawa, M.; Inubushi, H.; Moriyoshi, T.
J. Chem. Thermodyn. 1981, *13*, 171-8

VARIABLES:

One temperature: 298.15 K

Pressure: 0.1-142 MPa

PREPARED BY:

G.T. Hefter

EXPERIMENTAL VALUES:

No region of liquid-liquid immiscibility was observed over the entire pressure range in solutions containing up to 35 g(1)/100g sln. Isothermal compressibilities and partial molar volumes of the components are given.

AUXILIARY INFORMATION

METHOD/APPARATUS/PROCEDURE:

Compressibility studies

SOURCE AND PURITY OF MATERIALS: (ref 1)

(1) "best grade";
refluxed with freshly ignited calcium oxide and twice fractionally distilled;

density and refractive index agreed with literature to within 0.00015 g cm^{-3} and 0.0002 respectively.

(2) de-ionized, distilled from alkaline permanganate and redistilled.

ESTIMATED ERROR:

Temperature: control to ± 0.003 K

Pressure: ± 0.05%

REFERENCES:

1. Moriyoshi, T.; Morishita, T.; Inubushi, H. *J. Chem. Thermodyn.* 1977, *9*, 577

COMPONENTS:

(1) 1-Butanol (*n-butyl alcohol*); $C_4H_{10}O$; [71-36-3]

(2) Water; H_2O; [7732-18-5]

EVALUATOR:

G.T. Hefter, School of Mathematical and Physical Sciences, Murdoch University, Perth, Western Australia.
November 1982.

CRITICAL EVALUATION:

Solubilities in the system comprising 1-butanol (1) and water (2) have been reported in the following publications:

Reference	*T*/K	Solubility	Method
Reilly and Ralph (ref 1)	293	(1) in (2)	titration
Fuhner (ref 2)	273-383	(1) in (2)	synthetic
Drouillon (ref 3)	293-393	(1) in (2)	synthetic
Hill and Malisoff (ref 4)	278-398	mutual	volumetric
Jones (ref 5)	255-398	mutual	synthetic
Mueller *et al.* (ref 6)	273-378	mutual	densimetric
Stockhardt and Hull (ref 7)	298	(1) in (2)	gravimetric
Butler *et al.* (ref 8)	292-304	mutual	turbidimetric
Berkengeim (ref 9)	255-313	(2) in (1)	analytical
Reber *et al.* (ref 10)	364-397	mutual	synthetic
Othmer *et al.* (ref 11)	299,323	(1) in (2)	synthetic
Booth and Everson (ref 12)	298	(1) in (2)	titration
Hansen *et al.* (ref 13)	298	(1) in (2)	interferometric
Donahue *et al.* (ref 14)	298	mutual	analytical
Erichsen (ref 15)	273-293	(1) in (2)	synthetic
Erichsen (ref 16)	273-398	mutual	synthetic
McCants *et al.* (ref 17)	311	mutual	titration
Jones and McCants (ref 18)	311	mutual	titration
Skrzec and Murphy (ref 19)	300	mutual	titration
Hayashi and Sasaki (ref 20)	303	(1) in (2)	turbidimetric
Kakovskii (ref 21)	298	(1) in (2)	not stated
Rao and Rao (ref 22)	300	mutual	turbidimetric
Kinoshita *et al.* (ref 23)	298	(1) in (2)	surface tension
Venkataratnam and Rao (ref 24)	303	mutual	turbidimetric
Petriris and Geankopolis (ref 25)	298	mutual	titration
Ababi and Popa (ref 26)	298	mutual	turbidimetric
Smirnova and Morachevskii (ref 27)	293	mutual	densimetric
Ratouis and Dodé (ref 28)	298,303	(1) in (2)	analytical
Meussen and Huyskens (ref 29)	298	(1) in (2)	interferometric
Lesteva *et al.* (ref 30)	293,348	mutual	titration
Hanssens (ref 31)	298	(1) in (2)	interferometric
Mullens (ref 32)	298	mutual	interferometric
Vochten and Petre (ref 33)	288	(1) in (2)	surface tension
Korenman *et al.* (ref 34)	298	mutual	analytical
Prochazka *et al.* (ref 35)	396-398	mutual	turbidimetric
De Santis *et al.* (ref 36)	298	mutual	analytical
De Santis *et al.* (ref 37)	293-313	(1) in (2)	analytical
Lavrova and Lesteva (ref 38)	313-333	mutual	titration
Aoki and Moriyoshi (ref 39)	303-398	mutual	refractometric

(continued next page)

COMPONENTS:

(1) 1-Butanol (*n-butyl alcohol*); $C_4H_{10}O$; [71-36-3]

(2) Water; H_2O; [7732-18-5]

EVALUATOR:

G.T. Hefter, School of Mathematical and Physical Sciences, Murdoch University, Perth, Western Australia.
November 1982.

CRITICAL EVALUATION (continued)

Reference	T/K	Solubility	Method
Lyzlova (ref 40)	293,363	mutual	analytical
Singh and Haque (ref 41)	303	mutual	titration
Fu *et al.* (ref 42)	366	mutual	turbidimetric
Tokunaga *et al.* (ref 43)	288-308	(2) in (1)	analytical
Nishino and Nakamura (ref 44)	275-360	mutual	turbidimetric
Lutugina and Reshchetova (ref 47)	366	mutual	analytical

Apart from one other publication (ref 45), which did not contain sufficient information to justify its inclusion, all the original data are given in the data sheets following this Critical Evaluation. Solubilities in the system comprising 1-butanol-*d*, C_4H_9DO, and water-d_2, D_2O, (ref 46) are also included.

In the Critical Evaluation the data of Booth and Everson (ref 12), in volume fractions, and the data in ref 1, 7, 21, 23, 31-34 , in weight/volume fractions, are excluded from further consideration as density information was not given in the original references. The data of Nishino and Nakamura (ref 44) were also excluded as only a graphical presentation was given.

The data of Drouillon (ref 3), Jones (ref 5) (water-rich phase data only), Berkengeim (9), Othmer *et al.* (ref 11), Donahue and Bartell (ref 14) (alcohol-rich phase data only) Venkataratnam and Rao (ref 24), Meeussen and Huyskens (ref 29), Lesteva *et al.* (ref 30), Lyzlova (ref 40) and Singh and Haque (ref 41), disagree markedly from all other studies and are rejected.

The following individual points in otherwise satisfactory studies are also in marked disagreement with other studies and have been rejected: in the water-rich phase 303 K (ref 16), 323 and 333 K (ref 39), in the alcohol-rich phase 353 K (ref 4). All other data are included in the tables below.

Values obtained by the Evaluator by graphical interpolation or extrapolation from the data sheets are indicated by an asterisk (*). "Best" values have been obtained by simple averaging. The uncertainty limits (σ_n) attached to these "best" values do not have statistical significance and should be regarded only as a convenient representation of the spread of the values and not as error limits. The letter (*R*) designates "recommended" data. Data are "recommended" if two or more apparently reliable studies are in reasonable (± 5% relative) agreement. All other data are regarded as tentative only.

(continued next page)

COMPONENTS:

(1) 1-Butanol (*n-butyl alcohol*); $C_4H_{10}O$; [71-36-3]

(2) Water; H_2O; [7732-18-5]

EVALUATOR:

G.T. Hefter, School of Mathematical and Physical Sciences, Murdoch University, Perth, Western Australia.
November 1982.

CRITICAL EVALUATION (continued)

Tentative and recommended (*R*) values for the solubility of 1-butanol (1) in water (2)

T/K	Solubility g(1)/100g sln: Reported values	"Best" values ($\pm\sigma_n$)
273	10.45 (ref 2), 10.32 (ref 15), 10.28 (ref 16)	10.4 ± 0.1 (*R*)
278	9.55 (ref 4)	9.6
283	9.00 (ref 2), 8.91 (ref 4), 8.68 (ref 15), 8.83 (ref 16)	8.9 ± 0.1 (*R*)
288	8.21 (ref 4)	8.2
293	7.90 (ref 2), 7.81(ref 4), 7.72(ref 8), 7.56(ref 15), 7.63(ref 16), 7.8(ref 36)	7.8 ± 0.2 (*R*)
298	7.35(ref 4), 7.31(ref 8), 7.41(ref 13), 7.3(ref 14), 7.4(ref 23), 7.2(ref 26), 7.34(ref 27), 7.4(ref 35)	7.4 ± 0.1 (*R*)
303	7.10(ref 2), 7.08(ref 4), 7.06(ref 15), 7.01(ref 20), 6.99(ref 27), 7.1(ref 37), 7.0(ref 38)	7.1 ± 0.1 (*R*)
308	6.83 (ref 4)	6.8
313	6.55 (ref 2), 6.60(ref 4), 6.72(ref 15), 6.53(ref 16), 6.6(ref 37), 6.6(ref 38)	6.6 ± 0.1 (*R*)
323	6.35(ref 2), 6.46(ref 4), 6.1(ref 11), 6.55(ref 15), 6.38(ref 16)	6.4 ± 0.2
333	6.35(ref 2), 6.52(ref 4), 6.52(ref 6), 6.52(ref 15), 6.45(ref 38)	6.5 ± 0.1 (*R*)
343	6.55 (ref 2), 6.73 (ref 4), 6.67 (ref 15), 6.7* (ref 39)	6.7 ± 0.1 (*R*)
348	6.8 (ref 6), 6.9* (ref 39)	6.9 ± 0.1 (*R*)
353	7.00(ref 2), 6.89(ref 4), 6.90(ref 15), 7.1*(ref 35), 7.1*(ref 39)	7.0 ± 0.1 (*R*)
358	7.2*(ref 4), 7.5*(ref 35), 7.2*(ref 39)	7.3 ± 0.1 (*R*)
363	7.80(ref 2), 7.6*(ref 4), 7.8(ref 6), 7.50(ref 15), 7.9*(ref 35), 7.5*(ref 39)	7.7 ± 0.2 (*R*)
368	8.3*(ref 4), 8.4*(ref 35), 8.2*(ref 39)	8.3 ± 0.1 (*R*)
373	9.05(ref 2), 9.2*(ref 4), 9.1*(ref 10), 8.82(ref 15), 9.1*(ref 35), 9.2*(ref 39)	9.1 ± 0.1 (*R*)
378	10.3*(ref 4), 9.8(ref 6), 10.0*(ref 10), 9.9*(ref 35), 10.2*(ref 39)	10.0 ± 0.2 (*R*)
383	10.90(ref 2), 11.5*(ref 4), 11.3*(ref 10), 11.05(ref 15), 11.1*(ref 35), 11.0(ref 39)	11.1 ± 0.2 (*R*)
388	12.8*(ref 4), 13.2*(ref 10), 13.1*(ref 35), 12.7 (ref 39)	13.0 ± 0.2 (*R*)

(continued next page)

COMPONENTS:	EVALUATOR:
(1) 1-Butanol (*n-butyl alcohol*); $C_4H_{10}O$; [71-36-3] (2) Water; H_2O; [7732-18-5]	G.T. Hefter, School of Mathematical and Physical Sciences, Murdoch University, Perth, Western Australia. November 1982.

CRITICAL EVALUATION (continued)

Tentative and recommended (*R*) values for the solubility of water (2) in 1-butanol (1)

T/K	Solubility g(2)/100g sln Reported values	"Best values ($\pm\sigma_n$)
258	19.0 (ref 5)	19.0
268	19.18 (ref 5)	19.2
273	19.4 (ref 5), 18.95 (ref 15)	19.7 ± 0.3 (*R*)
278	19.62 (ref 4)	19.6
283	19.67 (ref 4), 19.5 (ref 15)	19.6 ± 0.1 (*R*)
288	19.86 (ref 4), 19.9 (ref 43)	19.9
293	20.07(ref 4),19.8(ref 5),20.08(ref 8),20.0(ref 15),20.1(ref 27), 20.0(ref 37),20.1(ref 43)	20.0 ± 0.1 (*R*)
298	20.07(ref 4),20.36(ref 8), 20.7*(ref 25),19.9*(ref 26), 20.49*(ref 36),20.4(ref 43)	20.3 ± 0.3 (*R*)
303	20.62(ref 4),20.68*(ref 8),20.63(ref 15),20.6 (ref 37)	20.6 ± 0.1 (*R*)
308	21.06 (ref 4), 21.1 (ref 43)	21.1 (*R*)
313	21.41(ref 4),21.4(ref 5),21.40(ref 15),21.4(ref 37),21.5(ref 38)	21.4 ± 0.1 (*R*)
323	22.42 (ref 4), 22.41 (ref 15), 22.4*(ref 39)	22.4 (*R*)
328	23.2*(ref 39)	23.2
333	23.62(ref 4),23.8(ref 5),23.71(ref 15),23.69(ref 38),23.9*(ref 39)	23.7 ± 0.1 (*R*)
338	24.5*(ref 39)	24.5
343	25.21 (ref 4), 25.43 (ref 15), 25.1*(ref 39)	25.2 ± 0.1 (*R*)
348	26.3 (ref 6), 26.4 (ref 39)	26.3 ± 0.1 (*R*)
353	27.3 (ref 5), 27.60 (ref 15), 27.7*(ref 39)	27.5 ± 0.2 (*R*)
358	28.3*(ref 4), 29.1*(ref 35), 29.1*(ref 39)	28.8 ± 0.4 (*R*)
363	30.1*(ref 4),30.2(ref 6),30.7*(ref 10),30.10(ref 15),30.2*(ref 35) 30.5 (ref 39)	30.3 ± 0.2 (*R*)
368	32.0*(ref 4), 32.3*(ref 10), 31.9*(ref 35), 31.9*(ref 39)	32.0 ± 0.2 (*R*)
373	33.9*(ref 4), 33.6(ref 5), 34.2*(ref 10),33.30(ref 15), 34.0 (ref 35)	33.8 ± 0.3 (*R*)
378	35.8*(ref 4), 35.7 (ref 6), 36.6*(ref 10), 36.5 (ref 35)	36.1 ± 0.4 (*R*)
383	38.2*(ref 4), 38.5 (ref 5), 39.2*(ref 10), 38.15(ref 15), 39.4*(ref 35),38.3*(ref 39)	38.6 ± 0.5 (*R*)
388	42*(ref 4), 42.9*(ref 10), 43.0*(ref 35),42*(ref 39)	42.5 ± 0.5 (*R*)
393	47*(ref 4), 47.5(ref 5), 48.5*(ref 10), 46.65(ref 15), 48.5*(ref 35),46.3*(ref 39)	47.4 ± 0.9 (*R*)

(continued next page)

COMPONENTS:	EVALUATOR:
(1) 1-Butanol (*n-butyl alcohol*); $C_4H_{10}O$; [71-36-3] (2) Water; H_2O; [7732-18-5]	G.T. Hefter, School of Mathematical and Physical Sciences, Murdoch University, Perth, Western Australia. November 1982.

CRITICAL EVALUATION (continued)

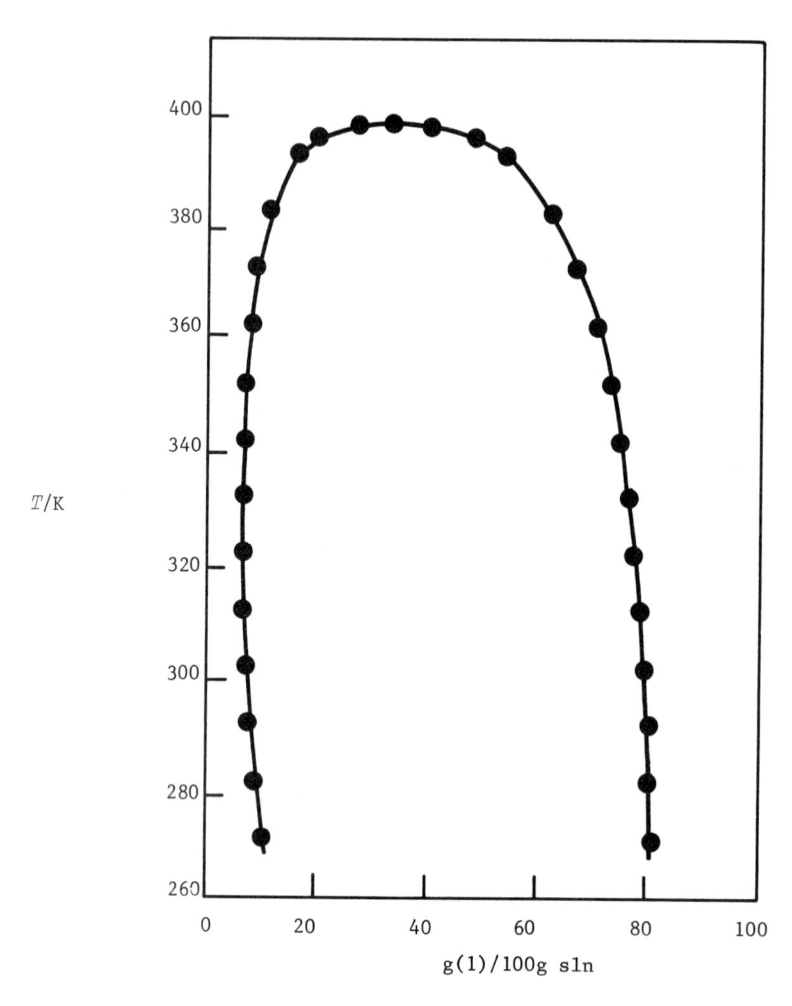

Figure 1. Mutual solubility of (1) and (2)

The "best" values from the above tables are plotted in Figure 1. The "best" values are in general in excellent agreement with the single most extensive determination (ref 4). Aoki and Moriyoshi (ref 39) have also determined mutual solubilities of (1) and (2) and the upper critical solution temperature at pressures of 200-2500 atm (20 - 250 MPa).

Excluding those data (ref 3,5) already rejected, the upper critical solution temperature has been reported as 397.55 (ref 10), 397.85 (ref 39), 398.30 (ref 4), and 398.5 K (ref 15). The corresponding critical solution compositions have been reported as 32.96 g(1)/100g sln (ref 10), x_1 = 0.110 (ref 39) and 32.5 g(1)/100g sln (ref 4).

(continued next page)

COMPONENTS:	EVALUATOR:
(1) 1-Butanol (*n-butyl alcohol*); $C_4H_{10}O$; [71-36-3] (2) Water; H_2O; [7732-18-5]	G.T. Hefter, School of Mathematical and Physical Sciences, Murdoch University, Perth, Western Australia. November 1982.

CRITICAL EVALUATION

References

1. Reilly, J.; Ralph, E.W. *Sci. Proc. Roy. Dublin Soc.* 1919, *15*, 597.
2. Fuhner, H. *Ber. Dtsch. Chem. Ges.* 1924, *57*, 510.
3. Drouillon, F. *J. Chim. Phys.* 1925, *22*, 149.
4. Hill, A.E.; Malisoff, W.M. *J. Am. Chem. Soc.* 1926, *48*, 918.
5. Jones, D.C. *J. Chem. Soc.* 1929, 799.
6. Mueller, A.J.; Pugsley, L.I.; Ferguson, J.B. *J. Phys. Chem.* 1931, *35*, 1314.
7. Stockhardt, J.S.; Hull, C.M. *Ind. Eng. Chem.* 1931, *23*, 1438.
8. Butler, J.A.V.; Thomson, D.W.; Maclennan, W.H. *J. Chem. Soc.* 1933, 674.
9. Berkengeim, T.I. *Zavod. Lab.* 1941, *10*, 952.
10. Reber, L.A.; McNabb, W.M.; Lucasse, W.A. *J. Phys. Chem.* 1942, *46*, 500.
11. Othmer, D.F.; Bergen, W.S.; Shlechter, N.; Bruins, P.F. *Ind. Eng. Chem.* 1945, *37*, 890.
12. Booth, H.S.; Everson, H.E. *Ind. Eng. Chem.* 1948, *40*, 1491.
13. Hansen, R.S.; Fu, Y.; Bartell, F.E. *J. Phys. Chem.* 1949, *53*, 769.
14. Donahue, D.J.; Bartell, F.E. *J. Phys. Chem.* 1952, *56*, 480.
15. Erichsen, L. von *Naturwissenschaften* 1952, *39*, 41.
16. Erichsen, L. von *Brennst. Chem.* 1952, *33*, 166.
17. McCants, J.F.; Jones, J.H.; Hopson, W.H. *Ind. Eng. Chem.* 1953, *45*, 454.
18. Jones, J.H.; McCants, J.F. *Ind. Eng. Chem.* 1954, *46*, 1956.
19. Skrzec, A.E.; Murphy, N.F. *Ind. Eng. Chem.* 1954, *46*, 2245.
20. Hayashi, M.; Sasaki, T. *Bull. Chem. Soc. Japan* 1956, *29*, 857.
21. Kakovskii, I.A. *Proc. Intern. Congr. Surface Activity, 2nd, London* 1957, *4*, 225.
22. Rao, R.M.; Rao, V.C. *J. Appl. Chem.* 1957, *7*, 659.
23. Kinoshita, K.; Ishikawa, H.; Shinoda, K. *Bull. Chem. Soc. Japan* 1958, *31*, 1081.
24. Venkataratnam, A.; Rao, R.I. *J. Sci. Ind. Res.* 1958, *17B*, 108.
25. Petriris, V.E.; Geankopolis, C.J. *J. Chem. Eng. Data* 1959, *4*, 197.
26. Ababi, V.; Popa, A. *An. Stiint. Univ. "Al. I. Cuza" Iasi* 1960, *6*, 929.
27. Smirnova, N.A.; Morachevskii, A.G. *Zh. Prikl. Khim. (Leningrad)* 1963, *36*, 2391.
28. Ratouis, M.; Dodé, M. *Bull. Soc. Chim. Fr.* 1965, 3318.
29. Meeussen, E.; Huyskens, P. *J. Chim. Phys.* 1966, *63*, 845.
30. Lesteva, T.M.; Ogorodnikov, S.K.; Tyvina, T.N. *Zh. Prikl. Khim. (Leningrad)* 1968, *41*, 1159.

(continued next page)

COMPONENTS:	EVALUATOR:
(1) 1-Butanol (*n-butyl alcohol*); $C_4H_{10}O$; [71-36-3] (2) Water; H_2O: [7732-18-5]	G.T. Hefter, School of Mathematical and Physical Sciences, Murdoch University, Perth, Western Australia. November 1982.

CRITICAL EVALUATION (continued)

31. Hanssens, I. *Associatie van normale alcoholen en hun affiniteit voor water en organische solventen*, Doctoraatsproefschrift, Leuven, 1969; Huyskens, P.; Mullens, J.; Gomez, A.; Tack, J. *Bull. Soc. Chim. Belg.* 1975, *84*, 253.

32. Mullens, J.; *Alcoholassociaten*, Doctoraatsproefschrift, Leuven, 1971; Huyskens, P.; Mullens, J.; Gomez, A.; Tack, J. *Bull. Soc. Chim. Belg.* 1975, *84*, 253.

33. Vochten, R.; Petre, G. *J. Colloid Interface Sci.* 1973, *42*, 320.

34. Korenman, I.M.; Gorokhov, A.A.; Polozenko, G.N. *Zh. Fiz. Khim.* 1974, *48*, 1810; 1975, *49*, 1490; *Russ. J. Phys. Chem.* 1974, *48*, 1065; *49*, 877.

35. Prochazka, O.; Sushka, J.; Pick, J. *Coll. Szech. Chem. Comm.* 1975, *40*, 781.

36. De Santis, R.; Marrelli, L.; Muscetta, P.N. *Chem. Eng. J.* 1976, *11*, 207.

37. De Santis, R.; Marrelli, L.; Muscetta, P.N. *J. Chem. Eng. Data* 1976, *21*, 324; Marrelli, L. *Chem. Eng. J.* 1979, *18*, 225.

38. Lavrova, O.A.; Lesteva, T.M. *Zh. Fiz. Khim.* 1976, *50*, 1617; *Dep. Doc. VINITI* 3813-75.

39. Aoki, Y.; Moriyoshi, T. *J. Chem. Thermodyn.* 1978, *10*, 1173.

40. Lyzlova, R.V. *Zh. Prikl. Khim. (Leningrad)* 1979, *52*, 545; *J. App. Chem. USSR* *52*, 509.

41. Singh, R.P.; Haque, M.M. *Indian J. Chem.* 1979, *17A*, 449.

42. Fu, C.F.; King, C.L.; Chang, Y.F.; Xeu, C.X. *Hua Kung Hseuh Pao* 1980 (3), 281.

43. Tokunaga, S.; Manabe, M.; Koda, M. *Niihama Kogyo Koto Semmon Gakko Kiyo, Rikogaku Hen (Memoirs Niihama Technical College, Sci and Eng.)* 1980, *16*, 96.

44. Nishino, N.; Nakamura, M. *Bull. Chem. Soc. Japan* 1978, *51*, 1617; 1981, *54*, 545.

45. Ito, K. *Kagaku Kenkyusha Hokoku* 1956, *32*, 207.

46. Rabinovich, I.B.; Fedorov, V.D.; Paskhin, N.P.; Avdesnyak, M.A.; Pimenov, N. Ya. *Dokl. Akad. Nauk SSSR* 1955, *105*, 108.

47. Lutugina, N.V.; Reshchetova, L.I. *Vestnik Leningr. Univ.* 1972, *16*, 75.

COMPONENTS:

(1) 1-Butanol; $C_4H_{10}O$; [71-36-3]

(2) Water; H_2O; [7732-18-5]

ORIGINAL MEASUREMENTS:

Reilly, J.; Ralph, E.W.

Sci. Proc. Roy. Dublin Soc. 1919, *15*, 597-608

VARIABLES:

One temperature: 20°C

PREPARED BY:

S.H. Yalkowsky; S.C. Valvani; A.F.M. Barton

EXPERIMENTAL VALUES:

The proportion of 1-butanol in the water-rich phase at 20°C was reported to be 79 g(1)/L sln.

The corresponding amount of substance concentration calculated by the compiler is 1.07 mol(1)/L sln.

AUXILIARY INFORMATION

METHOD/APPARATUS/PROCEDURE:

The behavior of three component 1-butanol/water/acetone systems was investigated. A mixture containing a slight excess of n-butanol over saturation was taken in a long-necked 50 mL flask fitted with a buret having a long-delivery tube reaching almost to the level of the liquid. The flask was immersed in a thermostat bath (20 ± 0.005°C) and acetone run in drop by drop with frequent shaking until the cloudy mixture just became clear on standing. From the increased weight the relative proportions of the components were calculated.

SOURCE AND PURITY OF MATERIALS:

(1) redistilled;
b.p. 117.6°C/763 mm Hg
d_4^{20} = 0.80953

(2) not stated

ESTIMATED ERROR:

Not stated

REFERENCES:

COMPONENTS:

(1) 1-Butanol; $C_4H_{10}O$; [71-36-3]

(2) Water; H_2O; [7732-18-5]

ORIGINAL MEASUREMENTS:

Führer, H.

Ber. Dtsch. Chem. Ges. 1924, *57*, 510-5.

VARIABLES:

Temperature: 0-110°C

PREPARED BY:

A. Maczynski; Z. Maczynska; Z. Szafranski

EXPERIMENTAL VALUES:

Solubility of 1-butanol (1) in water (2)

$t/^{\circ}C$	g(1)100g sln	x_1 (compiler)
0	10.45	0.0275
10	9.00	0.0235
20	7.90	0.0204
30	7.10	0.0182
40	6.55	0.0167
50	6.35	0.0162
60	6.35	0.0162
70	6.55	0.0167
80	7.00	0.0180
90	7.80	0.0201
100	9.05	0.0236
110	10.90	0.0289

AUXILIARY INFORMATION

METHOD/APPARATUS/PROCEDURE:

Rothmund's synthetic method (ref 1) was used.

Small amounts of (1) and (2) were sealed in a glass tube and heated with shaking in an oil bath to complete dissolution. The solution was cooled until a milky turbidity appeared and this temperature was adopted as the equilibrium temperature.

SOURCE AND PURITY OF MATERIALS:

(1) source not specified; specially purified but no details provided.

(2) not specified.

ESTIMATED ERROR:

Not specified.

REFERENCES:

1. Rothmund, V. *Z. Phys. Chem.* 1898, *26*, 433.

COMPONENTS:

(1) 1-Butanol; $C_4H_{10}O$; [71-36-3]

(2) Water; H_2O; [7732-18-5]

ORIGINAL MEASUREMENTS:

Drouillon, F.

J. Chim. Phys. 1925, *22*, 149-68.

VARIABLES:

Temperature: 20-120°C

PREPARED BY:

S.C. Valvani; S.H. Yalkowsky; A.F.M. Barton

EXPERIMENTAL VALUES:

Mutual solubility of 1-butanol (1) and water (2)

$t/^{\circ}C$	g(1)/100g sln		x_1 (compiler)	
	Alcohol-rich phase	Water-rich phase	Alcohol-rich phase	Water-rich phase
20.0[a]	81.6	6.6	0.519	0.169
23.0	-	6.5	-	0.0166
23.5	81.9	-	0.524	-
27.5	-	6.5	-	0.0166
32.5	-	6.2	-	0.0158
38.5	-	6.0	-	0.0153
40.0[a]	80.3	6.0	0.498	0.0153
45.0	-	5.7	-	0.0145
54.5	78.9	-	0.476	-
57.0	-	5.5	-	0.0140
60.0[a]	78.2	5.5	0.466	0.0140
60.5	78.2	-	0.466	-
67.0	-	5.7	-	0.0145
71.0	-	6.0	-	0.0153
72.0	76.3	-	0.439	-

(continued next page)

AUXILIARY INFORMATION

METHOD/APPARATUS/PROCEDURE:

The behavior of three-component 1-butanol/water/ethanol system was investigated. Mixtures of known composition were heated in sealed tubes to complete solution of alcohol and water, then cooled slowly, with shaking, until cloudiness appeared. For the water-rich mixtures, where a closed miscibility gap occurs, observations were made on both heating and cooling.

SOURCE AND PURITY OF MATERIALS:

Not stated.

ESTIMATED ERROR:

Solubility: about 2%

Temperature: ± 0.5°C

REFERENCES:

COMPONENTS:	ORIGINAL MEASUREMENTS:
(1) 1-Butanol; $C_4H_{10}O$; [71-36-3]	Drouillon, F.
(2) Water; H_2O; [7732-18-5]	*J. Chim. Phys.* 1925, *22*, 149-68.

EXPERIMENTAL VALUES (continued)

$t/^{o}C$	g(1)/100g sln		x_1 (compiler)	
	Alcohol-rich phase	Water-rich phase	Alcohol-rich phase	Water-rich phase
76.0	-	6.2	-	0.0158
80.0[a]	74.6	6.4	0.417	0.0164
81.5	-	6.5	-	0.0166
88.0	72.4	-	0.390	-
90.5	-	6.5	-	0.0166
94.0	-	7.2	-	0.0185
97.0	69.9	-	0.361	-
100.0[a]	68.8	7.5	0.349	0.0196
102.0	-	7.6	-	0.0196
105.5	-	8.4	-	0.0218
107.5	66.3	-	0.324	-
119.5	57.8	11.5	0.250	0.0306
120.0[a]	57.3	11.8	0.246	0.0315
124.5	53.2	-	0.216	-
125.0	50.8	-	0.200	-
126.5	-	16.4	-	0.0455
127.0	45.8	-	0.171	-
128.0	-	18.6	-	0.0526
128.5	39.6	-	0.137	-
129.0	34.0	22.5	0.111	0.0659
129.5	30.1	28.0	0.095	0.0864

[a] interpolated

The critical point was reported as 129.5°C, at 25 g(1)/100g sln (x_1 = 0.086)

COMPONENTS:

(1) 1-Butanol; $C_4H_{10}O$; [71-36-3]

(2) Water; H_2O; [7732-18-5]

ORIGINAL MEASUREMENTS:

Hill, A.E.; Malisoff, W.M.

J. Am. Chem. Soc. 1926, *48*, 918-27.

VARIABLES:

Temperature: 5-125°C

PREPARED BY:

A.F.M. Barton

EXPERIMENTAL VALUES:

Mutual Solubility of 1-butanol(1) and water(2)

$t/°C$	Alcohol-rich phase			Water-rich phase		
	g(1)/100g sln	x_1 (compiler)	density/g cm^{-3}	g(1)/100g sln	x_1 (compiler)	density/g cm^{-3}
5.0	80.38	0.499	0.8598	9.55	0.0250	0.9883
10.0	80.33	0.498	0.8567	8.91	0.0232	0.9877
15.0	80.14	0.495	0.8533	8.21	0.0213	0.9881
20.0	79.93	0.492	0.8484	7.81	0.0202	0.9873
25.0	79.73	0.489	0.8450	7.35	0.0189	0.9865
30.0	79.38	0.484	0.8424	7.08	0.0182	0.9851
35.0	78.94	0.477	0.8397	6.83	0.0175	0.9835
40.0	78.59	0.472	0.8345	6.60	0.0169	0.9841
50.0	77.58	0.457	0.8307	6.46	0.0165	0.9799
60.0	76.38	0.440	0.8253	6.52	0.0166	0.9766
70.0	74.79	0.419	0.8200	6.73	0.0172	0.9721
80.0	73.53	0.403	0.8159	6.89	0.0177	0.9675
92.0	69.24	0.354	-	-	-	-
97.9	-	-	-	8.74	0.0227	-
106.1	63.88	0.300	-	-	-	-

(continued next page)

AUXILIARY INFORMATION

METHOD/APPARATUS/PROCEDURE:

The volumetric method previously described (ref 1) was used, by measuring the volumes of the two phases which result when the two components are combined in two different but known ratios by weight. The determinations between 5°C and 40°C were by the volumetric method, with total volumes of 400 mL and a ratio of 3:1; those between 50°C and 80°C used 50 mL and a volume ratio of about 3:1. The remainder of the results are by the Alexejeff plethostatic method (ref 2).

SOURCE AND PURITY OF MATERIALS:

(1) Eastman Kodak Co.; refluxed with lime, dried with Na, repeatedly fractionated with 5-bulb still head; b.p. 117.70-117.80°C.

(2) not stated

ESTIMATED ERROR:

Not stated

REFERENCES:

1. Hill, A.E. *J. Am. Chem. Soc.* 1923, *45*, 1143.

2. Alexejeff, W. *Wied. Ann.* 1886, *28*, 305.

COMPONENTS:

(1) 1-Butanol; $C_4H_{10}O$; [71-36-3]

(2) Water; H_2O; [7732-18-5]

ORIGINAL MEASUREMENTS:

Hill, A.E.; Malisoff, W.M.

J. Am. Chem. Soc. 1926, *48*, 918-27.

EXPERIMENTAL VALUES (continued)

Mutual solubility of 1-butanol(1) and water(2) (continued)

$t/^{\circ}C$	Alcohol-rich phase			Water-rich phase		
	g(1)/100g sln	x_1 (compiler)	density/g cm^{-3}	g(1)/100g sln	x_1 (compiler)	density/g cm^{-3}
114.5	-	-	-	12.73	0.0343	-
116.9	-	-	-	13.46	0.0364	-
122.3	49.85	0.195	-	-	-	-
123.3	-	-	-	19.73	0.0564	-
124.33	42.02	0.150	-	-	-	-
124.83	-	-	-	27.26	0.0835	-
125.10	-	-	-	32.82	0.1062	-
125.15	32.82	0.106	-	30.44	0.0962	-

The two-liquid system was found to range from the quadruple point (-2.95°C) to 125.15°C, the consolute solution is 32.5 g(1)/100g sln (x_1 = 0.1048).

COMPONENTS:

(1) 1-Butanol; $C_4H_{10}O$; [71-36-3]

(2) Water; H_2O; [7732-18-5]

ORIGINAL MEASUREMENTS:

Jones, D.C.

J. Chem. Soc. 1929, 799-813.

VARIABLES:

Temperature: -15°C - 125°C

PREPARED BY:

S.H. Yalkowsky; S.C. Valvani; A.F.M. Barton

EXPERIMENTAL VALUES:

Mutual solubility of 1-butanol (1) and water (2)

t/°C	Alcohol-rich phase		Water-rich phase	
	g(1)/100g sln	x_1 (compiler)	g(1)/100g sln	x_1 (compiler)
-18.01[d]	-	-	12.72	0.0342
-15.0[l]	81.0	0.509	12.0	0.0321
- 5.0[k]	80.82	0.506	-	-
- 3.11	-	-	9.79	0.0257
0.0[l]	80.6	0.503	9.1	0.0238
13.0	80.46	0.500	-	-
19.3[b]	81.0	0.509	-	-
20.0[l]	80.2	0.496	6.4	0.0164
29.82[k]	79.51	0.486	-	-
40.0[a]	-	-	6.03	0.0154
40.0[l]	78.6	0.472	6.0	0.0153
58.50[d]	76.27	0.439	-	-
60.0[l]	76.2	0.437	6.0	0.0153
65.0[a]	-	-	6.03	0.0154
80.0[l]	72.7	0.393	6.4	0.0164

(continued next page)

AUXILIARY INFORMATION

METHOD/APPARATUS/PROCEDURE:

The synthetic method described by Jones and Betts (ref 2) was used. Well-steamed Pyrex glass gave consistent results identical with those in quartz glass tubes (soda glass did not give reproducible results). The thermometer was tested at the National Physical Laboratory.

SOURCE AND PURITY OF MATERIALS:

(1) prepared as in ref 1;
dried with calcium oxide, fractionated (the several fractions giving identical C.S.T.'s with a hydrochloric acid soln);
$d^{14.4}_{14.4}$ = 0.81417, n^{20}_{D} = 1.39711.

(2) conductivity water from Bousfield still freshly boiled.

ESTIMATED ERROR:

Temperature: given in footnote to experimental values; most accurate to 0.02°C.

REFERENCES:

1. Orton, K.J.P.; Jones, D.C. *J. Chem. Soc.* 1919, *115*, 1194.

2. Jones, D.C.; Betts, H.F. *J. Chem. Soc.* 1928, 1179.

COMPONENTS:

(1) 1-Butanol; $C_4H_{10}O$; [71-36-3]

(2) Water; H_2O; [7732-18-5]

ORIGINAL MEASUREMENTS:

Jones, D.C.

J. Chem. Soc. 1929, 799-813.

EXPERIMENTAL VALUES (continued)

Mutual solubility of 1-butanol (1) and water (2)

$t/^{\circ}C$	Alcohol-rich phase		Water-rich phase	
	g(1)/100g sln	x_1 (compiler)	g(1)/100g sln	x_1 (compiler)
81.0[b]	-	-	6.47	0.0166
100.0[l]	66.4	0.325	8.2	0.0213
106.05[d]	63.44	0.326	-	-
107.72[c]	-	-	9.79	0.0257
110.0[l]	61.5	0.279	10.2	0.0269
115.00[d]	57.8	0.250	-	-
117.40[d]	-	-	12.72	0.0342
120.0[l]	52.5	0.212	14.7	0.0402
120.30[e]	-	-	15.15	0.0416
122.45[d]	-	-	17.51	0.0490
122.60[d]	48.01	0.183	-	-
123.0[l]	46.8	0.176	19.0	0.0539
123.75[d]	44.03	0.160	-	-
124.05[d,j]	41.30	0.146	-	-
124.66[d,f]	38.05	0.130	-	-
124.72[d,e]	33.79	0.110	-	-
124.73[d,g]	30.39	0.0960	-	-
124.74[d,h]	32.49	0.105	-	-
124.74[d,f]	28.16	0.0870	-	-
124.74[d]	27.88	0.0858	-	-
124.75[d,i]	32.85	0.106	-	-
124.75[l]	32.4	0.1043	32.4	0.1043

[a] Accuracy ±2 K; slightly cloudy between these temperatures.

[b] Accuracy ±2 K.

[c] Rising temperature, 107.72°C; falling 107.60°C.

[d] Observation accurate to 0.02 K; rising and falling temperatures identical within these limits of accuracy.

[e] Repeated after 3 months.

[f] Slight critical opalescences; striations almost absent.

[g] Lower layer very large.

[h] Equal volumes; clearest critical phenomena.

[i] Upper layer greater than lower; quartz experimental tube.

[j] Determinations made at intervals up to 8 days; no change observed even after heating to 130°C.

[k] Accuracy 0.1 K; difference of 1 K between rising and falling temperatures.

[l] Interpolated.

COMPONENTS:

(1) 1-Butanol; $C_4H_{10}O$; [71-36-3]

(2) Water; H_2O; [7732-18-5]

ORIGINAL MEASUREMENTS:

Mueller, A.J.; Pugsley, L.I.; Ferguson, J.B.

J. Phys. Chem. 1931, *35*, 1314-27.

VARIABLES:

Temperature: 0-105°C

PREPARED BY:

S.H. Yalkowsky; S.C. Valvani; A.F.M. Barton

EXPERIMENTAL VALUES:

Mutual solubility of 1-butanol (1) and water (2)

t/°C	Alcohol-rich phase			Water-rich phase		
	g(1)100g sln	x_1 (compiler)	density/g cm^{-3}	g(1)100g sln	x_1 (compiler)	density/g cm^{-3}
0[a]	80.38	0.499		9.55	0.0251	
15[a]	80.2	0.496	0.853	8.30	0.0216	0.988
30[a]	79.4	0.484	0.842	7.08	0.0182	0.985
45[a]	78.2	0.466	0.833	6.50	0.0166	0.982
60[a]	76.4	0.440	0.825	6.52	0.0167	0.977
75	73.7	0.406		6.8	0.0174	
90	69.8	0.360		7.8	0.0202	
105	64.3	0.305		9.8	0.0257	

[a] Extrapolated or interpolated from ref 1 and described by the authors as being in good agreement with their results (not given).

AUXILIARY INFORMATION

METHOD/APPARATUS/PROCEDURE:

These data are from a study of the ternary 1 butanol/methanol/water system. Values for the binary system were determined by extrapolation of ternary binodal curves to zero methanol, giving results largely in agreement with those of ref 1.

At 0°C the amount of one component necessary to give rise to heterogeneity was found by weighing. Between 15°C and 60°C a sample was held in a calibrated Pyrex tube and the volume of the final homogeneous solution obtained from cathetometer readings. The densities of these solutions were calculated from the volumes and weights. The volumes of the two phases present in the sample selected for the detn of a tie line were found in a similar manner. The tie lines at 0°C were more directly determined, the two phases being separated and weighed. At 75°C and 90°C samples were sealed to prevent volatilization.

SOURCE AND PURITY OF MATERIALS:

(1) British Acetone Co. (1917); treated with lime, distilled; b.p. 117.6°C, d_4^{20} 0.8095-0.8097

(2) distilled

ESTIMATED ERROR:

Temperature: ±0.1°C (0-60°C); ±0.2°C (100°C); ±0.3°C (120°C).
Solubility ±0.3 wt% below 60°C; ±0.5 wt% above 60°C.

REFERENCES:

1. Hill, A.E.; Malisoff, W.M. *J. Am. Chem. Soc.* 1926, *48*, 918.

COMPONENTS:

(1) 1-Butanol; $C_4H_{10}O$; [71-36-3]

(2) Water; H_2O; [7732-18-5]

ORIGINAL MEASUREMENTS:

Stockhardt, J.S.; Hull, C.M.

Ind. Eng. Chem. 1931, *23*, 1438-1440

VARIABLES:

One temperature: 25°C

PREPARED BY:

S.H. Yalkowsky; S.C. Valvani; A.F.M. Barton

EXPERIMENTAL VALUES:

The proportion of 1-butanol in the water-rich phase at 25°C was reported to be 7.45 g(1)/100 cm^{-3} sln corresponding to 1.01 mol (1)/L sln.

AUXILIARY INFORMATION

METHOD/APPARATUS/PROCEDURE:

The method of ref 1 was used, except that the determinations were wholly gravimetric.

SOURCE AND PURITY OF MATERIALS:

(1) fractionated in a carborundum-packed column at 4:1 reflux;
b.p. 117.3-117.5°C, d_4^{20} 0.8097

ESTIMATED ERROR:

Not stated

REFERENCES:

1. Hill, A.E.; Malisoff, W.M. *J. Am. Chem. Soc.* 1926, *48*, 918.

COMPONENTS:

(1) 1-Butanol; $C_4H_{10}O$; [71-36-3]

(2) Water; H_2O; [7732-18-5]

ORIGINAL MEASUREMENTS:

Butler, J.A.V.; Thomson, D.W.; Maclennan W.H.

J. Chem. Soc. 1933, 674-86.

VARIABLES:

Temperature: 18-31°C

PREPARED BY:

S.H. Yalkowsky; S.C. Valvani; A.F.M. Barton

EXPERIMENTAL VALUES:

Mutual solubility of 1-butanol(1) and water(2)

$t/°C$	Alcohol-rich phase		Water-rich phase	
	(1)/100g sln	x_1(compiler)	g(1)/100g sln	x_1(compiler)
18.45	80.01	0.493	-	-
22.60	-	-	7.497	0.0193
23.40	79.73	0.489	-	-
23.70	-	-	7.407	0.0191
24.85	-	-	7.318	0.0188
25.00	79.64	0.488	7.31	0.0188
26.40	-	-	7.202	0.0185
27.45	79.50	0.486	-	-
28.06	-	-	7.090	0.0182
29.18	-	-	7.016	0.0180
30.83	79.28	0.482	-	-

AUXILIARY INFORMATION

METHOD/APPARATUS/PROCEDURE:

The method of ref 1 was used. Solutions of suitable composition by weight were placed in a soda-glass flask fitted with a thermometer. The flask was constantly shaken in a slowly-heated water bath, and the temperature of onset of cloudiness was noted. Disappearance of cloudiness was also studied. The determinations were repeated in Pyrex-glass flasks, but no difference was observed.

SOURCE AND PURITY OF MATERIALS:

(1) dried over CaO for a week, refluxed 8 h, fractionated; the whole distilled 118.19-118.35°C at 775.3 mm Hg; middle fraction b.p. 117-71°C (corr.) used d_4^{25} 0.8055

(2) not stated.

ESTIMATED ERROR:

Temperature: repeated observations within 0.050°C

REFERENCES:

1. Sidgwick, N.V.; Pickford, P.; Wilsdon, B.H. *J. Chem. Soc.* 1911, *99*, 1122.

COMPONENTS:

(1) 1-Butanol; $C_4H_{10}O$; [71-36-3]

(2) Water; H_2O; [7732-18-5]

ORIGINAL MEASUREMENTS:

Berkengeim, T.I.

Zavod. Lab. 1941, *10*, 952-4.

VARIABLES:

Temperature: (-18)-40°C

PREPARED BY:

A. Maczynski

EXPERIMENTAL VALUES:

Solubility of water (2) in 1-butanol (1)

$t/^{\circ}C$	g(2)/100g sln	x_2 (compiler)
-18	2.80	0.106
- 7	12.5	0.370
20	23.1	0.553
40	26.4	0.596

AUXILIARY INFORMATION

METHOD/APPARATUS/PROCEDURE:

The analytical method was used.

The solubility of (2) in (1) was determined by the Karl Fischer reagent method.

SOURCE AND PURITY OF MATERIALS:

(1) CP reagent, source not specified; used as received; b.p. range 116.5-117°C.

(2) not specified.

ESTIMATED ERROR:

Not specified.

REFERENCES:

COMPONENTS:

(1) 1-Butanol; $C_4H_{10}O$; [71-36-3]

(2) Water; H_2O; [7732-18-5]

ORIGINAL MEASUREMENTS:

Reber, L.A.; McNabb, W.M.; Lucasse, W.A.
J. Phys. Chem. 1942, *46*, 500-15.

VARIABLES:

Temperature: 91-125°C
Added sodium salts

PREPARED BY:

A.F.M. Barton

EXPERIMENTAL VALUES:

Mutual solubility of 1-butanol (1) and water (2)

	Alcohol-rich phase		Water-rich phase	
$t/^{\circ}C$	g(1)/100g sln	x_1 (compiler)	g(1)/100g sln	x_1 (compiler)
91.15	69.01	0.3513	-	-
98.35	-	-	8.89	0.0232
99.45	66.07	0.3213	-	-
103.50	64.22	0.3038	-	-
104.35	-	-	9.86	0.0259
110.90	60.21	0.2690	-	-
112.95	-	-	12.30	0.0330
115.15	57.03	0.2440	-	-
118.25	-	-	15.03	0.0412
119.40	52.33	0.2107	-	-
120.75	-	-	16.99	0.0474
121.60	48.86	0.1885	-	-
123.20	44.39	0.1625	21.12	0.0611
123.60	-	-	22.23	0.0650

(continued next page)

AUXILIARY INFORMATION

METHOD/APPARATUS/PROCEDURE:

The "synthetic" method was used. 5 mL ampoules were boiled with hydrochloric acid, washed, steamed and dried. (1) was added by hypodermic syringe, (2) by microburet, weighed, and sealed. The ampoules were rotated in an oil bath, the temperature by NBS-tested thermometer being read to ±0.05°C. In mixtures with 50-80% water, temperatures of appearance and disappearance of a second phase were essentially the same. Beyond these limits the clouding temperature was 0.1-0.5°C below the clearing temperature; the former were more precise and were reported. Results were also reported in the presence of sodium nitrate, halides, sulfate and thiocyanate.

SOURCE AND PURITY OF MATERIALS:

(1) Eastman Kodak Co. "best grade" purified according to ref 1 (refluxed with lime 4 h, Mg and iodine for 4 h, distilled in 15 bulb Snyder column);
b.p. 117.55 ± 0.05°C (corrected).
n_D^{25} 1.3974, d_4^{25} 0.8058,
negative iodoform test.

(2) distilled water, redistilled from alkaline potassium permanganate in Pyrex glass.

ESTIMATED ERROR:

Composition: within 0.01 wt %

Temperature: ±0.05°C

REFERENCES:

1. Lund, H.; Bjerrum, J. *Ber.* 1931, *64B*, 210.

COMPONENTS:	EVALUATOR:
(1) 1-Butanol; $C_4H_{10}O$; [71-36-3] (2) Water; H_2O; [7732-18-5]	Reber, L.A.; McNabb, W.M.; Lucasse, W.A. *J. Phys. Chem.* 1942, *46*, 500-15.

EXPERIMENTAL VALUES (continued)

Mutual solubility of 1-butanol (1) and water (2)

	Alcohol-rich phase		Water-rich phase	
$t/^{\circ}C$	g(1)/100g sln	x_1 (compiler)	g(1)/100g sln	x_1 (compiler)
124.10	39.62	0.1376	-	-
124.20	-	-	24.34	0.0725
124.30	36.83	0.1242	-	-
124.35	-	-	31.81	0.1019
124.40	-	-	30.52	0.0965
124.40	-	-	32.96	0.1068

The critical solution temperature was reported as 124.40°C.

The addition of all salts led to an increase in temperature of complete miscibility, and this effect was more marked the lower the proportion of water and the greater the salt concentration.

COMPONENTS:

(1) 1-Butanol; $C_4H_{10}O$; [71-36-3]

(2) Water; H_2O; [7732-18-5]

ORIGINAL MEASUREMENTS:

Othmer, D.F.; Bergen, W.S.; Shlechter, N.; Bruins, P.F.
Ind. Eng. Chem. 1945, *37*, 890-4.

VARIABLES:

Temperature: 26 and 50°C

PREPARED BY:

A. Maczynski

EXPERIMENTAL VALUES:

Mutual solubility of 1-butanol (1) and water (2)

$t/^\circ C$	g(1)/100g sln		x_1 (compiler)	
	(2)-rich phase	(1)-rich phase	(2)-rich phase	(1)-rich phase
26	6.5	80.2	0.017	0.496
50	6.1	79	0.015	0.48

AUXILIARY INFORMATION

METHOD/APPARATUS/PROCEDURE:

The indirect or synthetic method (ref 1,2) was used.

No details were reported in the paper.

SOURCE AND PURITY OF MATERIALS:

(1) commercial material;
carefully fractionated;
b.p. range 1°C, purity not specified.

(2) not specified.

ESTIMATED ERROR:

Not specified.

REFERENCES:

1. Othmer, D.F.; Tobias, P.E. *Ind. Eng. Chem.* 1942, *34*, 690.

2. Othmer, D.F.; White, R.E.; Trueger, E. *Ind. Eng. Chem.* 1942, *33*, 1240.

COMPONENTS:

(1) 1-Butanol; $C_4H_{10}O$; [71-36-3]

(2) Water; H_2O; [7732-18-5]

ORIGINAL MEASUREMENTS:

Booth, H.S.; Everson, H.E.

Ind. Eng. Chem. 1948, *40*, 1491-3.

VARIABLES:

One temperature: 25°C
Sodium xylene sulfonate

PREPARED BY:

S.H. Yalkowsky; S.C. Valvani; A.F.M. Barton

EXPERIMENTAL VALUES:

It was reported that the solubility of 1-butanol in water was 9.1 mL(1)/100 mL(2) at 25°C.

The corresponding figure in 40% sodium xylene sulfonate solution as solvent was > 400 mL(1)/100 mL solvent.

AUXILIARY INFORMATION

METHOD/APPARATUS/PROCEDURE:

A known volume of solvent (usually 50 mL) in a tightly stoppered calibrated Babcock tube was thermostatted. Successive measured quantities of solute were added and equilibrated until a slight excess of solute remained. The solution was centrifuged, returned to the thermostat bath for 10 min, and the volume of excess solute measured directly. This was a modification of the method described in ref 1.

SOURCE AND PURITY OF MATERIALS:

(1) "CP or highest grade commercial"

(2) distilled.

ESTIMATED ERROR:

Solubility within 0.1 mL/100 mL.

REFERENCES:

1. Hanslick, R.S. Dissertation, Columbia University, 1935.

COMPONENTS:

(1) 1-Butanol, $C_4H_{10}O$; [71-36-3]

(2) Water; H_2O; [7732-18-5]

ORIGINAL MEASUREMENTS:

Hansen, R.S.; Fu, Y.; Bartell, F.E.

J. Phys. Chem. 1949, *53*, 769-85.

VARIABLES:

One temperature: 25°C

PREPARED BY:

S.H. Yalkowsky; S.C. Valvani; A.F.M. Barton

EXPERIMENTAL VALUES:

At equilibrium at 25.0°C the proportion of 1-butanol in the water-rich phase was reported to be 7.41 g(1)/100g sln, a concentration of 0.985 mol (1)/L sln.

The corresponding mole fraction solubility calculated by the compiler is x_1 = 0.0191.

AUXILIARY INFORMATION

METHOD/APPARATUS/PROCEDURE:

An excess of the alcohol was added to redistilled water in a mercury-sealed flask which was shaken mechanically for 48 h in an air chamber thermostatted to 25.0 ± 0.1°C. The flask was then allowed to stand for 3 h in the airbath, after which a portion of the water-rich phase was removed by means of a hypodermic syringe and was compared interferometrically with the most concentrated alcohol solution which could be prepared conveniently. The solubility determination was associated with a study of multimolecular absorption from binary liquid solutions.

SOURCE AND PURITY OF MATERIALS:

(1) refluxed 4h over magnesium and iodine, distilled with 60cm glass packed reflux column;

b.p. 117°C/745 mm Hg

(2) distilled laboratory water, redistilled from alkaline permanganate solution.

ESTIMATED ERROR:

Temperature: 0.1°C

Solubility: deviation from mean of three determinations ±0.03 wt%.

REFERENCES:

COMPONENTS:

(1) 1-Butanol; C_4H_{10}; [71-36-3]

(2) Water; H_2O; [7732-18-5]

ORIGINAL MEASUREMENTS:

Donahue, D.J.; Bartell, F.E.

J. Phys. Chem. 1952, *56*, 480-4.

VARIABLES:

One temperature: 25°C

PREPARED BY:

A.F.M. Barton

EXPERIMENTAL VALUES:

	Density	Mutual solubility of 1-butanol (1) and water (2)	
	g mL^{-1}	x_1	g(1)/100g sln (compiler)
Alcohol-rich phase	0.8432	0.500	80.5
Water-rich phase	0.9860	0.0188[a]	7.3

[a]From ref 1 and 2

AUXILIARY INFORMATION

METHOD/APPARATUS/PROCEDURE:

Mixtures were placed in glass stoppered flasks and were shaken intermittently for at least 3 days in a water bath. The organic phase was analyzed for water content by the Karl Fischer method and the aqueous phase was analyzed interferometrically. The solubility measurements formed part of a study of water-organic liquid interfacial tensions.

SOURCE AND PURITY OF MATERIALS:

(1) "best reagent grade"; fractional distillation

(2) "purified"

ESTIMATED ERROR:

Temperature: ± 0.1°C

REFERENCES:

1. Butler, J.A.V.; Thomson, D.W.; Maclennan, W.H. *J. Chem. Soc.* 1933, 674.

2. Hansen, R.S.; Fu, Y.; Bartell, F.E. *J. Phys. Chem.* 1949, *53*, 769.

COMPONENTS:

(1) 1-Butanol; C_4H_{10}; [71-36-3]

(2) Water; H_2O; [7732-18-5]

ORIGINAL MEASUREMENTS:

Erichsen, L. von

Naturwissenschaften 1952, *39*, 41-2.

VARIABLES:

Temperature: 0-50°C

PREPARED BY:

A. Maczynski; Z. Maczynska

EXPERIMENTAL VALUES:

Solubility of 1-butanol (1) in water (2)

$t/°C$	x_1	g(1)/100g sln (compiler)
0	0.0271	10.28
10	0.0230	8.83
20	0.0197	7.63
30	0.0179	7.41
40	0.0167	6.53
50	0.0163	6.38

AUXILIARY INFORMATION

METHOD/APPARATUS/PROCEDURE:

The synthetic method was used.

No details were reported in the paper.

SOURCE AND PURITY OF MATERIALS:

(1) not specified.

(2) not specified.

ESTIMATED ERROR:

Not specified.

REFERENCES:

COMPONENTS:	ORIGINAL MEASUREMENTS:
(1) 1-Butanol; $C_4H_{10}O$; [71-36-3] (2) Water; H_2O; [7732-18-5]	Erichsen, L. von *Brennst. Chem.* 1952, *33*, 166-72.
VARIABLES:	**PREPARED BY:**
Temperature: 0-125°C	S.H. Yalkowsky ; Z. Maczynska

EXPERIMENTAL VALUES:

Mutual solubility of 1-butanol (1) and water (2)

t/°C	(2)-rich phase		(1)-rich phase	
	g(1)/100g sln	x_1	g(1)/100g sln	x_1
0	10.32	0.0271	81.05	0.5112
10	8.68	0.0230	80.50	0.5061
20	7.56	0.0197	80.00	0.4980
30	7.06	0.0179	79.37	0.4861
40	6.72	0.0167	78.60	0.4717
50	6.55	0.0163	77.59	0.4562
60	6.52	0.0162	76.29	0.4381
70	6.67	0.0166	74.57	0.4152
80	6.90	0.0175	72.40	0.3890
90	7.50	0.0195	69.90	0.3600
100	8.82	0.0232	66.70	0.3290
110	11.05	0.0295	61.85	0.2875
120	15.45	0.0440	53.35	0.2180
125	23.50	0.0710	43.10	0.1490

The UCST is 125.3°C

AUXILIARY INFORMATION

METHOD/APPARATUS/PROCEDURE:

The synthetic method was used.

The measurements were carried out in 2 ml glass ampoules which were placed in an aluminium block equipped with two glass windows. Cloud points were measured with a thermocouple wound up around the ampoule; each measurement being repeated twice.

SOURCE AND PURITY OF MATERIALS:

(1) Merck, or Ciba, or industrial product; distilled and chemically free from isomers; b.p. 116.6-116.7°C (755 mm Hg), n_D^{20} 1.3996.

(2) not specified.

ESTIMATED ERROR:

Not specified.

REFERENCES:

COMPONENTS:

(1) 1-Butanol; $C_4H_{10}O$; [71-36-3]

(2) Water; H_2O; [7732-18-5]

ORIGINAL MEASUREMENTS:

McCants, J.F.; Jones, J.H.; Hopson, W.H.

Ind. Eng. Chem. 1953, *45*, 454-6.

VARIABLES:

One temperature: 37.7°C

PREPARED BY:

A. Maczynski

EXPERIMENTAL VALUES:

The solubility of 1-butanol in water at 37.7°C was reported to be 6.7 g (1)/100g sln.

The corresponding mole fraction, x_1, calculated by the compiler is 0.017.

The solubility of water in 1-butanol at 37.7°C was reported to be 21.2 g(2)/100g sln.

The corresponding mole fraction, x_2, calculated by the compiler is 0.525.

AUXILIARY INFORMATION

METHOD/APPARATUS/PROCEDURE:

The titration method described by Washburn et al. (ref 1) was used.

SOURCE AND PURITY OF MATERIALS:

(1) Carbide and Carbon Co; technical product without further purification
n_D^{20} 1.3988;
purity not specified.

(2) distilled.

ESTIMATED ERROR:

Not specified.

REFERENCES:

1. Washburn, E.R.; Graham, C.L.; Arnold, G.R.; Trausue, L.F
J. Am. Chem. Soc. 1940, *62*, 1454.

COMPONENTS:	ORIGINAL MEASUREMENTS:
(1) 1-Butanol; $C_4H_{10}O$; [71-36-3] (2) Water; H_2O; [7732-18-5]	Jones, J.H.; McCants, J.F. *Ind. Eng. Chem.* 1954, *46*, 1956-8.
VARIABLES:	**PREPARED BY:**
One temperature: 100°F (37.8°C)	S.H. Yalkowsky; S.C. Valvani; A.F.M. Barton

EXPERIMENTAL VALUES:

At equilibrium at 100°F (37.8°C) the proportion of 1-butanol in the alcohol-rich phase was reported to be 79.7 g(1)/100g sln and the proportion of 1-butanol in the water-rich phase was reported to be 7.2 g(1)/100g sln.

The corresponding mole fraction solubility values calculated by the compiler are x_1 = 0.489 and x_1 = 0.0185 respectively. The refractive indexes, $n_D^{38.3}$, were reported as 1.3871 and 1.3393, respectively.

AUXILIARY INFORMATION

METHOD/APPARATUS/PROCEDURE:

The cloud point method by titrating, patterned after that described in ref 1, was used. The system under study was the ternary 1-butanol/1-hexanone/water.

SOURCE AND PURITY OF MATERIALS:

(1) U.S.I.C. "approved" grade; n_D^{20} 1.3991, d_4^{20} 0.810

(2) distilled; n_D^{20} 1.3330

ESTIMATED ERROR:

Temperature: controlled to within 0.1°C.

REFERENCES:

1. Washburn, E.R.; Hnizda, V.; Vold, R. *J. Am. Chem. Soc.* 1931, *53*, 3237.

COMPONENTS:

(1) 1-Butanol; $C_4H_{10}O$; [71-36-3]

(2) Water; H_2O; [7732-18-5]

ORIGINAL MEASUREMENTS:

Skrzec, A.E.; Murphy, N.F.

Ind. Eng. Chem. 1954, *46*, 2245-7.

VARIABLES:

One temperature: 26.7°C

PREPARED BY:

A. Maczynski

EXPERIMENTAL VALUES:

The solubility of 1-butanol in water at 26.7°C was reported to be 7.30 g(1)/100g sln.

The corresponding mole fraction, x_1, calculated by the compiler is 0.0188.

The solubility of water in 1-butanol at 26.7°C was reported to be 20.50 g(2)/100g sln.

The corresponding mole fraction, x_2, calculated by the compiler is 0.515.

AUXILIARY INFORMATION

METHOD/APPARATUS/PROCEDURE:

Probably the titration method was used. (This method was described for the determination of mutual solubilities in ternary systems but nothing is reported on the binary system determination).

SOURCE AND PURITY OF MATERIALS:

(1) Commercial Solvents Corp., technical grade; used as received; 99.5% pure.

(2) not specified.

ESTIMATED ERROR:

Temperature: ± 0.5°C.

REFERENCES:

COMPONENTS:

(1) 1-Butanol; $C_4H_{10}O$; [71-36-3]

(2) Water; H_2O [7732-18-5]

ORIGINAL MEASUREMENTS:

Hayashi, M.; Sasaki, T.

Bull. Chem. Soc. Jpn. 1956, *29*, 857-9.

VARIABLES:

One temperature: 30.0°C

Tween 80 concentration

PREPARED BY:

S.H. Yalkowsky; S.C. Valvani; A.F.M. Barton

EXPERIMENTAL VALUES:

The proportion of 1-butanol in the aqueous phase at 30.0°C was reported to be 7.01 g(1)/100g sln. The corresponding mole fraction calculated by the compiler is x_1 = 0.0180.

The solubility of the alcohol in dilute solutions of Tween 80, obtained by extrapolating to zero turbidity the linear relation between turbidity and solute concentration in the surfactant solution, was less than that in pure water but increased with increasing concentration of Tween 80.

AUXILIARY INFORMATION

METHOD/APPARATUS/PROCEDURE:

The mixture was well shaken at a temperature below 30.0°C and then stood in the thermostat for 24 h. After the excess solute particles cleared, a transparent saturated solution was obtained which was taken from the bottom of the vessel by a siphon. A known amount of this solution (about 20 g) was titrated with Tween 80 solution. Concentration was determined by comparison of turbidity to standard samples.

SOURCE AND PURITY OF MATERIALS:

(1) boiled with conc. sodium hydroxide sln, washed with water, dried with anhydrous potassium carbonate, distilled over calcium oxide, redistilled over calcium metal;
b.p. 118°C

(2) not stated.

ESTIMATED ERROR:

Solubility: ± 0.4% "possible error".

REFERENCES:

COMPONENTS:	ORIGINAL MEASUREMENTS:
(1) 1-Butanol; $C_4H_{10}O$; [71-36-3] (2) Water; H_2O; [7732-18-5]	Kakovskii, I.A. *Proc. Intern. Congr. Surface Activity, 2nd, London* 1957, *4*, 225-37.
VARIABLES:	PREPARED BY:
One temperature: 25°C	S.H. Yalkowsky and S.C. Valvani

EXPERIMENTAL VALUES:

The solubility of 1-butanol in water at 25°C was reported to be 1.07 mol L^{-1} (79.3 g(1)/L sln: compiler).

AUXILIARY INFORMATION

METHOD/APPARATUS/PROCEDURE:	SOURCE AND PURITY OF MATERIALS:
No experimental details given.	Not specified.
	ESTIMATED ERROR:
	Not specified.
	REFERENCES:

COMPONENTS:	ORIGINAL MEASUREMENTS:
(1) 1-Butanol; $C_4H_{10}O$; [71-36-3] (2) Water; H_2O; [7732-18-5]	Rao, R.M.; Rao, V.C. *J. Appl. Chem.* 1957, *7*, 659-66.
VARIABLES:	**PREPARED BY:**
One temperature: 27°C	A. Maczynski

EXPERIMENTAL VALUES:

The solubility of 1-butanol in water at 27°C was reported to be 7.0 g(1)/100g sln.

The corresponding mole fraction, x_1, calculated by the compiler is 0.018.

The solubility of water in 1-butanol at 27°C was reported to be 19.9 g(2)/100g sln.

The corresponding mole fraction, x_2, calculated by the compiler is 0.505.

AUXILIARY INFORMATION

METHOD/APPARATUS/PROCEDURE:

The method of appearance and disappearance of turbidity described in ref 1 was used.

No details were reported in the paper.

SOURCE AND PURITY OF MATERIALS:

(1) Merck reagent grade; distilled; b.p. 117.7°C, d^{30} 0.8018, n^{30} 1.3940.

(2) not specified.

ESTIMATED ERROR:

Not specified.

REFERENCES:

1. Othmer, D.F.; White, R.E.; Trueger, E. *Ind. Eng. Chem.* 1941, *33*, 1240.

COMPONENTS:	ORIGINAL MEASUREMENTS:
(1) 1-Butanol; $C_4H_{10}O$; [71-36-3] (2) Water; H_2O; [7732-18-5]	Kinoshita, K.; Ishikawa, H.; Shinoda, K. *Bull. Chem. Soc. Japan* 1958, *31*, 1081-4.
VARIABLES:	PREPARED BY:
One temperature: 25°C	S.H. Yalkowsky; S.C. Valvani; A.F.M. Barton.

EXPERIMENTAL VALUES:

At equilibrium at 25.0°C the concentration of 1-butanol in the water-rich phase was reported to be 0.97 mol(1) L^{-1}. The weight percentage solubility was reported as 7.4 g(1)/100g sln, and the corresponding mole fraction solubility calculated by the compiler is x_1 = 0.0191.

AUXILIARY INFORMATION

METHOD/APPARATUS/PROCEDURE:

The surface tension in aqueous solutions of alcohols monotonically decreases up to their saturation concentration and remains constant in the heterogeneous region (ref 1-4). Surface tension was measured by the drop weight method, using a tip 6 mm in diameter. The measurements were carried out in a water thermostat at 25 ± 0.05°C. From the (surface tension)-(logarithm of concentration) curves the saturation points were determined as the intersections of the curves with the horizontal straight lines passing through the lowest experimental points.

SOURCE AND PURITY OF MATERIALS:

(1) purified by vacuum distillation through 50-100cm column;
b.p. 11°C (pressure uncertain)

(2) not stated.

ESTIMATED ERROR:

Temperature: ±0.05°C.

Solubility: within 4%.

1. Motylewski, S. *Z. Anorg. Chem.* 1904, *38*, 410.
2. Taubamann, A. *Z. Physik. Chem.* 1932, *A161*, 141.
3. Zimmermann, H.K., Jr. *Chem. Rev.* 1952, *51*, 25.
4. Shinoda, K.; Yamanaka, T.; Kinoshita, K. *J. Phys. Chem.* 1959, *63*, 648.

COMPONENTS:

(1) 1-Butanol; $C_4H_{10}O$; [71-36-3]

(2) Water; H_2O; [7732-18-5]

ORIGINAL MEASUREMENTS:

Venkataratnam, A.; Rao, R.I.

J. Sci. Ind. Res. 1958, *17B*, 108-10.

VARIABLES:

One temperature: 30°C

PREPARED BY:

A. Maczynski

EXPERIMENTAL VALUES:

The solubility of 1-butanol in water at 30°C was reported to be 6.8 g(1)/100g sln.

The corresponding mole fraction, x_1, calculated by the compiler is 0.017.

The solubility of water in 1-butanol at 30°C was reported to be 20.0 g(2)/100g sln.

The corresponding mole fraction, x_2, calculated by the compiler is 0.507.

AUXILIARY INFORMATION

METHOD/APPARATUS/PROCEDURE:

The method of appearance and disappearance of turbidity described in ref 1 was used.

No details were reported in the paper.

SOURCE AND PURITY OF MATERIALS:

(1) Merck and Co.;
used as received;
b.p. 117.2°C, n^{30} 1.3930,
d^{30} 0.7996 g/ml.

(2) distilled;
free from carbon dioxide.

ESTIMATED ERROR:

Not specified.

REFERENCES:

1. Othmer, D.F.; White, R.E.; Trueger, E.
Ind. Eng. Chem. 1941, *33*, 1240.

COMPONENTS:

(1) 1-Butanol; $C_4H_{10}O$; [71-36-3]

(2) Water; H_2O; [7732-18-5]

ORIGINAL MEASUREMENTS:

Petriris, V.E.; Geankopolis, C.J.

J. Chem. Eng. Data 1959, *4*, 197-8.

VARIABLES:

One temperature: 25°C

PREPARED BY:

A. Maczynski

EXPERIMENTAL VALUES:

The solubility of 1-butanol in water at 25°C was reported to be 7.0 g(1)/100g sln.

The corresponding mole fraction, x_1, calculated by the compiler is 0.018.

The solubility of water in 1-butanol at 25°C was reported to be 20.7 g(2)/100g sln.

The corresponding mole fraction, x_2, calculated by the compiler is 0.518.

AUXILIARY INFORMATION

METHOD/APPARATUS/PROCEDURE:

The titration method was used.

No details were reported in the paper.

SOURCE AND PURITY OF MATERIALS:

(1) Baker, analytical reagent grade; used as received; purity not specified.

(2) not specified.

ESTIMATED ERROR:

Not specified.

REFERENCES:

COMPONENTS:	ORIGINAL MEASUREMENTS:
(1) 1-Butanol; $C_4H_{10}O$; [71-36-3] (2) Water; H_2O; [7732-18-5]	Ababi, V.; Popa, A. *An. Stiint. Univ."Al. I. Cuza"Iasi.* 1960, *6*, 929-42.
VARIABLES:	PREPARED BY:
One temperature: 25°C	A.Maczynski

EXPERIMENTAL VALUES:

The solubility of 1-butanol in water at 25°C was reported to be 7.2 g(1)/100g sln.

The corresponding mole fraction, x_1, calculated by the compiler is 0.018.

The solubility of water in 1-butanol at 25°C was reported to be 19.9 g(2)/100 sln.

The corresponding mole fraction, x_2, calculated by the compiler is 0.506.

AUXILIARY INFORMATION

METHOD/APPARATUS/PROCEDURE:	SOURCE AND PURITY OF MATERIALS:
The turbidimetric method was used. Ternary solubility methods were described in the paper but nothing was reported for binary solubilities.	(1) Merck analytical reagent; used as received. (2) not specified. ESTIMATED ERROR: Not specified. REFERENCES:

COMPONENTS:

(1) 1-Butanol; $C_4H_{10}O$; [77-36-3]

(2) Water; H_2O; [7732-18-5]

ORIGINAL MEASUREMENTS:

Smirnova, N.A.; Morachevskii, A.G.

Zh. Prikl. Khim. (Leningrad) 1963, *36*, 2391-7.

VARIABLES:

One temperature: 20°C

PREPARED BY:

A. Maczynski

EXPERIMENTAL VALUES:

The solubility of 1-butanol in water at 20°C was reported to be 8.1 g(1)/100g sln.

The corresponding mole fraction, x_1, calculated by the compiler is 0.021.

The solubility of water in 1-butanol at 20°C was reported to be 20.1 g(2)/100g sln.

The corresponding mole fraction, x_2, calculated by the compiler is 0.509.

AUXILIARY INFORMATION

METHOD/APPARATUS/PROCEDURE:

The density method was used.

The density was measured to an accuracy of 0.0001g cm^{-3}.

No additional details were reported in the paper.

SOURCE AND PURITY OF MATERIALS:

(1) CP reagent; treated with 5% solution of $KMnO_4$, dried over potassium and distilled; n_D^{20} 1.3993, d_4^{20} 0.8097.

(2) not specified.

ESTIMATED ERROR:

Not specified.

REFERENCES:

COMPONENTS:	ORIGINAL MEASUREMENTS:
(1) 1-Butanol; $C_4H_{10}O$; [71-36-3] (2) Water; H_2O; [7732-18-5]	Ratouis, M.; Dodé, M. *Bull. Soc. Chim. Fr.* 1965, 3318-22.
VARIABLES:	PREPARED BY:
Temperature: 25-30°C Ringer solution also studied	S.C. Valvani; S.H. Yalkowsky; A.F.M.Barton

EXPERIMENTAL VALUES:

Proportion of 1-butanol(1) in water-rich phase

$t/^{\circ}C$	g(1)/100g sln	x_1 (compiler)
25	7.34	0.0189
30	6.99	0.0180

Proportion of 1-butanol(1) in water-rich phase (Ringer solution)

$t/^{\circ}C$	g(1)/100g sln
25	6.87
30	6.70

AUXILIARY INFORMATION

METHOD/APPARATUS/PROCEDURE:

In a round bottomed flask, 50 mL of water and a sufficient quantity of alcohol were introduced until two separate layers were formed. The flask assembly was equilibrated by agitation for at least 3 h in a constant temp bath. Equilibrium solubility was attained by first supersaturating at a slightly lower temperature (solubility of alcohols in water is inversely proportional to temperature) and then equilibrating at the desired temperature. The aqueous layer was separated after an overnight storage in a bath. The alcohol content was determined by reacting the aqueous solution with potassium dichromate and titrating the excess dichromate with ferrous sulfate solution in the presence of phosphoric acid and diphenylamine barium sulfonate as an indicator.

SOURCE AND PURITY OF MATERIALS:

(1) Prolabo;
redistilled with 10:1 reflux;
b.p. 117.5°C/767 mm Hg
n_D^{25} 1.39745

(2) twice distilled from silica apparatus or ion exchanged with Sagei A20

ESTIMATED ERROR:

Solubility: relative error of 2 determinations less than 1%
Temperature: ± 0.05°C

REFERENCES:

COMPONENTS:

(1) 1-Butanol; $C_4H_{10}O$; [71-36-3]

(2) Water; H_2O; [7732-18-5]

ORIGINAL MEASUREMENTS:

Meeussen, E.; Huyskens, P.

J. Chim. Phys. 1966, *63*, 845-54.

VARIABLES:

One temperature: 25°C

PREPARED BY:

A. Maczynski

EXPERIMENTAL VALUES:

The mole fraction solubility of water in 1-butanol at 25°C was reported to be x_2 = 0.492.

The corresponding mass percentage calculated by the compiler is 19.05 g(2)/100g sln.

AUXILIARY INFORMATION

METHOD/APPARATUS/PROCEDURE:

The interferometric method of Brown and Bury (ref 1) modified by Guillerm (ref 2) was used.
A Rayleigh M75 interference refractometer with a M160 attachment was employed.
Pycnometric densities of the solutions were used to evaluate solubility concentrations.

SOURCE AND PURITY OF MATERIALS:

(1) commercial product for chromatography; purified; purity not specified.

(2) not specified.

ESTIMATED ERROR:

Temperature: ± 0.05°C
Solubility : ± 0.002 mol(2)dm^{-3} sln
(type of error not specified).

REFERENCES:

1. Brown, F.S.; Bury, C.R.; *J. Chem. Soc.* 1923, *123*, 2430.

2. Guillerm, S., *Doctoral thesis*, 1963.

COMPONENTS:

(1) 1-Butanol; $C_4H_{10}O$; [71-36-3]

(2) Water; H_2O; [7732-18-5]

ORIGINAL MEASUREMENTS:

Lesteva, T.M.; Ogorodnikov, S.K.; Tyvina, T.N.

Zh. Prikl. Khim. (Leningrad) 1968, *41*, 1159-63.

VARIABLES:

Temperature: 20 and 75°C

PREPARED BY:

A. Maczynski

EXPERIMENTAL VALUES:

Mutual solubility of 1-butanol (1) and water (2)

$t/^{\circ}C$	g(1)/100g sln		x_1	
	(2)-rich phase	(1)-rich phase	(2)-rich phase	(1)-rich phase
20.0	4.9	80.3	0.012	0.498
75.0	5.2	75.8	0.013	0.432

AUXILIARY INFORMATION

METHOD/APPARATUS/PROCEDURE:

The titration method was used.

No details were reported in the paper.

SOURCE AND PURITY OF MATERIALS:

(1) source not specified;
CP reagent;
purity not specified.

(2) not specified.

ESTIMATED ERROR:

Not specified.

REFERENCES:

COMPONENTS:

(1) 1-Butanol; $C_4H_{10}O$; [71-36-3]

(2) Water; H_2O; [7732-18-5]

ORIGINAL MEASUREMENTS:

*Hanssens, I.

Associatie van normale alcoholen en hun affiniteit voor water en organische solventen Doctoraatsproefschrift, Leuven, 1969.

Huyskens, P. Mullens, J.; Gomez, A.; Tack, J.

Bull. Soc. Chim. Belg. 1975, *84*, 253-62.

VARIABLES:

One temperature: 298K

PREPARED BY:

M.C. Haulait-Pirson; A.F.M. Barton

EXPERIMENTAL VALUES:

The solubility of 1-butanol(1) in the water-rich phase was reported as 0.926 mol(1)/L sln, and the solubility of water(2) in the alcohol-rich phase was reported as 9.261 mol(2)/L sln.

The corresponding values on a weight/volume basis are 68.6 g(1)/L sln, and 166.9 g(2)/L sln (compiler).

(The temperature was unspecified in the thesis, but reported as 298 K in the 1975 published paper).

AUXILIARY INFORMATION

METHOD/APPARATUS/PROCEDURE:

(1) and (2) were equilibrated using a cell described in ref 1. The Rayleigh M75 interference refractometer with the cell M160 for liquids was used for the determination of the concentrations. Cell thicknesses were 1, 3 and 10 cm depending on the concentration range. Standard solutions covering the whole range of concentrations investigated were used for the calibration.

SOURCE AND PURITY OF MATERIALS:

(1) Merck p.a.

(2) distilled

ESTIMATED ERROR:

soly ± 0.00036 - 0.05 mol/L sln, depending on the concentration

REFERENCES:

1. Meeussen, E.; Huyskens, P. *J. Chim. Phys.* 1966, *63*, 845.

COMPONENTS:	ORIGINAL MEASUREMENTS:
(1) 1-butanol; $C_4H_{10}O$; [71-36-3] (2) Water; H_2O; [7732-18-5]	*Mullens, J. *Alcoholassociaten,* Doctoraatsproefschrift, Leuven, 1971. Huyskens, P.; Mullens, J.; Gomez, A.; Tack,J. *Bull. Soc. Chim. Belg.* 1975, *84*, 253-62.
VARIABLES:	PREPARED BY:
One temperature: 25°C	M.C. Haulait-Pirson; A.F.M. Barton

EXPERIMENTAL VALUES:

At 25°C the solubility of 1-butanol(1) in the water-rich phase was reported as 0.926 mol(1)/L sln, and the solubility of water(2) in the alcohol-rich phase was reported as 9.261 mol(2)/L sln.

The corresponding values on a weight/volume basis are 68.6 g(1)/L sln, and 166.9 g(2)/L sln (compiler).

AUXILIARY INFORMATION

METHOD/APPARATUS/PROCEDURE:

The partition of the two components was made using a cell described in ref 1. The Rayleigh Interference Refractometer M154 was used for the determination of the concentrations.

Standard solutions covering the whole range of concentration investigated were used for the calibration.

SOURCE AND PURITY OF MATERIALS:

(1) Merck product (p.a.)

(2) distilled

ESTIMATED ERROR:

Soly ± 0.001 mol(1)/L sln.

REFERENCES:

1. Meeussen, E.; Huyskens,P. *J. Chim. Phys.* 1966, *63*, 845.

COMPONENTS:	ORIGINAL MEASUREMENTS:
(1) 1-Butanol; $C_4H_{10}O$; [71-36-3] (2) Water; H_2O; [7732-18-5]	Vochten, R.; Petre, G. *J. Colloid Interface Sci.* 1973, 42, 320-7.
VARIABLES:	PREPARED BY:
One temperature: 15°C	S.H. Yalkowsky; S.C. Valvani; A.F.M. Barton

EXPERIMENTAL VALUES:

The concentration of 1-butanol in the water-rich phase at equilibrium at 15°C was reported to be 1.1 ± 0.1 mol(1)/L sln. This corresponds to 81.5 g(1)/L sln (compiler).

AUXILIARY INFORMATION

METHOD/APPARATUS/PROCEDURE:

The solubilities were obtained from surface tensions, measured by the static method of Wilhelmy (platinum plate). The apparatus consisted of an electrobalance (R.G. Cahn) connected with a high impedance null detector (FLUKE type 845 AR). An all-Pyrex vessel was used.

SOURCE AND PURITY OF MATERIALS:

(1) purified by distillation and preparative gas chromatography; b.p. 117.5°C/760 mm Hg

(2) triply distilled from permanganate solution

ESTIMATED ERROR:

Temperature: ± 0.1°C
Solubility: (probably std deviation) ±0.1 mol(1)/L sln.

REFERENCES:

COMPONENTS:	ORIGINAL MEASUREMENTS:
1) 1-Butanol; $C_4H_{10}O$; [71-36-3] (2) Water; H_2O; [7732-18-5]	Korenman, I.M.; Gorokhov, A.A.; Polozenko,G.N. *Zh. Fiz. Khim.* 1974, *48*, 1810-2; **Russ. J. Phys. Chem.* 1974, *48*, 1065-7. *Zh. Fiz. Khim.* 1975, *49*, 1490-3; *Russ. J. Phys. Chem.* 1975, *49*, 877-8.
VARIABLES:	PREPARED BY:
One temperature: 25°C	A.F.M.Barton

EXPERIMENTAL VALUES:

At equilibrium at 25.0°C the concentration of 1-butanol in the water-rich phase was reported to be 1.05 mol (1)/L sln, and the concentration of water in the alcohol-rich phase was reported to be 9.70 mol (2)/L sln.

The corresponding solubilities on a mass/volume basis, calculated by the compiler, are 77.8 g(1)/L sln, and 174.8 g(2)/L sln respectively.

AUXILIARY INFORMATION

METHOD/APPARATUS/PROCEDURE:

The two liquids were shaken in a closed vessel at 25.0 ± 0.1°C until equilibrium was established. The solubility of the alcohol in the aqueous phase was determined on a Tsvet-1 chromatograph with a flame-ionization detector. The sorbent was a polyethylene glycol adipate deposited on Polychrom-1. The solubility of water in the alcohol was determined on a UKh-2 universal chromatograph under isothermal conditions. The study formed part of an investigation of salting out by alkali halides of higher alcohol-water systems.

SOURCE AND PURITY OF MATERIALS:

Not stated.

ESTIMATED ERROR:

Temperature: ± 0.1°C
Solubility: not stated; the results shown are the arithmetic mean of four experiments.

REFERENCES:

COMPONENTS:

(1) 1-Butanol; $C_4H_{10}O$; [71-36-3]

(2) Water; H_2O; [7732-18-5]

ORIGINAL MEASUREMENTS:

Prochazka, O.; Sushka, J.; Pick, J.

Coll. Czech. Chem. Comm. 1975, *40*, 781-6.

VARIABLES:

Temperature: 83-124°C

PREPARED BY:

A. Maczynski

EXPERIMENTAL VALUES:

Mutual solubility of 1-butanol (1) and water (2)

$t/^{\circ}C$	g(1)/100g sln		x_1 (compiler)	
	(2)-rich phase	(1)-rich phase	(2)-rich phase	(1)-rich phase
82.7	7.3	-	0.019	-
87.1	-	70.5	-	0.367
90.1	-	69.8	-	0.360
94.0	8.3	-	0.022	-
98.5	8.9	-	0.023	-
98.7	-	66.6	-	0.326
105.2	-	63.4	-	0.296
106.1	-	62.9	-	0.292
107.2	10.3	-	0.027	-
110.2	-	60.4	-	0.270
111.6	11.7	-	0.031	-
114.0	12.6	-	0.034	-
115.2	-	56.8	-	0.242
117.8	14.9	-	0.041	-

(continued next page)

AUXILIARY INFORMATION

METHOD/APPARATUS/PROCEDURE:

The turbidity method was used.

The apparatus with visual indication of turbidity was described in ref 1. No more additional details were reported in the paper.

SOURCE AND PURITY OF MATERIALS:

(1) Analytical grade reagent; distilled over forty-plate bubble cup column; b.p. 117.9°C/760 torr, n_D^{25} 1.3974, d_4^{25} 0.8057.

(2) redistilled.

ESTIMATED ERROR:

Not specified.

REFERENCES:

1. Matous, J.; Novak, J.P.; Sobr, J.; Pick, J. *Coll. Czech. Chem. Comm.* 1972, *37*, 2653.

COMPONENTS:	ORIGINAL MEASUREMENTS:
(1) 1-Butanol; $C_4H_{10}O$; [71-36-3] (2) Water; H_2O; [7732-18-5]	Prochazka, O.; Sushka, J.; Pick, J. *Coll. Czech. Chem. Comm.*, 1975, *40*, 781-6.

EXPERIMENTAL VALUES (continued)

$t/^{\circ}C$	g(1)/100g sln		x_1(compiler)	
	(2)-rich phase	(1)-rich phase	(2)-rich phase	(1)-rich phase
119.3	-	52.6	-	0.212
119.8	-	52.1	-	0.209
121.0	-	50.0	-	0.195
121.7	18.4	-	0.052	-
121.9	18.6	-	0.053	-
123.1	20.8	45.5	0.060	0.169
123.9	-	41.4	-	0.146
124.1	26.1	-	0.079	-
124.2	-	38.0	-	0.130
124.3	29.7	36.5	0.093	0.123
124.4	31.8	34.0	0.102	0.111
124.4	32.5	32.5	0.105	0.105

COMPONENTS:

(1) 1-Butanol; $C_4H_{10}O$; [71-36-3]

(2) Water; H_2O; [7732-18-5]

ORIGINAL MEASUREMENTS:

De Santis, R.; Marrelli, L.; Muscetta, P.N.

Chem. Eng. J. 1976, *11*, 207-14.

VARIABLES:

One temperature: 25°C

PREPARED BY:

A. Maczynski

EXPERIMENTAL VALUES:

The proportion of 1-butanol(1) in the water-rich phase at equilibrium at 25°C was reported to be 7.4 g(1)/100g sln.

The corresponding mole fraction solubility, x_1, calculated by the compiler, is 0.019.

The proportion of water(2) in the alcohol-rich phase at equilibrium at 25°C was reported to be 20.49 g(2)/100g sln.

The corresponding mole fraction solubility, x_2, calculated by the compiler, is 0.513.

AUXILIARY INFORMATION

METHOD/APPARATUS/PROCEDURE:

The determinations were carried out using a separating funnel with a thermostatic jacket. The extractor was loaded with (1) and (2) and after an extended period of mixing and quantitative gravity separation, samples were withdrawn from the aqueous phase. The concentration of (1) in (2) was determined by colorimetric analysis (double-beam Lange colorimeter) of the cerium complex. The concentration of (2) in (1) was derived from a material balance based upon starting quantities and compositions. Each of the determinations was carried out several times. The method is described also in ref 1.

SOURCE AND PURITY OF MATERIALS:

(1) Carlo Erba analytical purity; fractionated before use.

(2) doubly distilled.

ESTIMATED ERROR:

Temperature: ± 0.1°C.

REFERENCES:

1. De Santis, R.; Marrelli, L.; Muscetta, P.N. *J. Chem. Eng. Data*, 1976, *21*, 324.

COMPONENTS:

(1) 1-Butanol; $C_4H_{10}O$; [71-36-3]

(2) Water; H_2O; [7732-18-5]

ORIGINAL MEASUREMENTS:

De Santis, R.; Marrelli, L.; Muscetta, P.N.

J. Chem. Eng. Data 1976, *21*, 324-7.

Marrelli, L.

Chem. Eng. J., 1979, *18*, 225-32.

VARIABLES:

Temperature: 20-40°C

PREPARED BY:

A. Maczynski; S.H. Yalkowsky; S.C. Valvani

EXPERIMENTAL VALUES:

Mutual solubility of butanol (1) and water (2)

$t/^{\circ}C$	g(1)/100g sln		x_1 (compiler)	
	(2)-rich phase	(1)-rich phase	(2)-rich phase	(1)-rich phase
20	7.8	80.0	0.020	0.493
30	7.1	79.4	0.018	0.484
40	6.6	78.6	0.017	0.476

AUXILIARY INFORMATION

METHOD/APPARATUS/PROCEDURE:

The measurements of the solubility limits have been carried out at 20, 30, and 40°C using separator funnels with a thermostatic jacket for temperature control (±0.1 °C). Extractors were loaded with equal quantities of solution at concentrations between zero and the saturation value in the water. After an extended period of mixing and a quantitative gravity separation, samples were withdrawn from both phases. Equilibrium compositions were determined by analyzing the alcohol in the aqueous phase. Alcohol in the aqueous phase was determined by colorimetric analysis (double beam Lange colorimeter) of the cerium complex. Each of the determinations was carried out with several repetitions. The method is also described in ref 1.

SOURCE AND PURITY OF MATERIALS:

(1) Carlo Erba, analytical purity; refractionated.

(2) doubly distilled.

ESTIMATED ERROR:

Temperature: ±0.1 °C

Solubility standard deviation: ±0.5 wt percent

REFERENCES:

1. De Santis, R.; Marrelli, L.; Muscetta, P.N. *Chem. Eng. J.* 1976, *11*, 207-14.

COMPONENTS:

(1) 1-Butanol; $C_4H_{10}O$; [71-36-3]

(2) Water; H_2O; [7732-18-5]

ORIGINAL MEASUREMENTS:

Lavrova, O.A.; Lesteva, T.M.

Zh. Fiz. Khim. 1976, *50*, 1617; *Dep. Doc. VINITI* 3813-75.

VARIABLES:

Temperature: 40 and 60°C

PREPARED BY:

A. Maczynski

EXPERIMENTAL VALUES:

Mutual solubility of 1-butanol(1) and water (2)

t/°C	g(1)/100g sln		x_1 (compiler)	
	(2)-rich phase	(1)-rich phase	(2)-rich phase	(1)-rich phase
40	6.6	78.50	0.017	0.4701
60	6.45	76.31	0.016	0.4391

AUXILIARY INFORMATION

METHOD/APPARATUS/PROCEDURE:

The titration method was used.

No details were reported in the paper.

SOURCE AND PURITY OF MATERIALS:

(1) source not specified; distilled with heptane; purity 99.91 wt %, 0.09 wt % of water; n_D^{20} 1.3992, d_4^{20} 0.8099, b.p. 117.3°C.

(2) not specified.

ESTIMATED ERROR:

Not specified.

REFERENCES:

COMPONENTS:

(1) 1-Butanol; $C_4H_{10}O$; [71-36-3]

(2) Water; H_2O; [7732-18-5]

ORIGINAL MEASUREMENTS:

Aoki, Y.; Moriyoshi, T.

J. Chem. Thermodyn. 1978, *10*, 1173-9.

VARIABLES:

Temperature: 303-398K

Pressure: 1-2450 atm (0.1-25 MPa)

PREPARED BY:

S.C. Valvani; S.H. Yalkowsky; A.F.M.Barton; G.T. Hefter

EXPERIMENTAL VALUES:

Mutual solubility of 1-butanol (1) and water (2)

T/K	p/atm	g(1)/100g sln		x_1 (compiler)	
		(2)-rich phase	(1)-rich phase	(2)-rich phase	(1)-rich phase
302.95	1	7.0	79.3	0.0180	0.482
302.95	500	9.0	77.7	0.0235	0.459
302.95	970	9.6	76.9	0.0252	0.447
302.95	1150	10.2	76.3	0.0269	0.439
302.95	1240	11.0	76.1	0.0292	0.436
302.95	1500	11.2	75.0	0.0298	0.422
302.95	2000	12.7	74.0	0.0342	0.409
302.95	2250	13.9	72.5	0.0377	0.391
302.95	2450	14.3	72.2	0.0390	0.387
322.75	1	6.9	77.7	0.0177	0.459
322.75	500	8.6	76.0	0.0224	0.435
322.75	1000	9.8	74.4	0.0257	0.414
322.75	1240	10.6	73.5	0.0280	0.403
322.75	1550	11.0	73.2	0.0292	0.399
322.75	2040	12.6	72.2	0.0339	0.387

(continued next page)

AUXILIARY INFORMATION

METHOD/APPARATUS/PROCEDURE:

The method,described in ref 1, was that used for previous studies on 2-butanol and iso-butanol. Both components were placed in a cut-off glass syringe of about 20 cm^3 capacity used as a sample vessel, which was placed in a stainless steel pressure vessel mechanically shaken in an oil thermostat bath (± 0.02 K). After stirring for at least 12 h and then standing for another 6 h at the desired temperature and pressure a sample of the upper layer was withdrawn. Subsequently the pressure vessel was moved, the contents allowed to settle, and the lower layer sampled.

Methanol as a mixing agent was added to the samples before refractometric analysis until the mass ratio of sample to methanol became 5.000 ± 0.037.

SOURCE AND PURITY OF MATERIALS:

(1) "best grade reagent"; dried by refluxing over CaO and then distilling twice; n^{25} = 1.3973, d^{25} = 0.8062 g cm^{-3}

(2) de-ionized, distilled from alkaline $KMnO_4$ and then redistilled; n^{25} = 1.3327

ESTIMATED ERROR:

Solubility: mean error within ± 0.28 g(1)/100g sln

REFERENCES:

1. Moriyoshi, T.; Kaneshina, S.; Aihara, K.; Yabumoto, K. *J. Chem. Thermodyn.* 1975, 7, 537.

COMPONENTS:	ORIGINAL MEASUREMENTS:
(1) 1-Butanol; $C_4H_{10}O$; [71-36-3]	Aoki, Y.; Moriyoshi, T.
(2) Water; H_2O; [7732-18-5]	*J. Chem. Thermodyn.* 1978, *10*, 1173-9.

EXPERIMENTAL VALUES (continued)

T/K	p/atm	g(1)/100g sln		x_1 (compiler)	
		(2)-rich phase	(1)-rich phase	(2)-rich phase	(1)-rich phase
322.75	2250	12.7	71.5	0.0342	0.379
322.75	2450	13.2	70.7	0.0356	0.370
332.65	1	7.0	76.1	0.0180	0.436
332.65	500	7.5	74.5	0.0193	0.415
332.65	1000	8.6	73.2	0.0224	0.399
332.65	1500	9.3	71.7	0.0243	0.381
332.65	2010	9.5	70.5	0.0249	0.368
332.65	2180	10.8	70.0	0.0286	0.362
332.65	2250	10.8	69.2	0.0286	0.353
332.65	2400	11.3	69.0	0.0302	0.351
332.65	2450	11.5	69.0	0.0306	0.351
342.65	1	6.7	75.0	0.0171	0.422
342.65	670	7.8	72.5	0.0202	0.391
342.65	1700	10.2	69.5	0.0269	0.357
342.65	2040	10.7	68.5	0.0283	0.346
342.65	2450	11.9	66.6	0.0318	0.327
362.65	1	7.5	69.7	0.0193	0.359
362.65	1000	10.7	65.4	0.0283	0.315
362.65	1040	-	65.4	-	0.315
362.65	1290	11.3	64.9	0.0302	0.310
362.65	1700	13.2	63.5	0.0356	0.297
362.65	2000	14.5	62.6	0.0396	0.289
362.65	2180	15.5	62.3	0.0427	0.287
362.65	2450	16.1	61.4	0.0446	0.278
372.45	1	9.5	66.8	0.0249	0.328
372.45	1000	12.3	62.8	0.0330	0.291
372.45	1290	12.7	62.3	0.0342	0.287
372.45	1570	13.7	60.7	0.0371	0.273
372.45	1800	15.6	60.3	0.0430	0.270
372.45	1970	16.7	58.9	0.0465	0.258
372.45	2450	18.5	55.2	0.0522	0.231
380.05	1	10.6	62.7	0.0280	0.290
380.05	500	11.5	60.9	0.0306	0.274
380.05	1000	13.2	59.0	0.0356	0.259
380.05	1500	15.8	57.4	0.0436	0.247
380.05	2000	16.7	54.5	0.0465	0.226
380.05	2300	20.3	52.3	0.0583	0.210
380.05	2450	25.5	44.9	0.0768	0.165
382.95	1	11.0	61.7	0.0292	0.281
382.95	500	11.9	60.1	0.0318	0.268

(continued next page)

COMPONENTS:

(1) 1-Butanol; $C_4H_{10}O$; [71-36-3]

(2) Water; H_2O; [7732-18-5]

ORIGINAL MEASUREMENTS:

Aoki, Y.; Moriyoshi, T.

J. Chem. Thermodyn. 1978, *10*, 1173-9.

EXPERIMENTAL VALUES (continued)

T/K	p/atm	g(1)/100g sln		x_1(compiler)	
		(2)-rich phase	(1)-rich phase	(2)-rich phase	(1)-rich phase
382.95	1000	13.9	58.3	0.0377	0.254
382.95	1500	16.5	56.4	0.0458	0.239
382.95	1570	17.2	56.3	0.0481	0.238
382.95	1700	17.3	55.8	0.0484	0.235
382.95	1900	18.0	54.1	0.0506	0.223
382.95	2000	19.6	52.5	0.0559	0.212
382.95	2100	23.1	48.9	0.0680	0.189
382.95	2200	31.6	41.0	0.1009	0.145
384.45	1700	19.4	54.6	0.0552	0.227
384.45	1800	20.0	52.4	0.0573	0.211
384.45	1900	22.5	49.9	0.0659	0.195
384.45	2000	30.8	41.0	0.0976	0.227
386.15	1500	20.2	53.5	0.0579	0.218
386.15	1570	22.2	51.8	0.0649	0.207
386.15	1670	27.5	46.0	0.0844	0.172
387.15	1500	26.4	46.1	0.0802	0.172
388.15	1000	17.8	53.2	0.0500	0.216
388.15	1090	18.0	53.0	0.0506	0.215
388.15	1200	19.0	51.8	0.0539	0.207
388.15	1360	25.9	45.5	0.0783	0.169
388.15	1400	31.2	40.2	0.0992	0.141
390.35	800	18.0	52.9	0.0506	0.214
390.35	900	19.0	50.1	0.0539	0.196
390.35	1000	25.2	45.7	0.0757	0.170
391.25	1	14.5	55.6	0.0396	0.233
391.25	300	15.6	54.6	0.0430	0.227
391.25	500	16.7	53.7	0.0465	0.220
391.25	600	18.0	53.5	0.0506	0.218
391.25	720	19.7	51.8	0.0562	0.207
391.25	770	21.6	48.5	0.0627	0.187
391.25	790	23.5	44.1	0.0695	0.160
392.45	450	25.5	45.5	0.0768	0.169
392.45	500	30.0	39.8	0.0944	0.139
392.95	1	17.2	53.7	0.0481	0.220
392.95	270	21.2	49.7	0.0613	0.194
392.95	340	26.3	46.3	0.0798	0.173
392.95	350	27.1	44.7	0.0829	0.164
395.25	1	18.2	47.7	0.0513	0.182
395.25	150	25.1	41.9	0.0753	0.149
395.25	200	32.3	36.8	0.1039	0.124

(continued next page)

COMPONENTS:

(1) 1-Butanol; $C_4H_{10}O$; [71-36-3]

(2) Water; H_2O; [7732-18-5]

ORIGINAL MEASUREMENTS:

Aoki, Y.; Moriyoshi, T.

J. Chem. Thermodyn. 1978, *10*, 1173-9.

EXPERIMENTAL VALUES (continued)

T/K	p/atm	g(1)/100g sln		x_1(compiler)	
		(2)-rich phase	(1)-rich phase	(2)-rich phase	(1)-rich phase
395.85	1	18.9	48.0	0.0536	0.183
397.05	1	21.8	44.9	0.0635	0.165
397.45	1	24.7	41.8	0.0739	0.148
397.75	1	27.9	38.8	0.0859	0.134

Properties of the critical solutions

p_c/atm	T_c(UCST)/K	x_{1c}
1	397.85	0.110
204	395.25	0.113
380	392.95	0.117
500	392.45	0.115
820	391.25	0.118
1000	390.65	0.118
1030	390.35	0.118
1390	388.15	0.118
1500	387.65	0.120
1710	386.15	0.122
2000	384.35	0.122
2010	384.45	0.121
2200	382.95	0.121
2450	381.15	0.123
2500	380.05	0.125

$(dT_c/dp) = -(12.0 \pm 0.5) \times 10^{-3}$ K atm^{-1} at $p < 400$ atm

$-(7.0 \pm 0.7) \times 10^{-3}$ K atm^{-1} at $800 < p < 2500$ atm

$(dx_{1,c}/dT) = -(4.0 \pm 0.5) \times 10^{-4}$ K^{-1}

COMPONENTS:	ORIGINAL MEASUREMENTS:
(1) 1-Butanol; $C_4H_{10}O$; [71-36-3] (2) Water; H_2O; [7732-18-5]	Nishino, N.; Nakamura, M. *Bull. Chem. Soc. Japan* 1978, *51*, 1617-20; 1981, *54*, 545-8.
VARIABLES:	PREPARED BY:
Temperature: 275-360 K	G.T. Hefter

EXPERIMENTAL VALUES:

The mutual solubility of (1) and (2) in mole fractions are reported over the temperature range in graphical form. Graphical data are also presented for the heat of solution at infinite dilution of (1) in (2) and for the heat of evaporation of (1).

AUXILIARY INFORMATION

METHOD/APPARATUS/PROCEDURE:

The turbidimetric method was used. Twenty to thirty glass ampoules containing aqueous solutions of *ca.* 5 cm^3 of various concentrations near the solubility at room temperature were immersed in a water thermostat. The distinction between clear and turbid ampoules was made after equilibrium was established (*ca.* 2h). The smooth curve drawn to separate the clear and turbid regions was regarded as the solubility curve.

SOURCE AND PURITY OF MATERIALS:

(1) G.R. grade (various commercial sources given); dried over calcium oxide; kept in ampoules over magnesium powder.

(2) Deionized, refluxed for 15 h with potassium permanganate then distilled.

ESTIMATED ERROR:

Not stated.

REFERENCES:

COMPONENTS:

(1) 1-Butanol; $C_4H_{10}O$; [71-36-3]

(2) Water; H_2O; [7732-18-5]

ORIGINAL MEASUREMENTS:

Lyzlova, R.V.

Zhur. Prikl. Khim. (Leningrad) 1979, *52*, 545-50.

**J. App. Chem. USSR* 1979, *52*, 509-14.

VARIABLES:

Temperature: 20°C and 90°C

PREPARED BY:

A.F.M. Barton

EXPERIMENTAL VALUES:

Mutual solubility of 1-butanol(1) and water(2)

$t/°C$	x_1		g(1)/100g sln (compiler)	
	Alcohol-rich phase	Water-rich phase	Alcohol-rich phase	Water-rich phase
20	0.478	0.021	79.0	8.1
90	0.360	0.020	69.8	7.7

AUXILIARY INFORMATION

METHOD/APPARATUS/PROCEDURE:

The analytical method was used, samples being withdrawn from coexisting liquid phases at equilibrium for analysis by refractometry. An IRF-23 refractometer measured the refractive index correct to 0.000 02. The study was concerned with phase equilibria in the ternary system 1-butanol/2-methyl-1-propanol/water, and only a few experiments dealt with mutual solubilities in the binary systems.

SOURCE AND PURITY OF MATERIALS:

(1) CP grade; dried over freshly ignited K_2CO_3, distilled twice with 1.5 m glass-packed fractionating column; n_D^{20} 1.39920 d_4^{20} 0.8099 b.p. 117.8°C

(2) not specified

ESTIMATED ERROR:

Not specified for binary systems; error below ± 1% for water and ± 1.5% for alcohol ratios in ternary systems.

REFERENCES:

COMPONENTS:

(1) 1-Butanol; $C_4H_{10}O$; [71-36-3]

(2) Water; H_2O; [7732-18-5]

ORIGINAL MEASUREMENTS:

Singh, R.P.; Haque, M.M.

Indian J. Chem. 1979, *17A*, 449-51.

VARIABLES:

One temperature: 30°C

PREPARED BY:

A.F.M. Barton

EXPERIMENTAL VALUES:

Mutual solubility of 1-butanol(1) and water(2) at 30°C

	mol(2)/mol(1)	x_1	g(1)/100g sln (compiler)
alcohol-rich phase	1.0339	0.4920	79.9
water-rich phase	56.505	0.01677	6.56

AUXILIARY INFORMATION

METHOD/APPARATUS/PROCEDURE:

Titrations of one component with the other (ref 1) were carried out in well-stoppered volumetric flasks. The shaking after each addition was done ultrasonically for at least 30 min. These results formed part of a study of the ternary system 1-butanol/methanol/water.

SOURCE AND PURITY OF MATERIALS:

(1) B.D.H. AR; purified; density and refractive index checked

(2) conductivity water from all-glass still

ESTIMATED ERROR:

Temperature: ± 0.1°C
Solubility: each titration repeated at least three times.

REFERENCES:

1. Simonsen, D.R.; Washburn, E.R. *J. Am. Chem. Soc.* 1946, *68*, 235.

COMPONENTS:

(1) 1-Butanol; $C_4H_{10}O$; [71-36-3]

(2) Water; H_2O; [7732-18-5]

ORIGINAL MEASUREMENTS:

Fu, C.F.; King, C.L.; Chang, Y.F.; Xeu, C.X.

Hua Kung Hsueh Pao 1980 (3), 281-92.

VARIABLES:

One temperature: 93°C

PREPARED BY:

C.F. Fu

EXPERIMENTAL VALUES:

The proportion of 1-butanol in the water-rich phase at equilibrium at 92.8°C was reported to be 6.87 g(1)/100g sln. The corresponding mole fraction solubility, x_1, is 0.0176.

The proportion of water in the alcohol-rich phase at 92.8°C was reported to be 30.80 g(2)/100g sln. The corresponding mole fraction solubility, x_2, is 0.6467.

AUXILIARY INFORMATION

METHOD/APPARATUS/PROCEDURE:

The turbidimetric method was used. Homogenous solutions were prepared and boiled at 760 mm Hg in a specially designed flask attached to a condenser of negligible hold-up compared with the volume of liquid solution. The solution was stirred by a magnetic stirrer and titrated with (1) or (2). The end point of titration was judged both by cloudiness and constancy of boiling temperature.

SOURCE AND PURITY OF MATERIALS: (Some information not in the published paper has been supplied by the compiler)

(1) Beijing Chemicals reagent; used as received; b.p. 117.1°C (760 mm Hg) n_D^{15} 1.3992, d_4^{15} 0.8091.

(2) distilled.

ESTIMATED ERROR: (Supplied by compiler)

Temperature: ± 0.02°C
Solubility: 0.4%

REFERENCES:

COMPONENTS:

(1) 1-Butanol; $C_4H_{10}O$; [71-36-3]

(2) Water; H_2O; [7732-18-5]

ORIGINAL MEASUREMENTS:

Tokunaga, S.; Manabe, M.; Koda, M.

Niihama Kogyo Koto Semmon Gakko Kiyo, Rikogaku Hen (Memoirs Niihama Technical College, Sci. and Eng.) 1980, *16*, 96-101.

VARIABLES:

Temperature: 15-35°C

PREPARED BY:

A.F.M. Barton

EXPERIMENTAL VALUES:

Solubility of water in the alcohol-rich phase

$t/°C$	g(2)/100g sln	x_2	mol(1)/mol(2)
15	19.9	0.506	0.983
20	20.1	0.509	0.969
25	20.4	0.513	0.955
35	21.1	0.523	0.929

AUXILIARY INFORMATION

METHOD/APPARATUS/PROCEDURE:

The mixtures of 1-butanol (~ 5 mL) and water (~ 10 mL) were stirred magnetically in a stoppered vessel and allowed to stand for 10-12 h in a water thermostat. The alcohol phase was analysed for water by Karl Fischer titration.

SOURCE AND PURITY OF MATERIALS:

(1) distilled;
no impurities detectable by gas chromatography.

(2) deionized;
distilled prior to use.

ESTIMATED ERROR:

Temperature: ± 0.1°C
Solubility: each result is the mean of three determinations.

REFERENCES:

COMPONENTS:

(1) 1-Butanol; $C_4H_{10}O$; [71-36-3]

(2) Water; H_2O; [7732-18-5]

ORIGINAL MEASUREMENTS:

Lutugina, N.V., Reshchetova, L.I.

Vestnik Leningr. Univ. 1972, *16*, 75-81.

VARIABLES:

One temperature: 92.7°C

PREPARED BY:

C.F. Fu and G.T. Hefter

EXPERIMENTAL VALUES:

The solubility of 1-butanol in the water-rich phase at 92.7°C was reported to be 0.020 mole fraction.

The solubility of 1-butanol in the alcohol-rich phase at 92.7°C was reported to be 0.360 mole fraction.

The corresponding mass solubilities calculated by the compilers are 7.7 g(1)/100g sln and 69.8 g(1)/100g sln.

AUXILIARY INFORMATION

METHOD/APPARATUS/PROCEDURE:

Vapor-liquid-liquid equilibrium still described in ref 1 was used for liquid-liquid equilibrium determination at boiling point. Immediately after cutting off heating and stopping of violent boiling, samples were taken from both layers for analysis. Analytical method was described in detail.

SOURCE AND PURITY OF MATERIALS:

(1) Source not specified; distilled over a column of 20 theoretical plates (according to C_6H_6 - CCl_4); b.p. 117.5°C (760 mm Hg), n_D^{20} 1.3993, d_4^{20} 0.8098

(2) distilled.

ESTIMATED ERROR:

Not specified

REFERENCES:

1. Morachevskii, A.G.; Smirnova, N.A. *Zh. Prikl. Khim.* 1963, *36*, 2391.

COMPONENTS:

(1) 1-Butanol-d; C_4H_9DO; [4712-38-3]

(2) Water-d_2; D_2O; [7789-20-0]

ORIGINAL MEASUREMENTS:

Rabinovich, I.B., Fedorov, V.D., Pashkin, N.P., Avdesnyak, M.A., Pimenov, N. Ya.

Dokl. Adad. Nauk SSSR 1955, *105*, 108-11.

VARIABLES:

Temperature: 40-130°C

PREPARED BY:

G. Jancsó and G.T. Hefter.

EXPERIMENTAL VALUES:

Effect of deuteration on solubility of 1-butanol in water

$t/^{\circ}C$	$100(L_H - L_D)/L_H$ [a]
40	42
66	35
80	32
100	29
103	24
108	30
125	67 [b]

[a] L_H and L_D are the solubilities (in mol %) of 1-butanol in H_2O and 1-butanol-d in D_2O, respectively.

[b] This value refers to a temperature that corresponds to within ± 0.5 K to the UCST of the 1-butanol-H_2O system.

(continued next page)

AUXILIARY INFORMATION

METHOD/APPARATUS/PROCEDURE:

The synthetic method (cloud-clear points) of Alexejew (ref 1) was used which consists in observing the temperature at which a measured weight of one liquid is visually soluble in a measured weight of the other liquid.

SOURCE AND PURITY OF MATERIALS:

(1) 1-butanol-d was prepared by the hydrolysis of the aluminium salt of the ordinary butanol with D_2O. The OD group of the product had a deuterium content of > 99 at. %.

(2) D_2O, 99.0 - 99.8 at. % D, source not given.

ESTIMATED ERROR:

Solubility: not specified

Temperature: ±0.02°C

REFERENCES:

1. Alexejew, W. *J. Prakt. Chem.* 1882, *25*, 518.

COMPONENTS:

(1) 1-Butanol-d; C_4H_9DO; [4712-38-3]

(2) Water-d_2; D_2O; [7789-20-0]

ORIGINAL MEASUREMENTS:

Rabinovich, I.R., Fedorov, V.D., Pashkin, N.P., Avdesnyak, M.A., Pimenov, N. Ya.

Dokl. Akad. Nauk SSSR 1955, *105*, 108-11.

EXPERIMENTAL VALUES: (continued)

The upper critical solution temperature of 1-butanol and H_2O is 125.1°C, compared with 131.1°C for 1-butanol-d and D_2O.

Mutual solubilities for the 1-butanol-H_2O system were also determined but only graphical data were given (Δ in Figure 1).

The literature reference for the data of Hill and Malisov (o) is not given but is presumably that listed as ref 4 in the Critical Evaluation.

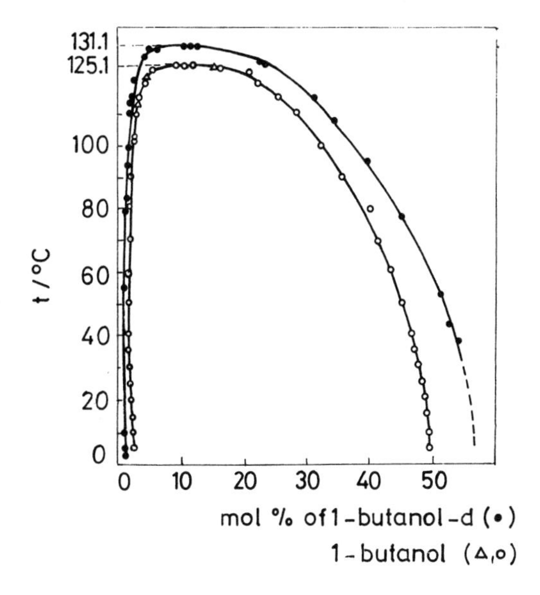

Fig. 1. Mutual solubilities of $CH_3CH_2CH_2CH_2OD(1)$ in $D_2O(2)$ (•) and $CH_3CH_2CH_2CH_2OH(1)$ in $H_2O(2)$ (Δ,o)

COMPONENTS:

(1) 2-Butanol (*sec-butanol*); $C_4H_{10}O$; [78-92-2]

(2) Water; H_2O; [7732-18-5]

EVALUATOR:

G.T. Hefter, School of Mathematical and Physical Sciences, Murdoch University, Perth, Western Australia.
November 1982.

CRITICAL EVALUATION:

The solubility in the 2-butanol (1) - water (2) system has been reported in a form suitable for critical evaluation in the works listed below.

Reference	T/K	Solubility	Method
Alexejew (ref 1)	283-376	mutual	synthetic/analytical
Dolgolenko (ref 3)	250-388	mutual	synthetic
Evans (ref 6)	293	mutual	volumetric
Altsybeeva *et al.* (ref 7)	293-353	mutual	analytical
Morachevskii and Popovich (ref 8)	283-368	(1) in (2)	refractometric
Ratouis and Dode (ref 9)	298 & 303	(1) in (2)	analytical
Shakhud *et al.* (ref 12)	293-333	mutual	titration
Moriyoshi *et al.* (ref 13)	283-383	mutual	refractometric
De Santis *et al.* (ref 14)	298	mutual	analytical
Becke and Quitzch (ref 15)	298	mutual	refractometric

The original data are given in the data sheets immediately following this Critical Evaluation.

Data sheets have not been prepared for the graphical data of Timmermans (ref 2), Schneider and Russo (ref 10) and Rabinovich *et al.* (ref 16), nor for Jones (ref 5) which does not contain original data.

Solubilities in the system 2-butanol-*d*, C_4H_9OD, and water-d_2, D_2O, are given in the data sheet for Rabinovich *et al.* (ref 16) following the 2-butanol/water system data sheets.

In the Critical Evaluation the data of Mullens *et al.*(ref 11) in weight/volume fractions are excluded from consideration because no density information was included.

Of the data obtained by Dolgolenko (ref 3) only the highest boiling fraction (b.p. 99.0 - 99.5^{o}C, see data sheet) has been considered in the Critical Evaluation as the physical properties of that fraction correspond most closely to those reported for 2-butanol by other workers (ref 8,12). This is consistent with the view of the original author (ref 3).

The data of Becke and Quitzch (ref 15), the water-rich phase data of Dolgolenko (ref 3), Altsybeeva *et al.* (ref 7) and De Santis *et al.* (ref 14), and the alcohol-rich phase data of Alexejew (ref 1) disagree markedly from all other studies and are rejected.

The data point at 333 K, water-rich phase, in the otherwise satisfactory study by Moriyoshi *et al.* (ref 13) is also in marked disagreement with other studies and has been rejected.

All other data are included in the tables following. Values obtained by the Evaluator by graphical interpolation or extrapolation from the data sheets are indicated by an asterisk (*). "Best" values have been obtained by simple averaging.

(continued next page)

COMPONENTS:	EVALUATOR:
(1) 2-Butanol (*sec-butanol*); $C_4H_{10}O$; [78-92-2] (2) Water; H_2O; [7732-18-5]	G.T. Hefter, School of Mathematical and Physical Sciences, Murdoch University, Perth, Western Australia. November 1982.

CRITICAL EVALUATION: (continued)

The uncertainty limits (σ_n) attached to these "best" values do not have statistical significance. They are to be regarded only as a convenient representation of the spread of the reported values and not as error limits.

It will be noted from the tables following that there are serious disagreements among the solubility data reported by independent workers for the system 2-butanol-water. This is reflected in the large σ_n values.

No one study or any of the averaged values can be singled out for recommendation and clearly this sytem requires a careful and thorough re-investigation over the entire temperature range. The values given in the following tables should be regarded as *tentative* only, until further independent studies can be made.

This situation is surprising since the 2-butanol-water system possesses some interesting features. At atmospheric pressures the solubility curve is strongly "waisted" at low temperatures (ref 2,3) and, depending on the purity of the 2-butanol, may even show a closed miscibility loop (ref 3)! This loop is unusually sensitive to pressure (ref 4,13) and disappears entirely at about 830 atm (the hypercritical point: see below).

The crucial effect of alcohol purity on the solubility curve (ref 3) may of course be the reason for the lack of agreement between independent workers already noted and supports the need for further precise studies of this sytem supported by modern methods of analysis and purity determination.

The solubility of 2-butanol (1) in water (2)

T/K	Solubility, g(1)/100g sln	
	Reported values	"Best" values ($\pm\sigma_n$)
283	23.9 (ref 13)	23.9
293	19.8 (ref 8), 19.62 (ref 12), 20.2 (ref 13)	19.9 ± 0.2
298	19.0* (ref 1), 18.5 (ref 6), 17.70 (ref 9), 17.0* (ref 13)	18.1 ± 0.8
303	17.0* (ref 1), 17.9 (ref 13)	17.5 ± 0.5
308	16.0* (ref 1), 16.5 (ref 9)	16.3 ± 0.3
313	15.3* (ref 1), 15.6 (ref 8), 16.24 (ref 12), 14.9 (ref 13)	15.5 ± 0.5
323	14.8* (ref 1), 13.2 (ref 13)	14.0 ± 0.8
333	14.5* (ref 1), 14.0 (ref 8), 13.98 (ref 12)	14.2 ± 0.2
343	14.6* (ref 1), 12.9 (ref 13)	13.8 ± 0.9
353	15.1* (ref 1), 14.0 (ref 8), 13.8 (ref 13)	14.3 ± 0.6
363	16.0* (ref 1), 15.5 (ref 13)	15.8 ± 0.3
373	18.9* (ref 1), 18.3 (ref 13)	18.6 ± 0.3
383	23.9 (ref 13)	23.9

(continued next page)

COMPONENTS:	EVALUATOR:
(1) 2-Butanol (*sec-butanol*); $C_4H_{10}O$; [78-92-2] (2) Water; H_2O; [7732-18-5]	G.T. Hefter, School of Mathematical and Physical Sciences, Murdoch University, Perth, Western Australia. November 1982.

CRITICAL EVALUATION (continued)

The solubility of water (2) in 2-butanol (1)

T/K	Solubility, g(2)/100g sln: Reported values	"Best" values ($\pm\sigma_n$)
268	37.5* (ref 3)	37.5
273	38.8* (ref 3)	38.8
283	38.4*(ref 3), 39.1 (ref 13)	38.8 ± 0.4
293	35.8*(ref 3), 35.6 (ref 6), 36.1 (ref 7), 36.8 (ref 8), 36.52 (ref 12) 36.2 (ref 13)	36.2 ± 0.4
298	34.7*(ref 3), 33.9*(ref 13), 33.1 (ref 14)	33.9 ± 0.7
303	34.0*(ref 3), 34.5 (ref 13)	34.3 ± 0.3
313	33.4*(ref 3), 33.0 (ref 7), 33.5 (ref 8), 33.0*(ref 12), 32.8 (ref 13)	33.1 ± 0.3
323	33.1*(ref 3), 32.5 (ref 13)	32.8 ± 0.3
333	33.3*(ref 3), 33.4 (ref 7) 33.8 (ref 8), 32.64 (ref 12), 33.1 (ref 13)	33.2 ± 0.4
343	33.8*(ref 3), 33.8 (ref 7), 32.8 (ref 13)	33.5 ± 0.5
353	35.0*(ref 3), 36.2 (ref 8), 34.7 (ref 13)	35.3 ± 0.6
363	37.7*(ref 3), 38.2 (ref 13)	38.0 ± 0.3
373	41.5*(ref 3), 42.3 (ref 13)	41.9 ± 0.4
383	49.0*(ref 3), 49.3 (ref 13)	49.2 ± 0.2

The "best" values from the above tables are plotted in Figures 1 and 2.

Timmermans (ref 4) and Moriyoshi *et al.* (ref 13) have determined the effects of pressure (1-800 atm, 0.1-80 MPa) on the lower and upper critical solution temperatures. Their results are in reasonable agreement (T_c differs by *ca.* 1-3K).

Moriyoshi *et al.* (ref 13) have also determined the mutual solubility of (1) and (2) at 1-800 atm (0.1-80 MPa) and the "hypercritical point", i.e. where the closed solubility loop disappears.

(continued next page)

COMPONENTS:

(1) 2-Butanol (*sec-butanol*); $C_4H_{10}O$; [78-92-2]

(2) Water; H_2O; [7732-18-5]

EVALUATOR:

G.T. Hefter, School of Mathematical and Physical Sciences, Murdoch University, Perth, Western Australia.
November 1982.

CRITICAL EVALUATION (continued)

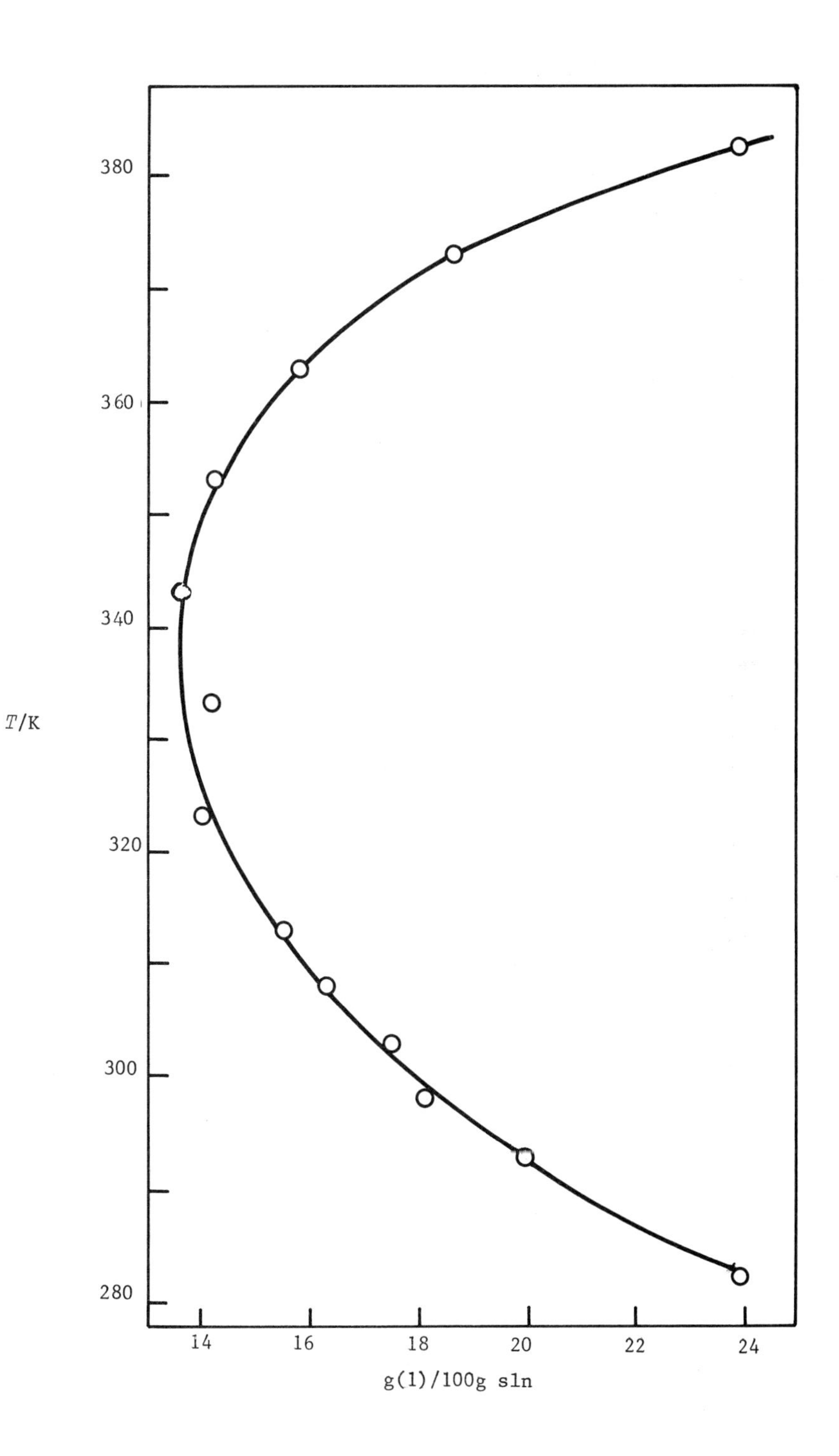

Figure 1. Solubility of 2-butanol in water, "best" values.

(continued next page)

COMPONENTS:	EVALUATOR:
(1) 2-Butanol (*sec-butanol*); $C_4H_{10}O$; [78-92-2] (2) Water; H_2O; [7732-18-5]	G.T. Hefter, School of Mathematical and Physical Sciences, Murdoch University, Perth, Western Australia. November 1982.

CRITICAL EVALUATION (continued)

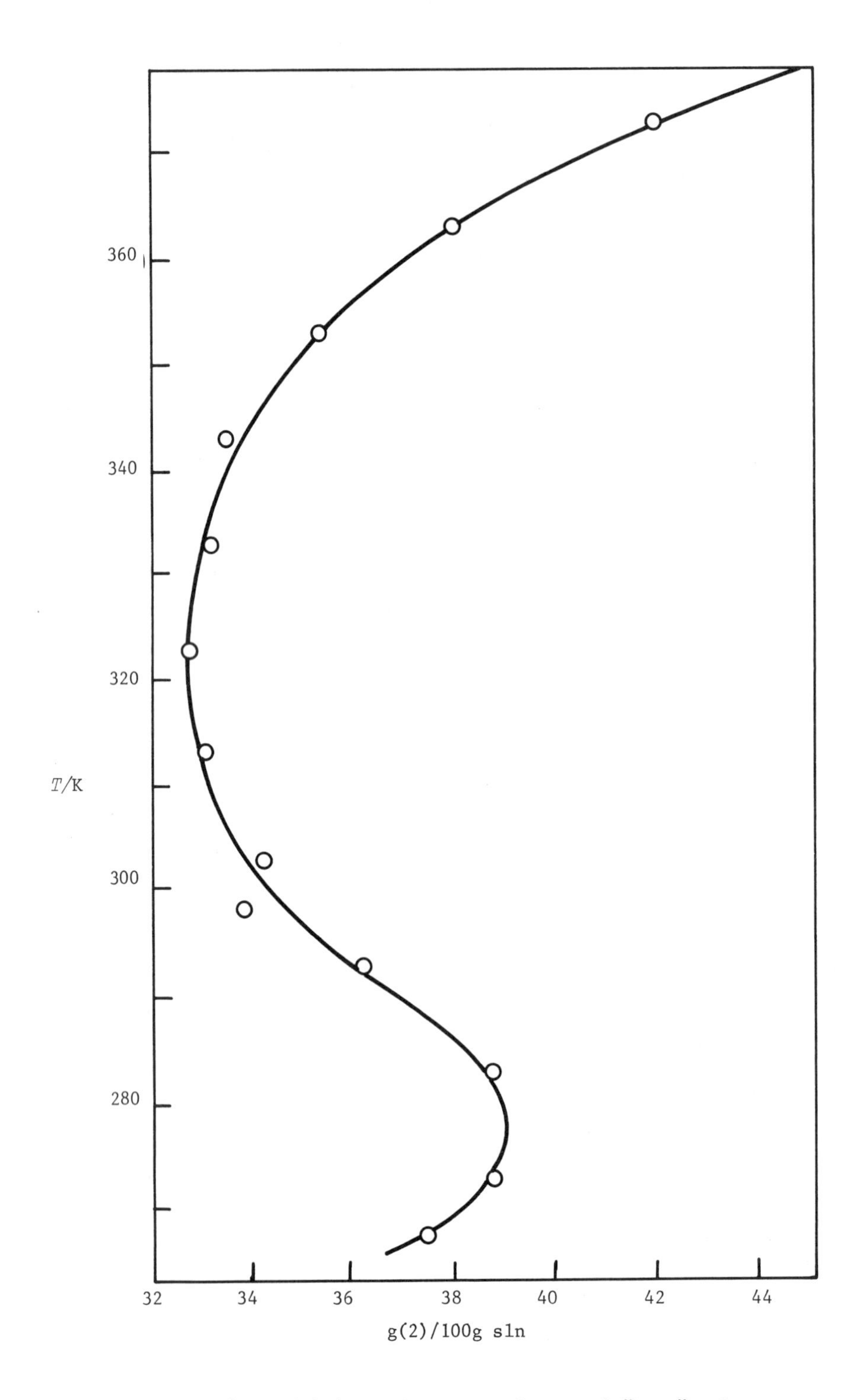

Figure 2. Solubility of water in 2-butanol, "best" values.

(continued next page)

COMPONENTS:	EVALUATOR:
(1) 2-Butanol (*sec-butanol*); $C_4H_{10}O$; [78-92-2] (2) Water; H_2O; [7732-18-5]	G.T. Hefter, School of Mathematical and Physical Sciences, Murdoch University, Perth, Western Australia. November 1982.

CRITICAL EVALUATION (continued)

References

1. Alexejew, W. *Ann. Phys. Chem.* 1886, *28*, 305.
2. Timmermans, J. *Z. Phys. Chem.* 1907, *58*, 129.
3. Dolgolenko, W. *Z. Phys. Chem.* 1908, *62*, 499.
4. Timmermans, J. *Arch. Neerl. Sci. Exactes Nat.* 1922, *A6(3)*, 147; *J. Phys. Chim. Physicochim. Biol.* 1923, *20*, 491.
5. Jones, D.C. *J. Chem. Soc.* 1929, 799.
6. Evans, T.W. *Anal. Chem.* 1936, *8*, 206.
7. Altsybeeva, A.N.; Belousov, V.N.; Ovtraht, H.V.; Morachevskii, A.G. *Zh. Fiz. Khim.* 1964, *38*, 1242; Altsybeeva, A.N.; Morachevskii, A.G. *Zh. Fiz. Khim.* 1964, *38*, 1574.
8. Morachevskii, A.G.; Popovich, Z.P. *Zh. Prikl. Khim.* 1965, *38*, 2129.
9. Ratouis, M.; Dodé, M. *Bull. Soc. Chim. Fr.* 1965, 3318.
10. Schneider, G.; Russo, C. *Ber. Bunsenges.* 1966, *70*, 1008.
11. Mullens, J. *Alcoholassociaten*, doctoraatsproefschrift Leuven, 1971.; Huyskens, P.; Mullens, J.; Gomez, A.; Tack, J. *Bull. Soc. Chim. Belg.* 1975, *84*, 253.
12. Shakhud, Zh. N.; Markuzin, N.P.; Storonkin, A.V. *Vest. Leningrad Univ. Ser. Fiz. Khim* 1972, *(10)*, 85; 89.
13. Moriyoshi, T.; Kaneshina, S.; Aihara, K.; Yabumoto, K. *J. Chem. Thermodyn.* 1975, *7*, 537.
14. De Santis, R.; Marrelli, L.; Muscetta, P.N. *Chem. Eng. J.* 1976, *11*, 207.
15. Becke, A.; Quitzch, G. *Chem. Tech.* 1977, *29*, 49.
16. Rabinovich, I.B.; Fedorov, V.D.; Pashkin, N.P.; Avdesnyak, M.A.; Pimenov, N, Ya. *Dokl. Akad/Nauk SSSR* 1955, *105*, 108.

COMPONENTS:

(1) 2-Butanol; *(sec-butanol)*; $C_4H_{10}O$; [78-92-2]

(2) Water; H_2O; [7732-18-5]

ORIGINAL MEASUREMENTS:

Alexejew, W.

Ann. Phys. Chem. 1886, *28*, 305-38.

VARIABLES:

Temperature: 10-103°C

PREPARED BY:

A. Maczynski; Z. Maczynska; A. Szafranski

EXPERIMENTAL VALUES:

Mutual solubility of 2-butanol (1) and water (2)

t/°C	g(1)/100g sln		x_1 (compiler)	
	(2)-rich phase	(1)-rich phase	(2)-rich phase	(1)-rich phase
10.1	-	55.2	-	0.230
20.9	-	60.3	-	0.270
24	20.1	-	0.0576	-
28.5	-	62.4	-	0.287
32	17.1	-	0.0477	-
34	15.9	-	0.0439	-
56	14.7	-	0.0402	-
72	14.7	-	0.0402	-
77.1	-	62.4	-	0.287
84.5	-	60.3	-	0.270
88.3	15.9	-	0.0439	-
96.1	-	55.2	-	0.230
97.5	-	54.9	-	0.228
97.6	17.1	-	0.0477	-
101.5	20.1	-	0.0576	-
102.5	-	48.8	-	0.188

AUXILIARY INFORMATION

METHOD/APPARATUS/PROCEDURE:

The synthetic and/or analytical method was used, the latter only when the solubility diminished with temperature.

Into a tared glass tube (1) was introduced and weighed, then (2) was added through a capillary funnel. The tube was sealed, reweighed, fastened to the bulk of a mercury thermometer and repeatedly heated and cooled in a water (or glycerol) bath until the mixture became respectively homogeneous and turbid.

SOURCE AND PURITY OF MATERIALS:

(1) not specified.

(2) not specified.

ESTIMATED ERROR:

Not specified.

REFERENCES:

COMPONENTS:

(1) 2-Butanol (*sec-butanol*); $C_4H_{10}O$; [78-92-2]

(2) Water; H_2O; [7732-18-5]

ORIGINAL MEASUREMENTS:

Dolgolenko, W.

Z. Phys. Chem. 1908, *62*, 499-511.

VARIABLES:

Temperature: -23 - 115 $^{\circ}$C

Distillation fractions of (1): 98.0-99.5 $^{\circ}$C

PREPARED BY:

G.T. Hefter

EXPERIMENTAL VALUES:

The mutual solubility of 2-butanol and water was determined for three boiling range fractions of 2-butanol. As quite considerable differences were observed, data reported by the author for all three fractions are given in Tables a to c below. Table d contains mutual solubility data for a separate sample of 2-butanol supplied to the author by N.P. Ipatiew; the data are similar to those in Table a.

Note that the third distillation fraction (Table c) shows a closed solubility loop with a lower critical solution temperature of *ca* 70 $^{\circ}$C. This is attributed by the author to contamination of the 2-butanol by *tertiary*-butanol or *iso*-butanol. The upper critical solution temperature varies from *ca.* 106-115 $^{\circ}$C depending on the fraction.

$t/^{\circ}C$	g(1)/100g sln		x_1	
	(2)-rich phase	(1)-rich phase	(2)-rich phase	(1)-rich phase
a. First distillation fraction of 2-butanol				
B.p. 98.0-97.6 $^{\circ}$C at 760.8 mm Hg; d_4^{20} 0.80596				
7.0	-	35.76	-	0.119
8.6	-	43.36	-	0.157
9.9	29.76	-	0.0934	-

(continued next page)

AUXILIARY INFORMATION

METHOD/APPARATUS/PROCEDURE:

The synthetic method of Alexejew (ref 1) was used. The two components were carefully weighed into glass tubes which were fitted into a wooden holder and placed in a large beaker or, for low temperature work, a Dewar flask. The temperature was then lowered and raised whilst the tubes were shaken and the turbidity temperature was noted until consistent results were obtained. See also comments above.

SOURCE AND PURITY OF MATERIALS:

(1) Kahlbaum; dried with barium oxide then fractionated.

(2) not specified.
Other details given in the data tables.

ESTIMATED ERROR:

Temperature: 0.05 $^{\circ}$C

Composition: See data

REFERENCES:

1. Alexejew, W. *Russ. Berg-J.* 1879; 1885.

COMPONENTS:	ORIGINAL MEASUREMENTS:
(1) 2-Butanol (*sec-butanol*); $C_4H_{10}O$; [78-92-2]	Dolgolenko, W.
(2) Water; H_2O; [7732-18-5]	*Z. Phys. Chem.* 1908, *62*, 499-511.

EXPERIMENTAL VALUES (continued)

$t/^{\circ}C$	g(1)/100g sln		x_1	
	(2)-rich phase	(1)-rich phase	(2)-rich phase	(1)-rich phase
11.2	28.55	-	0.0885	-
11.3	-	48.92	-	0.189
15.7	-	53.77	-	0.220
17.2	-	55.02	-	0.229
21.9	21.77	-	0.0634	-
22.4	-	58.19	-	0.253
26.5	-	59.89	-	0.261
29.9	18.77	-	0.0532	-
32.5	18.15	-	0.0512	-
37.3	-	62.96	-	0.292
38.3	-	63.18	-	0.294
46.3	15.51	-	0.0427	-
53.5	14.83	-	0.0406	-
58.0	14.59	-	0.0399	-
71.7	-	63.18	-	0.294
73.0	14.59	62.96	0.0399	0.292
77.6	14.83	-	0.0406	-
83.8	15.51	-	0.0427	-
85.7	-	59.89	-	0.266
90.3	-	58.19	-	0.253
95.2	18.15	-	0.0512	-
96.0	-	55.02	-	0.229
96.9	18.77	-	0.0532	-
97.7	-	53.77	-	0.220
103.0	21.77	-	0.0634	-
103.2	-	48.92	-	0.189
105.7	-	43.36	-	0.157
106.8	28.55	-	0.0885	-
106.9	-	35.76	-	0.119
107.0	29.76	-	0.0934	-

(continued next page)

COMPONENTS:	ORIGINAL MEASUREMENTS:
(1) 2-Butanol (*sec-butanol*); $C_4H_{10}O$; [78-92-2]	Dolgolenko, W.
(2) Water; H_2O; [7732-18-5]	*Z. Phys. Chem.* 1908, *62*, 499-511.

EXPERIMENTAL VALUES (continued)

b. Second distillation fraction of 2-butanol:
b.p. 98.6-99.0°C at 760.8 mm Hg, d_4^{20} 0.80619.

t/°C	g(1)/100g sln		x_1	
	(2)-rich phase	(1)-rich phase	(2)-rich phase	(1)-rich phase
-17.2	25.54	-	0.0770	-
-8.8	27.10	-	0.0829	-
-6.2	-	58.75	-	0.257
-5.4	27.70	-	0.0852	-
-3.0	-	56.90	-	0.243
1.3	-	55.73	-	0.234
2.3	27.70	-	0.0852	-
4.3	-	55.73	-	0.234
4.7	27.10	-	0.0829	-
9.0	25.54	56.90	0.0770	0.243
13.9	-	58.75	-	0.257
15.0	22.89	-	0.0673	-
19.8	20.87	-	0.0603	-
21.2	-	61.49	-	0.280
24.7	-	62.50	-	0.288
29.4	17.90	-	0.0503	-
30.1	-	63.81	-	0.300
37.5	16.16	-	0.0448	-
41.3	15.50	-	0.0426	-
77.2	-	63.81	-	0.300
83.1	-	62.50	-	0.288
85.8	-	61.49	-	0.280
88.1	15.50	-	0.0426	-
91.9	16.16	-	0.0448	-
93.5	-	58.75	-	0.257
97.3	-	56.90	-	0.243
97.9	17.90	-	0.0503	-
99.4	-	55.73	-	0.234
100.0	-	55.24	-	0.231
100.1	-	55.20	-	0.231
103.0	-	52.99	-	0.215
104.1	20.87	-	0.0603	-
105.0	-	51.01	-	0.202
106.5	22.89	-	0.0673	-
107.0	-	47.97	-	0.183
108.2	25.54	-	0.0770	-
108.9	-	42.49	-	0.152

(continued next page)

COMPONENTS:

(1) 2-Butanol (*sec-butanol*); $C_4H_{10}O$; [78-92-2]

(2) Water; H_2O; [7732-18-5]

ORIGINAL MEASUREMENTS:

Dolgolenko, W.

Z. Phys. Chem. 1908, *62*, 499-511.

EXPERIMENTAL VALUES (continued)

$t/^{\circ}C$	g(1)/100g sln		x_1	
	(2)-rich phase	(1)-rich phase	(2)-rich phase	(1)-rich phase
109.0	27.70	-	0.0829	-
109.2	-	40.08	-	0.140
109.3	29.20	-	0.0911	-
109.4	30.20	36.16	0.0951	0.121

c. Third distillation fraction of 2-butanol:
b.p. 99.0-99.5 $^{\circ}C$ at 760.8 mm Hg, d_4^{20} 0.80663

$t/^{\circ}C$	g(1)/100g sln (2)-rich phase	g(1)/100g sln (1)-rich phase	x_1 (2)-rich phase	x_1 (1)-rich phase
-23.4	20.28	-	0.0582	-
-21.0	22.81	-	0.0670	-
-14.9	23.89	-	0.0709	-
-7.5	-	63.65	-	0.299
-2.3	-	61.63	-	0.281
0.7	-	61.09	-	0.276
1.2	23.89	-	0.0709	-
5.9	-	61.09	-	0.276
6.2	22.81	-	0.0670	-
10.0	-	61.63	-	0.281
14.4	20.28	-	0.0582	-
18.0	19.02	-	0.0540	-
18.2	-	63.65	-	0.299
27.6	-	65.64	-	0.317
33.1	15.32	-	0.0421	-
40.2	-	66.68	-	0.327
44.9	13.60	-	0.0369	-
62.0	-	66.68	-	0.327
76.2	-	65.64	-	0.317
85.5	13.60	-	0.0369	-
85.8	-	63.65	-	0.299
92.0	-	61.63	-	0.281
93.4	-	61.09	-	0.276
95.5	15.32	-	0.0421	-
97.0	-	59.73	-	0.265
105.1	-	55.18	-	0.230
106.4	19.02	-	0.0540	-
106.9	-	53.79	-	0.221
109.1	20.28	-	0.0582	-
111.4	-	49.71	-	0.194
112.1	22.81	48.45	0.0670	0.186

(continued next page)

COMPONENTS:

(1) 2-Butanol (*sec-butanol*); $C_4H_{10}O$; [78-92-2]

(2) Water; H_2O; [7732-18-5]

ORIGINAL MEASUREMENTS:

Dolgolenko, W.

Z. Phys. Chem. 1908, *62*, 499-511.

EXPERIMENTAL VALUES (Continued)

$t/^{\circ}C$	g(1)/100g sln		x_1	
	(2)-rich phase	(1)-rich phase	(2)-rich phase	(1)-rich phase
112.9	23.89	-	0.0709	-
113.4	-	45.27	-	0.167
113.5	25.18	-	0.0756	-
114.0	-	43.31	-	0.157
114.3	27.60	-	0.0848	-
114.5	-	40.09	-	0.140
114.8	31.23	-	0.0994	-
114.9	35.20	-	0.117	-

d. 2-Butanol supplied by N.P. Ipatiew:

b.p. 98.5-98.8°C at 750.4 mm Hg; d_4^{20} 0.80656

$t/^{\circ}C$	g(1)/100g sln (2)-rich phase	g(1)/100g sln (1)-rich phase	x_1 (2)-rich phase	x_1 (1)-rich phase
-8.4	25.95	-	0.0786	-
-5.7	-	61.46	-	0.280
-4.3	26.33	-	0.0800	-
-3.5	-	60.34	-	0.270
-2.6	26.33	-	0.0800	-
-2.3	-	59.88	-	0.266
1.02	25.95	-	0.0786	-
2.22	-	59.16	-	0.261
2.32	-	59.16	-	0.261
8.02	-	59.88	-	0.266
8.07	23.96	-	0.0712	-
9.72	-	60.34	-	0.270
13.22	-	61.46	-	0.280
14.97	21.46	-	0.0623	-
20.30	19.59	-	0.0560	-
28.47	17.2	-	0.0482	-
91.6	-	61.46	-	0.280
94.5	-	60.34	-	0.270
95.4	-	59.88	-	0.266
97.5	-	59.16	-	0.261
99.5	17.2	-	0.0482	-
99.9	-	57.89	-	0.251
103.0	-	56.20	-	0.238
105.3	19.59	-	0.0560	-
108.1	21.46	-	0.0623	-
111.4	23.96	-	0.0712	-
112.5	25.95	-	0.0786	-
112.7	26.33	-	0.0800	-

COMPONENTS:

(1) 2-Butanol; *(sec-butanol)*; $C_4H_{10}O$; [78-92-2]

(2) Water; H_2O; [7732-18-5]

ORIGINAL MEASUREMENTS:

Timmermans, J.

Arch. Neerl. Sci. Exactes Nat. 1922, *A6(3)*, 147; *J. Chim. Phys. Physicocim. Biol.* 1923, *20*, 491.

VARIABLES:

Temperature: (-9) - 114 °C
Pressure: 0.1-81 MPa (1-800 atm)

PREPARED BY:

A. Maczynski; G.T. Hefter

EXPERIMENTAL VALUES:

p/atm	p/MPa (compiler)	lower critical temperature/°C	upper critical temperature/°C
1	0.101	-8.45	113.8
100	10.1	5.3	105.3
120	12.1	-	-
200	20.3	17.8	-
300	30.4	24.8	-
400	40.5	30.8	-
500	50.7	36.7	-
600	60.8	42.7	85.3
700	71.0	49.6	81.3
800	81.1	58.6	73.5
830	84.1	Homogeneous at all temperatures	

dT_c/dp (LCST)[a] = + 0.064 K atm^{-1}

dT_c/dp (UCST)[a] = - 0.045 K atm^{-1}

[a] Not reported in original papers but calculated in ref 1.

AUXILIARY INFORMATION

METHOD/APPARATUS/PROCEDURE:

The method and apparatus were described in an earlier reference which is not identified in the paper.

SOURCE AND PURITY OF MATERIALS:

(1) source not specified; b.p. 99.50 ± 0.01 °C, d_4^0 0.82263 ± 0.00002.

(2) not specified.

ESTIMATED ERROR:

Not specified.

REFERENCES:

1. Moriyoshi, T.; Kaneshina, S.; Aihara, K.; Yabumoto, K. *J. Chem. Thermodynamics* 1975, *7*, 537.

COMPONENTS:

(1) 2-Butanol; *(sec-butanol)*; $C_4H_{10}O$; [78-92-2]

(2) Water; H_2O; [7732-18-5]

ORIGINAL MEASUREMENTS:

Evans, T.W.

Anal. Chem. 1936, *8*, 206-8.

VARIABLES:

One temperature: 20°C

PREPARED BY:

A. Maczynski

EXPERIMENTAL VALUES:

The solubility of 2-butanol in water at 20°C was reported to be 18.5 g(1)/100g sln.

The corresponding mole fraction, x_1, calculated by the compiler is 5.23×10^{-2}.

The solubility of water in 2-butanol at 20°C was reported to be 35.6 g(2)/100g sln.

The corresponding mole fraction, x_1, calculated by the compiler is 0.695.

AUXILIARY INFORMATION

METHOD/APPARATUS/PROCEDURE:

Hill's method (ref 1) was used.

Weighed amounts of (1) and (2) were introduced into centrifuge tubes and shaken until equilibrium was reached. The phases were allowed to separate and the volumes of the upper and lower layers were read. The experiment was then repeated with a different ratio of the starting materials. The mutual solubility data were calculated from phase volumes and weights.

SOURCE AND PURITY OF MATERIALS:

(1) not specified.

(2) not specified.

ESTIMATED ERROR:

Solubility: 0.1-0.2% (type of error not specified).

REFERENCES:

1. Hill. A.E. *J. Am. Chem. Soc.* 1923, *45*, 1143.

COMPONENTS:

(1) 2-Butanol; *(sec-butanol)*; $C_4H_{10}O$; [78-92-2]

(2) Water; H_2O; [7732-18-5]

ORIGINAL MEASUREMENTS:

Altsybeeva, A.I.; Morachevskii, A.G. *Zh. Fiz. Khim.* 1964, *38*, 1574-9.

Altsybeeva, A.I.; Belousov, V.N.; Ovtraht, H.V.; Morachevskii, A.G. *Zh. Fiz. Khim.* 1964, *38*, 1242-7.

VARIABLES:

Temperature: 20-70°C

PREPARED BY:

A. Maczynski

EXPERIMENTAL VALUES:

Mutual solubility in the system 2-butanol (1) and water (2)

$t/^{\circ}C$	x_1		g(1)/100g sln(compiler)	
	(2)-rich phase	(1)-rich phase	(2)-rich phase	(1)-rich phase
20	0.058	0.301	20	63.9
40	0.046	0.330	17	67.0
60	0.039	0.326	14	66.6
70	0.039	0.322	14	66.2

AUXILIARY INFORMATION

METHOD/APPARATUS/PROCEDURE:

Probably the analytical method was used.

The refractometric and densimetric methods were described for the analysis of the ternary system 2-butanol/water/2-butanone, but nothing was specified for the determination of the solubility in the binary systems.

SOURCE AND PURITY OF MATERIALS:

(1) not specified;

(2) not specified.

ESTIMATED ERROR:

Not specified.

REFERENCES:

COMPONENTS:

(1) 2-Butanol; *(sec-butanol)*; $C_4H_{10}O$; [78-92-2]

(2) Water; H_2O; [7732-18-5]

ORIGINAL MEASUREMENTS:

Morachevskii, A.G.; Popovich, Z.P.

Zh. Prikl.Khim. 1965, *38*, 2129-31.

VARIABLES:

Temperature: 20-85 °C

PREPARED BY:

A. Maczynski; Z. Maczynska

EXPERIMENTAL VALUES:

Mutual solubility of 2-butanol (1) and water (2)

$t/^{\circ}C$	g(1)/100g sln		x_1 (compiler)	
	(2)-rich phase	(1)-rich phase	(2)-rich phase	(1)-rich phase
20	19.8	63.2	0.0566	0.294
40	15.6	66.5	0.0430	0.325
60	14.0	66.2	0.0381	0.332
80	14.0	63.8	0.0381	0.300
85	15.0	62.0	0.0411	0.284

AUXILIARY INFORMATION

METHOD/APPARATUS/PROCEDURE:

The analytical method was used.

A mixture of (1) and (2) was placed in a thermostatted bath and stirred for an hour. After some time (a few minutes to two hours) each phase was sampled. The compositions of the phases were determined by refraction using an IRF-22 refractometer.

SOURCE AND PURITY OF MATERIALS:

(1) source not specified; purified; b.p. 99.52 °C, n_D^{20} 1.3970, d_4^{20} 0.8067, 0.1-0.15% of water by Karl Fischer analysis.

(2) not specified.

ESTIMATED ERROR:

Temperature: ± 0.05 °C

Solubility : ± 0.3 g(1)/100 sln (type of error not specified.)

REFERENCES:

AWW-E

COMPONENTS:	ORIGINAL MEASUREMENTS:
(1) 2-Butanol (*sec-butanol*); $C_4H_{10}O$; [78-92-2] (2) Water; H_2O; [7732-18-5]	Ratouis, M.; Dodé, M. *Bull. Soc. Chim. Fr.* 1965, 3318-22.
VARIABLES:	PREPARED BY:
Temperature: 25-30°C Ringer solution also studied	S.C. Valvani; S.H. Yalkowsky; A.F.M. Barton

EXPERIMENTAL VALUES:

Solubility of 2-butanol in water-rich phase

$t/^{\circ}C$	g(1)/100g sln	x_1 (compiler)
25	17.70	0.0497
35	16.50	0.0458

Solubility of 2-butanol in Ringer solution

$t/^{\circ}C$	g(1)/100g sln
25	15.79
30	14.90

AUXILIARY INFORMATION

METHOD/APPARATUS/PROCEDURE:

In a round bottom flask, 50 mL of water and a sufficient quantity of alcohol were introduced until two separate layers were formed. The flask assembly was equilibrated by agitation for at least 3 h in a constant temperature bath. Equilibrium solubility was attained by first supersaturation at a slighly lower tempature (solubility of alcohols in water is inversely proportional to temperature) and then equilibriating at the desired temperature. The aqueous layer was separated after an overnight storage in a bath. The alcohol content was determined by reacting the aqueous solution with potassium dichromate and titrating the excess dichromate with ferrous sulfate solution in the presence of phosphoric acid and diphenylamine barium sulfonate as an indicator.

SOURCE AND PURITY OF MATERIALS:

(1) Prolabo, Paris; redistilled with 10:1 reflux. b.p. 99.1°C/762.9 mm Hg n_D^{25} 1.39534

(2) twice distilled from silica apparatus or ion exchanged with Sagei A20.

ESTIMATED ERROR:

Solubility: relative error of 2 determinations less than 1%
Temperature: ± 0.05°C

REFERENCES:

COMPONENTS:

(1) 2-Butanol (*sec-butanol*); $C_4H_{10}O$; [78-92-2]

(2) Water; H_2O; [7732-18-5]

ORIGINAL MEASUREMENTS:

*Mullens, J.

Alcoholassociaten, Doctoraatsproefschrift, Leuven, 1971.

Huyskens, P.; Mullens, J.; Gomez, A.; Tack, J. *Bull. Soc. Chim. Belg.* 1975, *84*, 253-62.

VARIABLES:

One temperature: 25°C

PREPARED BY:

M.C. Haulait-Pirson; A.F.M. Barton

EXPERIMENTAL VALUES:

At equlibrium at 25°C the concentration of 2-butanol(1) in the water-rich phase was reported as 2.449 mol(1)/L sln, and the concentration of water(2) in the alcohol-rich phase was reported as 7.825 mol(2)/L sln.

The corresponding solubilities on a mass/volume basis are 181.5 g(1)/L sln and 141.0 g(2)/L sln (compiler).

AUXILIARY INFORMATION

METHOD/APPARATUS/PROCEDURE:

The partition of the two components was made using a cell described in ref 1. The Rayleigh Interference Refractometer M154 was used to determine the concentrations. Standard solutions covering the whole range of concentration investigated were used for the calibration.

SOURCE AND PURITY OF MATERIALS:

(1) Merck (p.a.)

(2) distilled

ESTIMATED ERROR:

Solubility ± 0.001 mol(1)/L sln.

REFERENCES:

1. Meeussen, E.; Huyskens, P. *J. Chim. Phys.* 1966, *63*, 845.

COMPONENTS:

(1) 2-Butanol; *(sec-butanol)*; $C_4H_{10}O$; [78-92-2]

(2) Water; H_2O; [7732-18-5]

ORIGINAL MEASUREMENTS:

Shakhud, Zh.N.; Markuzin, N.P.; Storonkin, A.V.

Vestn. Leningr. Univ. Ser. Fiz. Khim. 1972 *(10)* 85-8; 89-92.

VARIABLES:

Temperature: 20-60°C

PREPARED BY:

A. Maczynski

EXPERIMENTAL VALUES:

Mutual solubility of 2-butanol (1) and water (2)

$t/°C$	x_1		g(1)/100g sln (compiler)	
	(2)-rich phase	(1)-rich phase	(2)-rich phase	(1)-rich phase
20	0.056	0.297	19.62	63.48
40	0.045	0.340	16.24	67.94
40	0.045	0.321	16.24	66.05
60	0.038	0.334	13.98	67.36

AUXILIARY INFORMATION

METHOD/APPARATUS/PROCEDURE:

The titration method was used. Measurements were carried out in a round-bottom flask equipped with mercury seal and immersed in a thermostat. No more details were reported in the papers.

SOURCE AND PURITY OF MATERIALS:

(1) source not specified; twice distilled;

d_4^{20} 0.8069, n_D^{20} 1.3972.

(2) twice distilled, d_D^{20} 1.3332.

ESTIMATED ERROR:

Temperature: ± 0.02°C

Solubility: ± 1% (max. dev.)

REFERENCES:

COMPONENTS:

(1) 2-Butanol; *(sec-butanol)*; $C_4H_{10}O$; [78-92-2]

(2) Water; H_2O; [7732-18-5]

ORIGINAL MEASUREMENTS:

Moriyoshi, T.; Kaneshina, S.; Aihara, K.; Yabumoto, K.

J. Chem. Thermodynamics 1975, *7*, 537-45.

VARIABLES:

Temperature: 283-381 K
Pressure: 0.1-81 MPa (1-800 atm)

PREPARED BY:

A. Maczynski; Z. Maczynska; G.T. Hefter

EXPERIMENTAL VALUES:

Mutual solubility of 2-butanol(1) and water (2)

p/atm	p/MPa (compiler)	T/K	(2)-rich phase g(1)/100g sln	x_1 (compiler)	(1)-rich phase g(1)/100g sln	x_1 (compiler)
1	0.101	283.15	23.9*	0.0709	60.9*	0.274
		293.15	20.2*	0.0579	63.8*	0.300
		300.15	17.9*	0.0503	65.5*	0.316
		313.15	14.9*	0.0498	67.2*	0.332
		323.15	13.2*	0.0356	67.5*	0.335
		333.15	12.8*	0.0344	66.9*	0.329
		343.15	12.9*	0.0347	67.2*	0.332
		353.15	13.8*	0.0374	65.3*	0.314
		363.15	15.5*	0.0427	61.8*	0.282
		373.15	18.3*	0.0516	57.7*	0.249
		383.15	23.9*	0.0709	50.7*	0.200
100	10.1	283.15	26.9	0.0821	56.0	0.236
		293.15	22.8	0.0670	59.4	0.262
		300.15	20.0	0.0573	62.1	0.285
		313.15	16.5	0.0458	64.4	0.305
		323.15	14.6	0.0399	65.3	0.314
		333.15	13.9	0.0377	64.9	0.310
		343.15	14.1	0.0383	65.5	0.316
		353.15	15.1	0.0414	64.0	0.302
		363.15	17.0	0.0474	60.1	0.268
		373.15	20.8*	0.0600	53.5*	0.218
		380.55	25.6*	0.0772	48.6*	0.187

(continued next page)

AUXILIARY INFORMATION

METHOD/APPARATUS/PROCEDURE:

Both components were placed in a cut-off glass syringe of about 20 cm^3 capacity used as a sample vessel, which was placed in a thick-walled stainless-steel cylinder of about 56 cm^3 capacity as a pressure vessel and connected to a high pressure pump. The pressure vessel was mechanically shaken in an oil thermostat bath. After stirring for at least 6 h and then standing for another 6 h at the desired condition of temperature and pressure a sample of the upper layer was taken. Subsequently, the pressure vessel was moved, the contents allowed to settle, and the lower layer sampled. The analysis of samples was made refractometrically.

SOURCE AND PURITY OF MATERIALS:

(1) source not specified; dried by refluxing with freshly burned lime, distilled; n_D^{25} 1.3951

(2) de-ionized, distilled from alkaline potassium permanganate, redistilled.

ESTIMATED ERROR:

Temperature: ± 0.02 K
Solubility : ± 0.35 g(1)/100g sln (type of error not specified)

REFERENCES:

COMPONENTS:	ORIGINAL MEASUREMENTS:
(1) 2-Butanol; *(sec-butanol)*; $C_4H_{10}O$; [78-92-2]	Moriyoshi, T.; Kaneshina, S.; Aihara, K.; Yabumoto, K.
(2) Water; H_2O; [7732-18-5]	*J. Chem. Thermodynamics* 1975, *7*, 537-45.

EXPERIMENTAL VALUES (continued)

p/atm	p/MPa (compiler)	T/K	(2)-rich phase g(1)/100g sln	(2)-rich phase x_1 (compiler)	(1)-rich phase g(1)/100g sln	(1)-rich phase x_1 (compiler)
200	20.3	289.15	35.2*	0.1166	45.9*	0.171
		293.15	27.9	0.0859	53.4	0.218
		300.15	23.4	0.0691	58.2	0.253
		313.15	18.4	0.0519	62.0	0.284
		323.15	16.2	0.0449	63.1	0.294
		333.15	15.0	0.0411	63.2	0.294
		343.15	15.3	0.0420	64.1	0.303
		353.15	16.4	0.0455	62.7	0.290
		363.15	18.6	0.0526	58.2	0.253
		373.15	22.6*	0.0662	51.0*	0.202
		378.15	29.3*	0.0915	46.1*	0.172
300	30.4	298.15	33.0*	0.1069	47.5*	0.180
		300.15	28.9	0.0899	51.2	0.203
		313.15	20.5	0.0590	59.9	0.266
		323.15	17.9	0.0503	61.2	0.277
		333.15	16.3	0.0452	61.7	0.281
		343.15	16.5	0.0458	62.5	0.288
		353.15	17.8	0.0500	61.1	0.276
		363.15	20.4	0.0586	55.7	0.234
		373.15	33.2	0.1077	41.6	0.147
400	40.5	304.15	31.4*	0.1001	46.6*	0.175
		313.15	23.0	0.0677	57.8	0.250
		323.15	19.5	0.0556	59.4	0.262
		333.15	17.8	0.0500	60.1	0.268
		343.15	17.7	0.0497	60.7	0.273
		353.15	19.1	0.0543	59.2	0.261
		363.15	22.9*	0.0673	52.3*	0.210
		368.15	28.9*	0.0899	46.1*	0.172
500	50.7	309.15	39.6*	0.1374	40.5*	0.142
		131.15	26.8	0.0817	52.8	0.214
		323.15	21.4	0.0620	57.0	0.244
		333.15	19.4	0.0552	58.2	0.253
		343.15	19.0	0.0539	58.5	0.255
		353.15	20.6	0.0593	56.4	0.239
		363.15	27.3	0.0836	47.1	0.178
600	60.8	316.15	32.4*	0.1043	45.9*	0.171
		323.15	24.1	0.0716	53.0	0.215
		333.15	21.7	0.0631	55.4	0.232
		343.15	21.0	0.0607	55.9	0.235
		353.15	22.8*	0.0670	52.6*	0.212
		360.15	31.2*	0.0992	43.1*	0.155
700	71.0	323.15	31.6	0.1009	44.0	0.160
		333.15	24.9	0.0746	51.4	0.204
		343.15	23.9	0.0709	52.5	0.212
		353.15	26.9	0.0821	46.9	0.177
800	81.1	333.15	30.8*	0.0976	44.4*	0.162
		338.15	29.1*	0.0907	46.3*	0.173
		343.15	29.1*	0.0907	46.1*	0.172
		348.15	31.8*	0.1018	43.3*	0.156

*from direct measurements; all other values obtained by interpolation (by the original authors) of direct measurements.

(continued next page)

COMPONENTS:

(1) 2-Butanol (*sec-butanol*); $C_4H_{10}O$; [78-92-2]

(2) Water; H_2O: [7732-18-5]

ORIGINAL MEASUREMENTS:

Moriyoshi, T.; Kaneshina, S.; Aihara, M.; Yabumoto, K.

J. Chem. Thermodynamics 1975, *7*, 537-45

EXPERIMENTAL VALUES (continued)

Properties of the critical solutions

p/atm	T_c(LCST)/K	x_{1c}	T_c(UCST)/K	x_{1c}
1	-	-	390.35	-
100	-	-	387.15	-
160[a]	283.15[a]	0.149[a]	-	-
200	288.35	-	380.45	-
250[a]	293.15[a]	0.140[a]	-	-
300	297.05	-	373.45	-
303[a]	-	-	373.15[a]	0.131[a]
358[a]	300.15[a]	0.139[a]	-	-
400	302.45	-	369.35	-
500	309.15	-	365.55	-
552[a]	-	-	363.15[a]	0.129[a]
580[a]	313.15[a]	0.143[a]	-	-
600	314.95	-	360.95	-
700	322.05	-	355.45	-
714[a]	323.15[a]	0.134[a]	-	-
751[a]	-	-	353.15[a]	0.133[a]
800	330.65	-	350.15	-
830[a]	333.15[a]	0.130[a]	-	-
840[a]	-	-	343.15[a]	0.131[a]

[a] Calculated by the authors using the "rectilinear diameter law". All other values obtained (by the authors) by graphical interpolation of the experimental results.

The hypercritical point, where the closed solubility loop disappears, was estimated to occur at (340.0 ± 1.5) K, (845 ± 5) atm, and x_1 = 0.131 ± 0.002

dT_c/dp(LCST) = + 0.060 K atm^{-1}

dT_c/dp(UCST) = – 0.046 K atm^{-1}

$dx_{1c}/dT \simeq 0\ K^{-1}$

COMPONENTS:

(1) 2-Butanol *(sec-butanol)*; $C_4H_{10}O$; [78-92-2]

(2) Water; H_2O; [7732-18-5]

ORIGINAL MEASUREMENTS:

De Santis, R.; Marrelli, L.; Muscetta, P.N.

Chem. Eng. J., 1976, *11*, 207-14.

VARIABLES:

One temperature: 25°C

PREPARED BY:

A. Maczynski

EXPERIMENTAL VALUES:

The proportion of 2-butanol(1) in the water-rich phase at equilibrium at 25°C was reported to be 22.5 g(1)/100g sln.

The corresponding mole fraction solubility, x_1, calculated by the compiler, is 0.0659.

The proportion of water(2) in the alcohol-rich phase at equilibrium at 25°C was reported to be 33.1 g(2)/100g sln.

The corresponding mole fraction solubility, x_2, calculated by the compiler, is 0.671.

AUXILIARY INFORMATION

METHOD/APPARATUS/PROCEDURE:

The determinations were carried out using a separating funnel with a thermostatic jacket. The extractor was loaded with (1) and (2) and after an extended period of mixing and quantitative gravity separation, samples were withdrawn from the aqueous phase. The concentration of (1) in (2) was determined by colorimetric analysis (double-beam Lange colorimeter) of the cerium complex. The concentration of (2) in (1) was derived from a material balance based upon starting quantities and compositions. Each of the determinations was carried out with several repetitions. The method is described in ref 1.

SOURCE AND PURITY OF MATERIALS:

(1) Merck, analytical purity; fractionated before use

(2) doubly distilled

ESTIMATED ERROR:

Temperature: ± 0.1°C

REFERENCES:

1. De Santis, R.; Marelli, L.; Muscetta, P.N. *J. Chem. Eng. Data*, 1976, 21, 324.

COMPONENTS:

(1) 2-Butanol (*sec-butanol*); $C_4H_{10}O$; [78-92-2]

(2) Water; H_2O; [7732-18-5]

ORIGINAL MEASUREMENTS:

Becke, A.; Quitzch, C.

Chem. Tech., 1977, *29*, 49-51.

VARIABLES:

One temperature: 25°C

PREPARED BY:

A. Maczynski

EXPERIMENTAL VALUES:

The mole fraction of 2-butanol in the water-rich phase at equilibrium at 25°C was reported to be x_1 = 0.035, corresponding to 13 g(1)/100g sln (compiler).

The mole fraction of water in the alcohol-rich phase at equilibrium at 25°C was reported to be x_2 = 0.301, corresponding to 9.47 g(2)/100g sln (compiler).

Graphical data at 60°C were also presented.

AUXILIARY INFORMATION

METHOD/APPARATUS/PROCEDURE:

The refractometric and the Karl-Fischer dead-stop titration methods were used.

No more details are given in the paper.

SOURCE AND PURITY OF MATERIALS:

(1) not specified

(2) not specified

ESTIMATED ERROR:

Solubility: 0.05-1% for (1) and (2)
0.3 - 1.3% for (2) in (1)
(relative error)

REFERENCES:

COMPONENTS:

(1) 2-Butanol-d; C_4H_9DO; [4712-39-4]

(2) Water-d_2; D_2O; [7789-20-0]

ORIGINAL MEASUREMENTS:

Rabinovich, I.B., Fedorov, V.D., Pashkin, N.P., Avdesnyak, M.A., Pimenov, N. Ya.

Dokl. Akad. Nauk SSSR 1955, *105*, 108-11.

VARIABLES:

Temperature: 10-120°C

PREPARED BY:

G. Jancso and G.T. Hefter.

EXPERIMENTAL VALUES:

Effect of deuteration on solubility of 2-butanol in water

t/°C	$100(L_H - L_D)L_H$ [a]
10	9.0
20	7.8
26	7.0
40	5.5
66	4.7
80	6.3
100	14
103	17
108	22

[a] L_H and L_D are the solubilities (in mol %) of 2-butanol in water and 2-butanol-d in D_2O respectively.

(continued next page)

AUXILIARY INFORMATION

METHOD/APPARATUS/PROCEDURE:

The synthetic method (cloud-clear points) of Alexejew (ref 1) was used which consists in observing the temperature at which a measured weight of one liquid is visually soluble in a measured weight of the other liquid.

SOURCE AND PURITY OF MATERIALS:

(1) 2-butanol-d was prepared by the hydrolysis of the aluminium salt of the ordinary butanol with D_2O.
The OD group of the product had a deuterium content of > 99 at %.

(2) D_2O, 99.0 - 99.8 at %, source not given.

ESTIMATED ERROR:

Solubility: not specified

Temperature: ± 0.2°C

REFERENCES:

1. Alexejew, W.J. *J. Prakt. Chem.* 1882, *25*, 518.

2. Rabinovich, I.B. *Influence of Isotopy on the Physicochemical Properties of Liquids.* (In Russian). Nauka. Moscow 1968, p.261.

COMPONENTS:	ORIGINAL MEASUREMENTS:
(1) 2-Butanol-d; C_4H_9DO; [4712-39-4] (2) Water-d_2; D_2O; [7789-20-0]	Rabinovich, I.B., Fedorov, V.D., Pashkin, N.P., Avdesnyak, M.A., Pimenov, N. Ya. *Dokl. Akad. Nauk SSSR* 1955, *105*, 108-11.

EXPERIMENTAL VALUES (continued)

The upper critical solution temperature of 2-butanol - H_2O was reported to be 113.1°C, compared with 119.3°C for 2-butanol-d - D_2O.

The solubility diagram (Figure 1) was taken from ref 2. Mutual solubilities for the 2-butanol - H_2O system were also determined (●), but only given in graphical form. The source of the literature data (o in Figure 1) is not given.

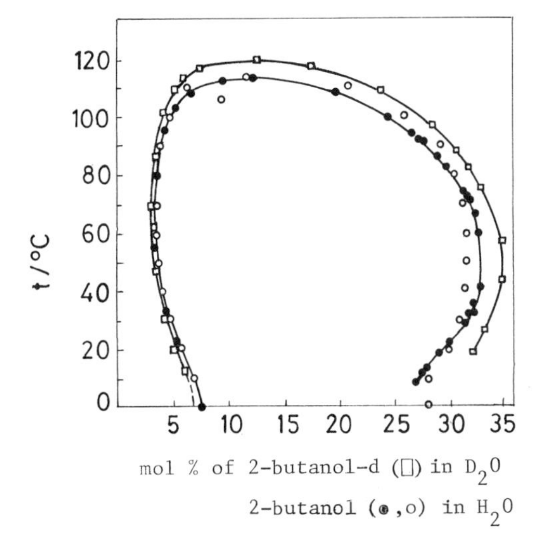

Fig. 1. Mutual solubilities of $C_4H_9OD(1)$ in $D_2O(2)$ (□) and of $C_4H_9OH(1)$ in $H_2O(2)$ (●,o).

COMPONENTS:	ORIGINAL MEASUREMENTS:
(1) 1-Penten-3-ol; $C_5H_{10}O$; [616-25-1] (2) Water; H_2O; [7732-18-5]	Ginnings, P.M.; Herring, E.; Coltrane, D. *J. Am. Chem. Soc.* 1939, *61*, 807-8.
VARIABLES:	**PREPARED BY:**
Temperature: 20-30°C	A. Maczynski; Z Maczynska

EXPERIMENTAL VALUES:

Mutual solubility of 1-penten-3-ol(1) and water(2)

t/°C	g(1)/100g sln (2)-rich phase	g(1)/100g sln (1)-rich phase	x_1(compiler) (2)-rich phase	x_1(compiler) (1)-rich phase
20	8.72	87.12	0.0196	0.5802
25	8.20	86.88	0.0183	0.5750
30	7.74	86.68	0.0172	0.5707

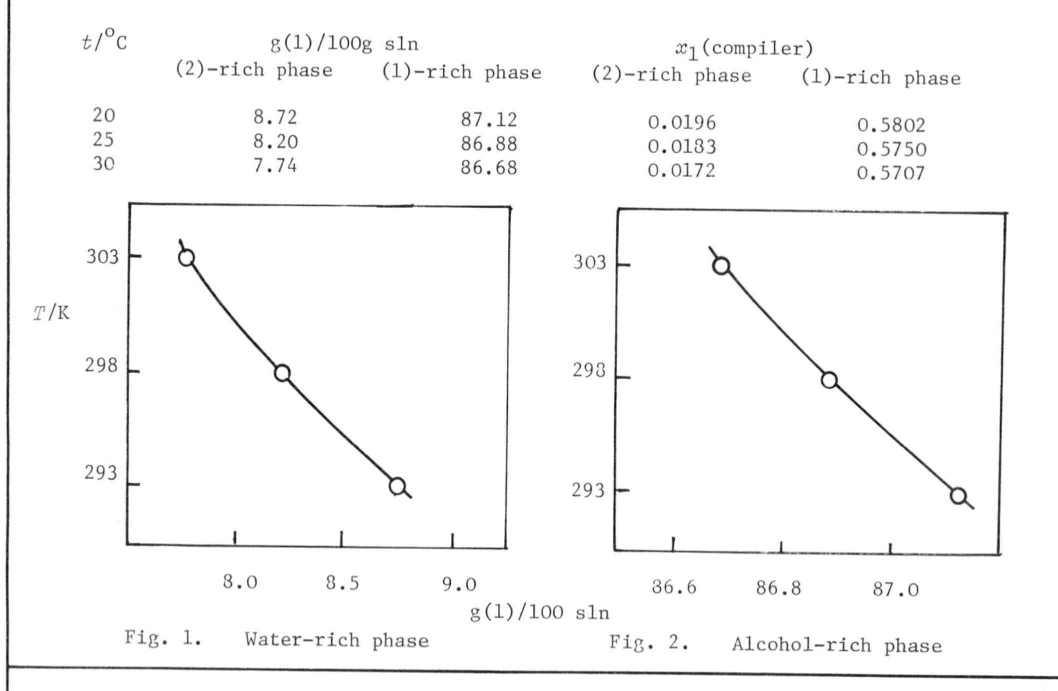

Fig. 1. Water-rich phase

Fig. 2. Alcohol-rich phase

AUXILIARY INFORMATION

METHOD/APPARATUS/PROCEDURE:

The volumetric method was used as described in ref 1.

Both components were introduced in known amounts into a two-bulb graduated and calibrated flask and shaken mechanically in a water-bath at constant temperature. After sufficient time the liquids were allowed to separate and the total volume was measured. Upon centrifugation, the phase separation line was read, and phase volumes were calculated. From the total weights of the components, the total volume, individual phase volumes, and component concentrations in either phase were evaluated.

Densities of the two phases were also determined.

SOURCE AND PURITY OF MATERIALS:

(1) prepared by Grignard synthesis; distilled; b.p. range 114.0-114.6°C, d_4^{25} 0.8347; purity not specified.

(2) not specified.

ESTIMATED ERROR:

Solubility: 0.3 wt % (type of error not specified)

REFERENCES:

1. Ginnings, P.M.; Baum, R.J. *J. Am. Chem. Soc.* 1937, *59*, 1111.

COMPONENTS:

(1) 3-Penten-2-ol; $C_5H_{10}O$; [1569-50-2]

(2) Water; H_2O; [7732-18-5]

ORIGINAL MEASUREMENTS:

Ginnings, P.M.; Herring, E.; Coltrane, D.

J. Am. Chem. Soc. 1939, *61*, 807-8.

VARIABLES:

Temperature: 20-30°C

PREPARED BY:

A. Maczynski and Z. Maczynska

EXPERIMENTAL VALUES:

Mutual solubility of 3-penten-2-ol(1) and water(2)

$t/°C$	g(1)/100g sln		x_1(compiler)	
	(2)-rich phase	(1)-rich phase	(2)-rich phase	(1)-rich phase
20	9.46	87.71	0.0214	0.5932
25	8.92	87.65	0.0201	0.5918
30	8.48	87.57	0.0190	0.5901

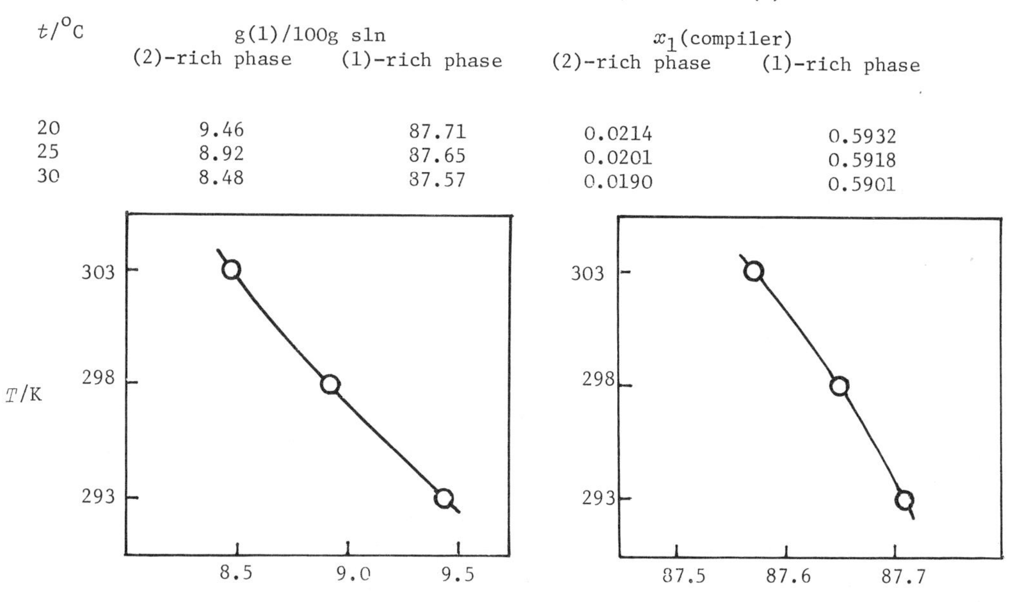

Fig.1. Water-rich phase Fig.2. Alcohol-rich phase

AUXILIARY INFORMATION

METHOD/APPARATUS/PROCEDURE:

The volumetric method was used as described in ref 1.

Both components were introduced in known amounts into a two-bulb graduated and calibrated flask and shaken mechanically in a water-bath at constant temperature. After sufficient time the liquids were allowed to separate and the total volume was measured. Upon centrifugation, the phase separation line was read, and phase volumes were calculated. From the total weights of the components, the total volume, individual phase volumes, and component concentrations in either phase were evaluated.

The density of each phase was also determined.

SOURCE AND PURITY OF MATERIALS:

(1) prepared by Grignard synthesis; distilled; b.p. range 121.7-124.2°C, d_4^{25} 0.8328; purity not specified.

(2) not specified.

ESTIMATED ERROR:

Solubility: 0.3 wt% (type of error not specified)

REFERENCES:

1. Ginnings, P.M.; Baum, R.J. *J. Am. Chem. Soc.* 1937, *59*, 1111.

COMPONENTS:

(1) 4-Penten-1-ol; $C_5H_{10}O$; [821-09-0]

(2) Water; H_2O; [7732-18-5]

ORIGINAL MEASUREMENTS:

Ginnings, P.M.; Herring, E.; Coltrane, D.

J. Am. Chem. Soc. 1939, *61*, 807-8.

VARIABLES:

Temperature: 20-30°C

PREPARED BY:

A. Maczynski and Z. Maczynska

EXPERIMENTAL VALUES:

Mutual solubility of 4-penten-1-ol(1) and water(2)

t/°C	g(1)/100g sln		x_1(compiler)	
	(2)-rich phase	(1)-rich phase	(2)-rich phase	(1)-rich phase
20	5.89	86.83	0.0129	0.5739
25	5.70	86.57	0.0125	0.5684
30	5.56	86.19	0.0122	0.5605

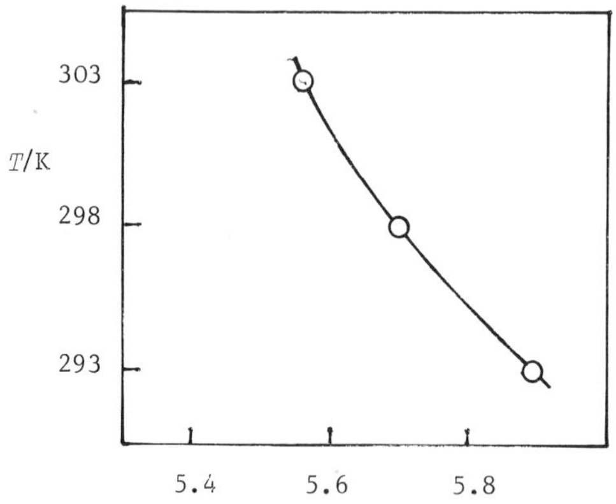

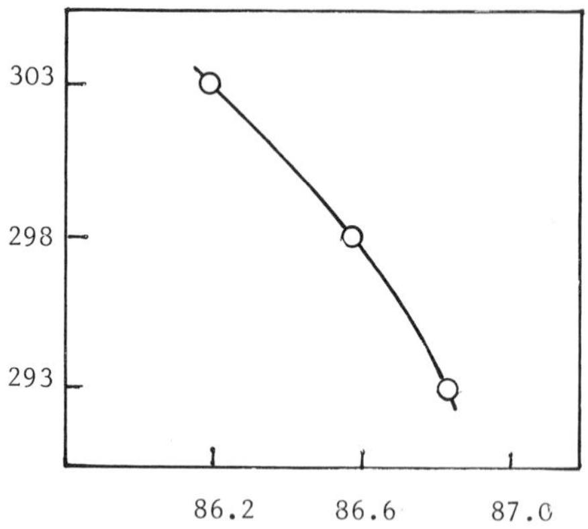

Fig.1. Water-rich phase

Fig.2. Alcohol-rich phase

AUXILIARY INFORMATION

METHOD/APPARATUS/PROCEDURE:

The volumetric method was used as described in ref 1.

Both components were introduced in known amounts into a two-bulb graduated and calibrated flask and shaken mechanically in a water-bath at constant temperature. After sufficient time the liquids were allowed to separate and the total volume was measured. Upon centrifugation, the phase separation line was read, and phase volumes were calculated. From the total weights of the components, the total volume, individual phase volumes, and component concentrations in either phase were evaluated.

The density of each phase was also determined.

SOURCE AND PURITY OF MATERIALS:

(1) prepared by Paul's synthesis; distilled; b.p.range 138.8-139.3°C, d_4^{25} 0.8578; purity not specified.

(2) not specified.

ESTIMATED ERROR:

Not specified

REFERENCES:

1. Ginnings, P.M.; Baum, R.J. *J. Am. Chem. Soc.* 1937, *59*, 1111.

COMPONENTS:	EVALUATOR:
(1) 2,2-Dimethyl-1-propanol (*neopentyl alcohol, tert-butyl carbinol*); $C_5H_{12}O$; [75-84-3] (2) Water; H_2O; [7732-18-5]	A. Maczynski, Institute of Physical Chemistry of the Polish Academy of Sciences, Warsaw, Poland; and A.F.M. Barton, Murdoch University, Perth, Western Australia. November 1982

CRITICAL EVALUATION:

Solubilities in the system comprising 2,2-dimethyl-1-propanol (1) and water (2) have been reported in two publications (Figures 1 and 2). Ginnings and Baum (ref 1) carried out measurements of the mutual solubilities of the components at 293, 298 and 303 K, by the volumetric method. Ratouis and Dodé (ref 2) determined the solubility of (1) in (2) at 298 and 303 K by an analytical method and their value 3.22 ± 0.1 g(1)/100g sln at 298 K is in reasonable agreement with the value 3.50 ± 0.1 g(1)/100g sln of ref 1. Although the 303 K value of 3.19 ± 0.1 g(1)/100g sln is in agreement with the value 3.28 ± 0.1 g(1)/100g sln given in ref 1, information is so limited that all values are regarded as tentative.

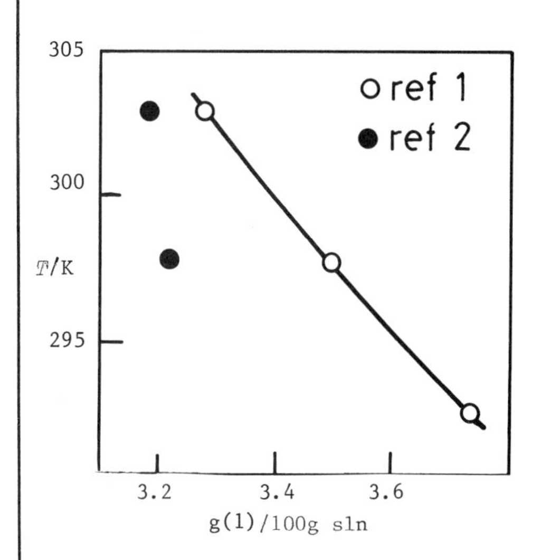

Fig. 1. Water-rich phase

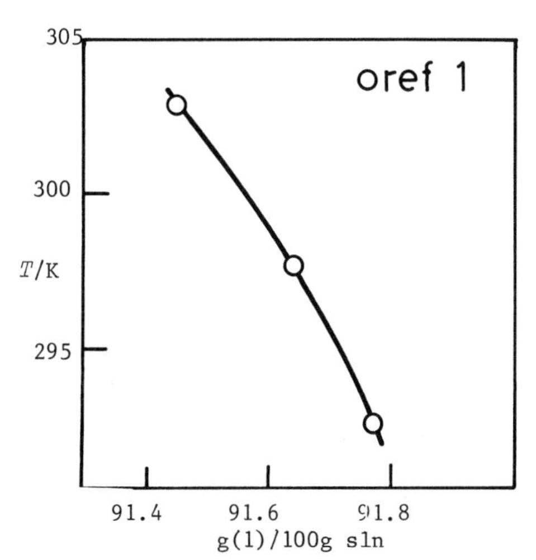

Fig. 2. Alcohol-rich phase

Tentative values for the mutual solubilities of 2,2-dimethyl-1-propanol (1) and water (2)

T/K	Water-rich phase		Alcohol-rich phase	
	g(1)/100g sln	x_1	g(2)/100g sln	x_2
293	3.7	0.0079	8.2	0.31
298	3.5	0.0074	8.4	0.31
303	3.3	0.0069	8.5	0.31

References

1. Ginnings, P.M.; Baum, R. *J. Am. Chem. Soc.* 1937, *59*, 1111.

2. Ratouis, M.; Dodé, M. *Bull. Soc. Chim. Fr.* 1965, 3318.

COMPONENTS:

(1) 2,2-Dimethyl-1-propanol *(neopentyl alcohol, tert-butylcarbinol)*; $C_5H_{12}O$; [75-84-3]

(2) Water; H_2O; [7732-18-5]

ORIGINAL MEASUREMENTS:

Ginnings, P.M.; Baum, R.

J. Am. Chem. Soc. 1937, *59*, 1111-3.

VARIABLES:

Temperature: 20-30°C

PREPARED BY:

A. Maczynski and Z. Maczynska

EXPERIMENTAL VALUES:

Mutual solubility of 2,2-dimethyl-1-propanol(1) and water(2)

$t/^\circ C$	g(1)/100g sln		x_1 (compiler)	
	(2)-rich phase	(1)-rich phase	(2)-rich phase	(1)-rich phase
20	3.74	91.77	0.00788	0.6950
25	3.50	91.64	0.00736	0.6913
30	3.28	91.46	0.00688	0.6863

Relative density d_4

$t/^\circ C$	Water-rich phase	Alcohol-rich phase
20	0.9936	0.8243
25	0.9930	0.8216
30	0.9925	0.8178

AUXILIARY INFORMATION

METHOD/APPARATUS/PROCEDURE:

The volumetric method was used.

Both components were introduced in known amounts into a two-bulb graduated and calibrated flask and shaken mechanically in a water-bath at constant temperature. After sufficient time the liquids were allowed to separate and the total volume was measured. Upon centrifugation, the phase separation line was read, and phase volumes were calculated. From the total weights of the components, the total volume individual phase volumes, and component concentrations in either phase were evaluated.

SOURCE AND PURITY OF MATERIALS:

(1) from the Grignard synthesis; distilled from metallic calcium; b.p. range 113.0-114.0°C, m.p. range 48-49°C; purity not specified.

(2) Not specified.

ESTIMATED ERROR:

Temperature: ± 0.1°C

Solubility : better than 0.1 wt % (type of error not specified)

REFERENCES:

COMPONENTS:

(1) 2,2-Dimethyl-1-propanol *(neopentyl alcohol, tert-butylcarbinol)*; $C_5H_{12}O$; [75-84-3]

(2) Water, H_2O; [7732-18-5]

ORIGINAL MEASUREMENTS:

Ratouis, M., Dodé, M.

Bull. Soc. Chim. Fr. 1965, 3318-22.

VARIABLES:

Temperature: 25-30 °C

Ringer solution also studied

PREPARED BY:

S.C. Valvani; S.H. Yalkowsky; A.F.M. Barton.

EXPERIMENTAL VALUES:

Proportion of (1) in water-rich phase

t/°C	g(1)/100g sln	x_1 (compiler)
25	3.22	0.00675
30	3.19	0.00669
30 (Ringer solution)	2.84	

AUXILIARY INFORMATION

METHOD/APPARATUS/PROCEDURE:

In a round bottomed flask, 50mL of water and a sufficient quantity of alcohol were introduced until two separate layers were formed. The flask assembly was equilibrated by agitation for at least 3 h in a constant temp bath. Equilibrium solubility was attained by first supersaturating at a slightly lower temperature (solubility of alcohols in water is inversely proportional to temperature) and then equilibrating at the desired temperature. The aqueous layer was separated after an overnight storage in a bath. The alcohol content was determined by reacting the aqueous solution with potassium dichromate and titrating the excess dichromate with ferrous sulfate solution in the presence of phosphoric acid and diphenylamine barium sulfonate as an indicator.

SOURCE AND PURITY OF MATERIALS:

(1) Fluka A.G.; Buchs S.G.; redistilled with 10:1 reflux; b.p. 112 °C/755.3 mm Hg

(2) twice distilled from silica apparatus or ion-exchanged with Sagei A20

ESTIMATED ERROR:

Solubility relative error of 2 determinations less than 1%

Temperature: ± 0.05 °C

REFERENCES:

COMPONENTS:	EVALUATOR:
(1) 2-Methyl-1-butanol (*"active" amyl alcohol, sec-butylcarbinol*); $C_5H_{12}O$; [137-32-6] (2) Water; H_2O; [7732-18-5]	A. Maczynski, Institute of Physical Chemistry of the Polish Academy of Sciences, Warsaw, Poland; and A.F.M. Barton, Murdoch University, Perth, Western Australia November 1982

CRITICAL EVALUATION:

Solubilities in the system comprising 2-methyl-1-butanol (1) and water (2) have been reported in three publications. Ginnings and Baum (ref 1) carried out measurements of the mutual solubilities of the components at 293, 298 and 303 K by the volumetric method. Crittenden and Hixon (ref 2), determined the mutual solubilities at 298 K, presumably by the titration method, while Ratouis and Dodé (ref 3) analyzed the water-rich phase at 303 K.

For the water-rich phase (Figure 1) the value 3.0 g(1)/100g sln given in ref 2 at 298 K is in good agreement with ref 1, 2.97 g(1)/100g sln, and is recommended. The value 2.61 g(1)/100g sln for the solubility of (1) in (2) given in ref 3 at 303 is not in good agreement with ref 1 and 2, and is excluded from consideration.

In the alcohol-rich phase (Figure 2), the value 90.8 g(1)/100g sln given in ref 2 at 298 K is in very good agreement with the value 90.81 g(1)/100g sln of ref 1, and is recommended. The temperature dependence reported in ref 1 is inconsistent with that of related systems, and this information is to be considered as tentative.

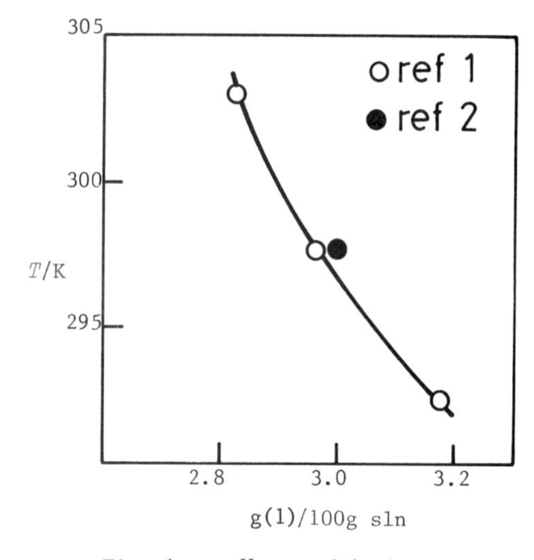

Fig. 1. Water-rich phase

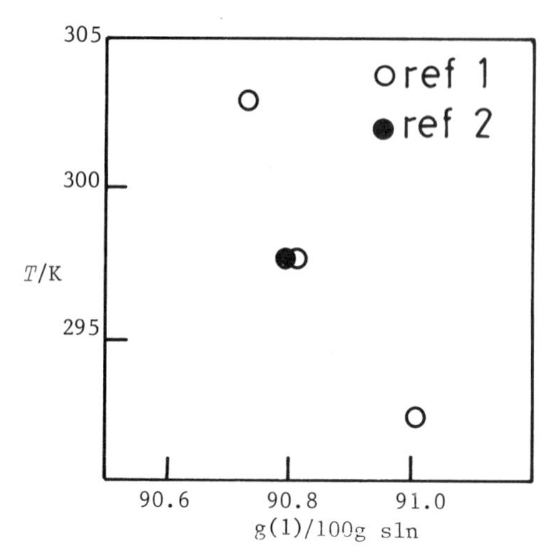

Fig. 2. Alcohol-rich phase

Recommended and tentative values for the mutual solubilities of 2-methyl-1-butanol (1) and water (2)

T/K	Water-rich phase		Alcohol-rich phase	
	g(1)/100g sln	10^3x_1	g(2)/100g sln	x_2
293	3.2 (tentative)	6.7	9.0 (tentative)	0.33
298	3.0 (recommended)	6.2	9.2 (recommended)	0.33
303	2.8 (tentative)	5.9	9.3 (tentative)	0.33

(continued next page)

COMPONENTS:	EVALUATOR:
(1) 2-Methyl-1-butanol (*"active" amyl alcohol, sec-butylcarbinol*); $C_5H_{12}O$; [137-32-6] (2) Water; H_2O; [7732-18-5]	A. Maczynski, Institute of Physical Chemistry of the Polish Academy of Sciences, Warsaw, Poland; and A.F.M. Barton, Murdoch University, Perth, Western Australia November 1982

CRITICAL EVALUATION (continued)

References

1. Ginnings, P.M.; Baum, R. *J. Am. Chem. Soc.* 1937, *59*, 1111.
2. Crittenden, E.D., Jr.; Hixon, A.N. *Ind. Eng. Chem.* 1954, *46*, 265.
3. Ratouis, M.; Dodé, M. *Bull. Soc. Chim. Fr.* 1965, 3318.

COMPONENTS:

(1) 2-Methyl-1-butanol *("active" amyl alcohol, sec-butylcarbinol)*; $C_5H_{12}O$; [137-32-6]

(2) Water; H_2O; [7732-18-5]

ORIGINAL MEASUREMENTS:

Ginnings, P.M.; Baum, R.

J. Am. Chem. Soc. 1937, *59*, 1111.

VARIABLES:

Temperature: 20-30°C

PREPARED BY:

A. Maczynski; Z Maczynski.

EXPERIMENTAL VALUES:

Mutual solubility of 2-methyl-1-butanol(1) and water(2)

t/°C	g(1)/100g sln		x_1(compiler)	
	(2)-rich phase	(1)-rich phase	(2)-rich phase	(1)-rich phase
20	3.18	91.05	0.00667	0.6752
25	2.97	90.81	0.00621	0.6688
30	2.83	90.74	0.00592	0.6669

Relative density, d_4

t/°C	Water-rich phase	Alcohol-rich phase
20	0.9943	0.8311
25	0.9930	0.8288
30	0.9928	0.8239

AUXILIARY INFORMATION

METHOD/APPARATUS/PROCEDURE:

Hill's volumetric method was adopted.

Both components were introduced in known amounts into a two-bulb graduated and calibrated flask and shaken mechanically in a water-bath at constant temperature. After sufficient time the liquids were allowed to separate and the total volume was measured. Upon centrifugation, the phase separation line was read, and phase volumes were calculated. From the total weights of the components, the total volume, individual phase volumes, and component concentrations in either phase were evaluated.

SOURCE AND PURITY OF MATERIALS:

(1) Eastman best grade; distilled from metallic calcium; b.p. range 128.4-129.1°C, d_4^{25} 0.8103.

(2) not specified.

ESTIMATED ERROR:

Temperature: ± 0.1°C

Solubility: better than 0.1 wt % (type of error not specified).

REFERENCES:

COMPONENTS:

(1) 2-Methyl-1-butanol (*"active" amyl alcohol, sec-butylcarbinol*); $C_5H_{12}O$; [137-32-6]

(2) Water; H_2O; [7732-18-5]

ORIGINAL MEASUREMENTS:

Crittenden, E.D.,Jr.; Hixon, A.N.

Ind. Eng. Chem. 1954, *46*, 265-8.

VARIABLES:

One temperature: 25°C

PREPARED BY:

A. Maczynski

EXPERIMENTAL VALUES:

The solubility of 2-methyl-1-butanol in water at 25°C was reported to be 3.0 g(1)/100 sln.

The corresponding mole fraction, x_1, calculated by the compiler is 0.0063.

The solubility of water in 2-methyl-1-butanol at 25°C was reported to be 9.2 g(2)/100g sln.

The corresponding mole fraction x_2, calculated by the compiler is 0.33.

AUXILIARY INFORMATION

METHOD/APPARATUS/PROCEDURE:

Presumably the titration method described for ternary systems containing HCl was used. In this method the solubility was determined by bringing 100 ml samples of (1) or (2) to a temperature 25° ± 0.10°C and the second component was then added from a calibrated buret, with vigorous stirring, until the solution became permanently cloudy.

SOURCE AND PURITY OF MATERIALS:

(1) source not specified;
purified;
purity not specified.

(2) not specified.

ESTIMATED ERROR:

Solubility: 2% (alcohol-rich)-10% (water-rich)

Temperature: ± 0.10°C (no further details)

REFERENCES:

COMPONENTS:	ORIGINAL MEASUREMENTS:
(1) 2-Methyl-1-butanol (*"active" amyl alcohol; sec-butylcarbinol*); $C_5H_{12}O$; [137-32-6] (2) Water, H_2O; [7732-18-5]	Ratouis, M., Dodé, M. *Bull. Soc. Chim. Fr.* 1965, 3318-22.
VARIABLES:	PREPARED BY:
One temperature: 30°C Ringer solution also studied	S.C. Valvani; S.H. Yalkowsky; A.F.M. Barton

EXPERIMENTAL VALUES:

The proportion of (1) in the water-rich phase at equilibrium at 30°C was reported to be 2.61 g(1)/100g sln.

The corresponding mole fraction, calculated by the compiler, is x_1 = 0.00545.

The proportion of (1) in the water-rich phase of a mixture with Ringer solution at equilibrium at 30°C was reported to be 2.46 g(1)/100g sln.

AUXILIARY INFORMATION

METHOD/APPARATUS/PROCEDURE:

In a round bottomed flask, 50 mL of water and a sufficient quantity of alcohol were introduced until two separate layers were formed. The flask assembly was equilibrated by agitation for a least 3 h in a constant temp bath. Equilibrium solubility was attained by first supersaturating at a slightly lower temperature (solubility of alcohols in water is inversely proportional to temperature and then equilibrating at the desired temperature. The aqueous layer was separated after an overnight storage in a bath. The alcohol content was determined by reacting the aqueous solution with potassium dichromate and titrating the excess dichromatewith ferrous sulfate solution in the presence of phosphoric acid and diphenylamine barium sulfonate as an indicator.

SOURCE AND PURITY OF MATERIALS:

(1) Prolabo, Paris; redistilled with 10:1 reflux; n_D^{25} 1.40780

(2) twice distilled from silica apparatus or ion-exchanged with Sagei A20

ESTIMATED ERROR:

Solubility: relative error of 2 determinations less than 1%

Temperature: ± 0.05°C

REFERENCES:

COMPONENTS:	EVALUATOR:
(1) 2-Methyl-2-butanol (*tert-pentanol, tert-amyl alcohol, ethyldimethylcarbinol*) $C_5H_{12}O$; [75-85-4] (2) Water; H_2O; [7732-18-5]	A. Maczynski, Institute of Physical Chemistry of the Polish Academy of Sciences, Warsaw, Poland;and A.F.M. Barton, Murdoch University, Perth, Western Australia. November 1982

CRITICAL EVALUATION:

Solubilities in the system comprising 2-methyl-2-butanol (1) and water (2) have been reported in six publications. Ginnings and Baum (ref 1) carried out measurements of mutual solubility of the components at 293, 298 and 303 K by the volumetric method. Krupatkin (ref 2) determined the solubility of (1) in (2) at 293 and 323 K by the titration method. The water-rich phase was studied also by Ratouis and Dode (ref 3) (at 303 K by an analytical method) and Mullens (ref 4) (an interferometric method at one temperature). Moriyoshi and Aoki (ref 5) determined mutual solubilities between 333 and 402 K and up to 2450 atm (248 MPa) by refractometry. Hyde *et al.* (ref 6) determined the upper critical solution temperature and a "room temperature" solubility.

Data for the water-rich phase at atmospheric pressure are collected in Figure 1. The results of Krupatkin (ref 2) disagree with all others, and are not considered further. At 303 K the value 9.90 ± 0.1 g(1)/100g sln of ref 3 is in good agreement with the value 10.10 ± 0.1 g(1)/100g sln of ref 1, although in poor agreement with the values at 298 K of ref 1 and 4. The values of ref 1 at 293, 298 and 303 K are selected as tentative values. Above 332 K the graphically smoothed data of Moriyoshi and Aoki (ref 5), all that is available, are also regarded as tentative.

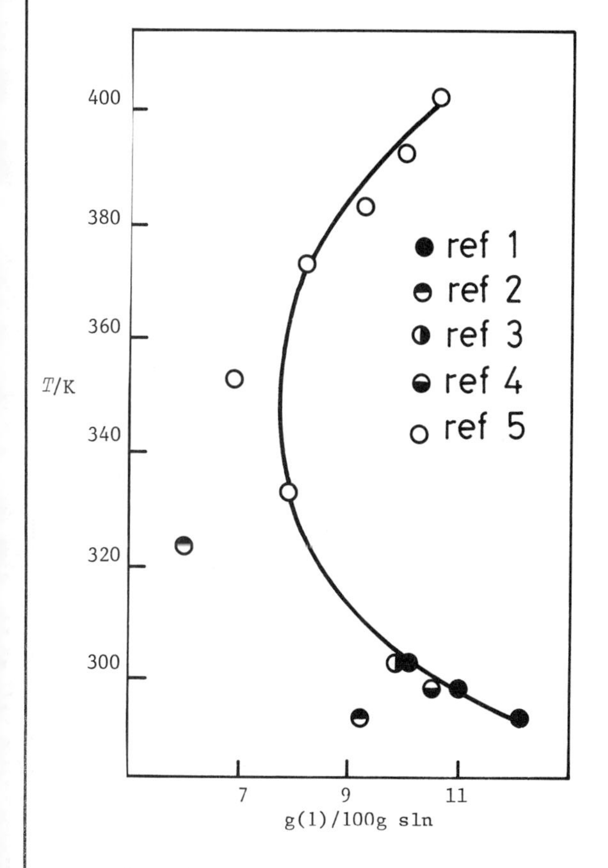

Fig. 1. Water-rich phase

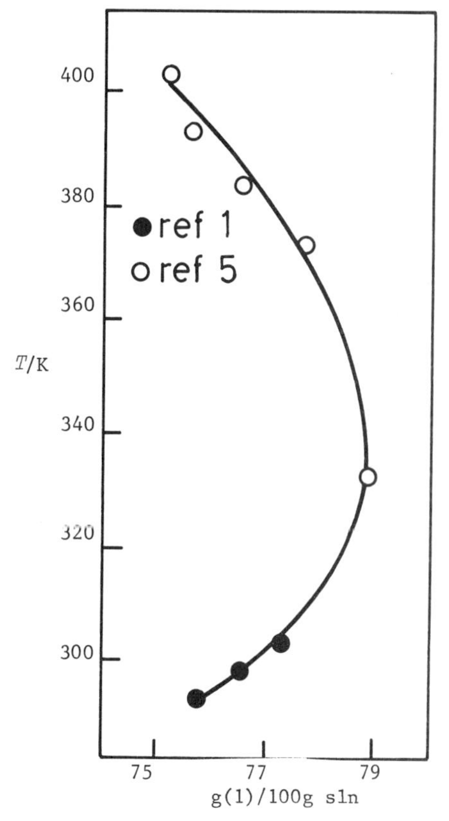

Fig. 2. Alcohol-rich phase

(continued next page)

COMPONENTS	EVALUATOR
(1) 2-Methyl-2-butanol (*tert-pentanol, tert-amyl alcohol, ethyldimethylcarbinol*) $C_5H_{12}O$: [75-85-4] (2) Water; H_2O; [7732-18-5]	A.Maczynski, Institute of Physical Chemistry of the Polish Academy of Sciences, Sciences, Warsaw, Poland and A.F.M. Barton, Murdoch University, Perth, Western Australia. November 1982

CRITICAL EVALUATION (continued)

The information on the alcohol-rich phase is collected in Figure 2. At 293, 298 and 303 K only the data of Ginnings and Baum (ref 1) are available, and are presented tentatively. Above 332 K the graphically smoothed data of Moriyoshi and Aoki (ref 5) are regarded as tentative.

Mutual solubilities of (1) and (2) as a function of pressure are given in the compilation sheet of ref 5. Information on this system is inadequate, and further study is required.

Tentative values for the mutual solubilities of 2-methyl-2-butanol (1) and water (2)

T/K	Water-rich phase		Alcohol-rich phase	
	g(1)/100g sln	x_1	g(2)/100g sln	x_2
293	12.1	0.027	24.3	0.61
298	11.0	0.025	23.7	0.60
303	10.1	0.024	22.7	0.59
313	9.0	0.020	22.1	0.58
333	7.8	0.019	21.2	0.57
353	7.7	0.017	21.4	0.57
373	8.2	0.018	22.5	0.59
393	9.9	0.022	24.1	0.61

References

1. Ginnings, P.M.; Baum, R. *J. Am. Chem. Soc.* 1937, *59*, 1111.

2. Krupatkin, I.L. *Zh. Obshch. Khim.* 1955, *25*, 1871; *J. Gen. Chem. USSR* 1955, *25*, 1815.

3. Ratouis, M.; Dodé, M. *Bull. Soc. Chim. Fr.* 1965, 3318.

4. Mullens, J. *Alcoholassociaten,* doctoraatsproefschrift, Leuven, 1971; Huyskens, P.; Mullen, J.; Gomez, A.; Tack, J. *Bull. Soc. Chim. Belg.* 1975, *84*, 253.

5. Moriyoshi, T.; Aoki, Y. *J. Chem. Eng. Japan,* 1978, *11*, 341.

6. Hyde, A.J.; Langbridge, D.M.; Lawrence, A.S.C. *Disc. Faraday Soc.* 1954, *18*, 239.

COMPONENTS:

(1) 2-Methyl-2-butanol (*tert-pentanol, tert-amyl alcohol, ethyldimethylcarbinol*)
$C_5H_{12}O$; [75-85-4]

(2) Water; H_2O; [7732-18-5]

ORIGINAL MEASUREMENTS:

Ginnings, P.M.; Baum, R.

J. Am. Chem. Soc. 1937, 59, 1111-13.

VARIABLES:

Temperature: 20-30°C

PREPARED BY:

A. Maczynski

EXPERIMENTAL VALUES:

Mutual solubility of 2-methyl-2-butanol and water

t/°C	g(1)/100g sln		x_1(compiler)	
	(2)-rich phase	(1)-rich phase	(2)-rich phase	(1)-rich phase
20	12.15	75.74	0.02748	0.3894
25	11.00	76.53	0.02463	0.3998
30	10.10	77.31	0.02244	0.4104

Relative density, d_4

t/°C	Water-rich phase	Alcohol-rich phase
20	0.9837	0.8662
25	0.9829	0.8552
30	0.9828	0.8498

AUXILIARY INFORMATION

METHOD/APPARATUS/PROCEDURE:

The volumetric method was used.

Both components were introduced in known amounts into a two-bulb graduated and calibrated flask and shaken mechanically in a water-bath at constant temperature. After sufficient time the liquids were allowed to separate and the total volume was measured. Upon centrifugation, the phase separation line was read, and phase volumes were calculated. From the total weights of the components, the total volume, individual phase volumes, and component concentrations in either phase were evaluated.

SOURCE AND PURITY OF MATERIALS:

(1) from Grignard synthesis; distilled from metallic calcium; b.p. range 101.9-102.1°C, d_4^{25} 0.8055; purity not specified.

(2) not specified.

ESTIMATED ERROR:

Temperature: ± 0.1°C
Solubility : better than 0.1 wt %
(type of error not specified)

REFERENCES:

COMPONENTS:	ORIGINAL MEASUREMENTS:
(1) 2-Methyl-2-butanol (*tert-pentanol, tert-amyl alcohol, ethyldimethylcarbinol*) $C_5H_{12}O$; [75-85-4] (2) Water; H_2O; [7732-18-5]	Hyde, A.J; Langbridge, D.M.; Lawrence, A.S.C. *Disc. Faraday Soc.* 1954, *18*, 239-58.
VARIABLES:	PREPARED BY:
One temperature; ca 260.0°C	A. Maczynski; A.F.M. Barton

EXPERIMENTAL VALUES:

The upper critical solution temperature was reported to be ca. 260.0°C 45.0 g(1)/100g sln (x_1 = 0.143, compiler).

The solubility of (1) in (2) at room temperature was reported as 11.0%.

AUXILIARY INFORMATION

METHOD/APPARATUS/PROCEDURE:	SOURCE AND PURITY OF MATERIALS:
Not specified	(1) not specified (2) not specified
	ESTIMATED ERROR:
	Not specified
	REFERENCES:

COMPONENTS:

(1) 2-Methyl-2-butanol *(tert-pentanol, tert-amyl alcohol, ethyldimethylcarbinol)*; $C_5H_{12}O$; [75-85-4]

(2) Water; H_2O; [7732-18-5]

ORIGINAL MEASUREMENTS:

Krupatkin, I.L.

Zh. Obshch. Khim. 1955, 25, 1871-6;

**J. Gen. Chem. USSR* 1955, 25, 1815-9.

VARIABLES:

Temperature: 20°C and 50°C

PREPARED BY:

S.H. Yalkowsky; S.C. Valvani; A.F.M. Barton

EXPERIMENTAL VALUES:

Mutual solubility of 2-methyl-2-butanol (1) and water (2)

t/°C	Water-rich phase		Alcohol-rich phase	
	g(1)/100g sln	x_1 (compilers)	g(1)/100g sln	x_1 (compilers)
20	9.23	0.0204	85.20	0.541
50	6	0.013	78.88	0.433

AUXILIARY INFORMATION

METHOD/APPARATUS/PROCEDURE:

The data above formed part of the results of a study of the ternary system including 3-methyl-1-butanol. The investigation used the isothermal method, in ampoules with ground-glass stoppers in a water thermostat. Samples of one component at the specified constant temperature were titrated with the other component until turbidity developed.

SOURCE AND PURITY OF MATERIALS:

(1) distilled;
b.p. 102°C

(2) distilled twice

ESTIMATED ERROR:

Not stated

REFERENCES:

COMPONENTS:

(1) 2-Methyl-2-butanol (*tert-pentanol, tert-amyl alcohol, ethyldimethylcarbinol*) $C_5H_{12}O$; [75-85-4]

(2) Water; H_2O; [7732-18-5]

ORIGINAL MEASUREMENTS:

Ratouis, M.; Dodé, M.

Bull. Soc. Chim. Fr. 1965, 3318.

VARIABLES:

One temperature: 30°C

Ringer solution also studied.

PREPARED BY:

S.C. Valvani; S.H. Yalkowsky; A.F.M. Barton.

EXPERIMENTAL VALUES:

The proportion of 2-methyl-2-butanol (1) in the water-rich phase at equilibrium at 30°C was reported to be 9.90 g(1)/100g sln.

The corresponding mole fraction solubility calculated by the compiler is x_1 = 0.0220.

The proportion of (1) in the water-rich phase of a mixture with Ringer solution at equilibrium at 30°C was reported to be 9.17 g(1)/100g sln.

AUXILIARY INFORMATION

METHOD/APPARATUS/PROCEDURE:

In a round bottomed flask, 50 mL of water and a sufficient quantity of alcohol were introduced until two separate layers were formed. The flask assembly was equilibrated by agitation for at least 3 h in a constant temp bath. Equilibrium solubility was attained by first supersaturating at a slightly lower temperature (solubility of alcohols in water is inversely proportional to temperature and then equilibrating at the desired temperature. The aqueous layer was separated after an overnight storage in a bath. The alcohol content was determined by reacting the aqueous solution with potassium dichromate and titrating the excess dichromate with ferrous sulfate solution in the presence of phosphoric acid and diphenylamine barium sulfonate as an indicator.

SOURCE AND PURITY OF MATERIALS:

(1) Prolabo, Paris;
redistilled with 10:1 reflux ratio;
b.p. 102.4-102.5°C/757.7 mm Hg
n_D^{25} 1.40258

(2) twice distilled from silica apparatus or ion-exchanged with Sagei A20

ESTIMATED ERROR:

Solubility: relative error of 2 determinations less than 1%

Temperature: ± 0.05°C

REFERENCES:

COMPONENTS:

(1) 2-Methyl-2-butanol; (*tert-pentanol, tert-amyl alcohol, ethyl dimethylcarbinol*), $C_5H_{12}O$; [75-85-4]

(2) Water; H_2O; [7732-18-5]

ORIGINAL MEASUREMENTS:

* Mullens, J.
Alcoholassociation, doctoraatsproetschrift, Leuven, 1971.

Huyskens, P.; Mullens, J.; Gomez, A.; Tack, J.
Bull. Soc. Chim. Belg. 1975, *84*, 253-62.

VARIABLES:

One temperature: 25°C

PREPARED BY:

M.C. Haulait-Pirson; A.F.M. Barton.

EXPERIMENTAL VALUES:

At equilibrium at 25°C the concentration of 2-methyl-2-butanol (1) in the water-rich phase was reported as 1.198 mol(1)/L sln, and the concentration of water (2) in the alcohol-rich phase was reported as 7.308 mol(2)/L sln.

The corresponding solubilities on a mass/volume basis are 105.6 g(1)/L sln and 131.7 g(2)/L sln (compiler).

AUXILIARY INFORMATION

METHOD/APPARATUS/PROCEDURE:

The partition of the two components was made using a cell described in ref 1. The Rayleigh Interference Refractometer M154 was used for the determination of the concentrations.
Standard solutions covering the whole range of concentration investigated were used for the calibration.

SOURCE AND PURITY OF MATERIALS:

(1) Merck (p.a.)

(2) distilled

ESTIMATED ERROR:

Solubility: ±0.001 mol(1)/L sln.

REFERENCES:

1. Meeussen, E.; Huyskens, P.
J. Chim. Phys. 1966, *63*, 845.

COMPONENTS:

(1) 2-Methyl-2-butanol (*tert-pentanol, tert-amyl alcohol, ethyldimethylcarbinol*); $C_5H_{12}O$; [75-85-4]

(2) Water; H_2O; [7732-18-5]

ORIGINAL MEASUREMENTS:

Moriyoshi, T.; Aoki, Y.

J. Chem. Eng. Jpn. 1978, *11*, 341-5.

VARIABLES:

Temperature: 333-402 K
Pressure: 0.1 (1-2450 atm)

PREPARED BY:

A. Maczynski; Z. Maczynski; A.F.M. Barton

EXPERIMENTAL VALUES:

Mutual solubility of 2-methyl-2-butanol (1) and water (2)

T/K	p/atm	p/MPa (compiler)	g(1)/100g sln (2)-rich phase	g(1)/100g sln (1)-rich phase	x_1 (compiler) (2)-rich phase	x_1 (compiler) (1)-rich phase
332.65	1	0.1	7.9	78.9	0.0172	0.433
	500	50.7	9.0	77.5	0.0198	0.413
	1000	101.3	9.3	75.8	0.0205	0.390
	1510	153.0	10.3	74.2	0.0229	0.370
	2000	202.7	11.2	72.7	0.0251	0.352
	2450	248.3	12.4	71.3	0.0281	0.337
352.55	1	0.1	6.9	82.4	0.0149	0.489
	500	50.7	7.5	80.1	0.0163	0.451
	816	82.7	8.1	78.9	0.0177	0.433
	1000	101.3	8.5	78.0	0.0186	0.420
	1530	155.0	9.5	75.8	0.0210	0.390
	2000	202.7	9.8	74.2	0.0217	0.370
	2450	249.3	10.8	72.4	0.0241	0.349
372.55	1	0.1	8.2	77.7	0.0179	0.416
	500	50.7	9.6	76.5	0.0212	0.399
	748	75.8	10.0	75.4	0.0222	0.376
	965	97.8	10.2	74.9	0.0227	0.379
	1020	103.4	10.6	74.7	0.0236	0.376
	1034	104.8	10.6	-	0.0236	-
	1224	124.0	-	74.5	-	0.374
	1361	137.9	11.2	74.0	0.0251	0.368
	1565	158.6	11.5	73.6	0.0258	0.363
	1864	188.9	12.5	72.9	0.0284	0.355

(continued next page)

AUXILIARY INFORMATION

METHOD/APPARATUS/PROCEDURE:

The refractometric method was used.

Under the desired conditions of temperature and pressure, mixtures of (1) and (2) were stirred for at least 24 h and then allowed to settle for another 6 h. Next a known mass of methanol was added to the samples by weight. After mixing, the refractive indices of the resulting solutions were measured at 298.15 K, and their compositions were determined from a calibration curve.

Details of the apparatus and experimental procedure are described in ref 1.

SOURCE AND PURITY OF MATERIALS:

(1) source not specified, best grade reagent;
dried by refluxing over freshly ignited calcium oxide and then distilled twice;
n_D^{25} 1.4025, d^{25} 0.8042 g cm^{-3}.

(2) not specified;
described in ref 1.

ESTIMATED ERROR:

Temperature: ± 0.05 K
Pressure: ± 7 atm
Solubility: ± 0.32 g(1)/100 sln (mean error)

REFERENCES:

1. Moriyoshi, T.; Aoki, Y.; Kamiyama, H. *J. Chem. Thermodyn.* 1977, *9*, 495.

COMPONENTS:	ORIGINAL MEASUREMENTS
(1) 2-Methyl-2-butanol *(tert-pentanol, tert-amyl alcohol, ethyldimethylcarbinol)*; $C_5H_{12}O$; [75-85-4] (2) Water; H_2O; [7732-18-5]	Moriyoshi, T.; Aoki, Y. *J. Chem. Eng. Jpn.* 1978, *11*, 341-5.

EXPERIMENTAL VALUES (continued)

T/K	p/atm	p/MPa (compiler)	g(1)/100g sln (2)-rich phase	g(1)/100g sln (1)-rich phase	x_1 (compiler) (2)-rich phase	x_1 (compiler) (1)-rich phase
372.55	2000	202.7	12.7	72.0	0.0289	0.344
	2245	227.5	13.5	70.7	0.0309	0.330
	2450	248.3	13.8	69.6	0.0317	0.319
382.05	1	0.1	9.3	76.5	0.0205	0.399
	500	50.7	9.8	75.4	0.0217	0.385
	816	82.7	10.6	74.5	0.0236	0.374
	1000	101.3	11.2	74.2	0.0251	0.370
	1225	124.1	11.5	73.6	0.0258	0.363
	1510	153.0	11.8	72.2	0.0266	0.347
	1864	188.9	12.7	71.3	0.0289	0.337
	2245	227.5	13.9	69.4	0.0319	0.317
	2450	248.3	14.5	68.5	0.0324	0.308
391.85	1	0.1	10.0	75.6	0.0222	0.388
	748	75.8	11.5	74.0	0.0258	0.368
	1000	101.3	11.6	73.6	0.0261	0.363
	1500	152.0	12.7	71.8	0.0289	0.342
	2245	227.5	14.5	69.0	0.0335	0.313
	2450	248.3	14.8	69.1	0.0343	0.304
401.75	1	0.1	10.7	75.2	0.0239	0.383
	500	50.7	11.6	74.0	0.0261	0.368
	1000	101.3	12.6	72.9	0.0286	0.355
	2000	202.7	14.4	69.4	0.0332	0.317
	2272	230.2	14.8	68.3	0.0343	0.306
	2450	248.3	15.6	67.7	0.0364	0.300

The existence of an upper critical solution temperature was expected on the basis of the behaviour of some butanol-water systems, but was not observed under the conditions used in this study.

COMPONENTS:	EVALUATOR:
(1) 3-Methyl-1-butanol (*isopentanol, isoamyl alcohol, isobutylcarbinol*); $C_5H_{12}O$; [123-51-3] (2) Water; H_2O; [7732-18-5]	A. Maczynski, Institute of Physical Chemistry of the Polish Academy of Sciences, Warsaw, Poland; and A.F.M. Barton, Murdoch University, Perth, Western Australia. November 1982

CRITICAL EVALUATION

Solubilities in the water-rich phase of the system comprising 3-methyl-1-butanol (1) and water (2) have been reported in the following publications:

Reference	T/K	Method
Coull and Hope (ref 5)	298	titration
Mitsui and Sasaki (ref 7)	283	indirect from 3-component system
Addison (ref 8)	293	surface tension
Booth and Everson (ref 9)	298	titration
Krupatkin (ref 12)	293, 333	titration
Hayashi and Sasaki (ref 14)	303	turbidity
Ratouis and Dodé (ref 17)	303	analytical

The mutual solubilities of (1) and (2) were measured in the following works:

Reference	T/K	Method
Brun (ref 3)	273	synthetic
Kablukov and Malischeva (ref 4)	288-303	volumetric
Ginnings and Baum (ref 6)	293-303	volumetric
Crittenden and Hixon (ref 10)	298	titration
Weiser and Geankopolis (ref 13)	298-323	titration
Arnold and Washburn (ref 15)	283-313	synthetic and analytical
Lavrova and Lesteva (ref 18)	313, 333	titration

Hyde *et al.* (ref 11) determined the upper critical solution temperature and a "room temperature" solubility.

The remaining data have not been compiled: Fontein (ref 1) used mixed amyl alcohols; Fuhner (ref 2) carried out only preliminary experiments on this sytem; and Ito (ref 16) reported the critical solution temperature. The results of Coull and Hope (ref 5), Mitsui and Sasaki (ref 7) and Booth and Everson (ref 9) are in volume units without densities provided, and have not been compared with the other data. The data of Krupatkin (ref 12) are in disagreement with those of all other references, and are rejected.

Values for the solubility of (1) in (2) over the temperature range 283 to 333 K are listed as follows.

(continued next page)

COMPONENTS:	EVALUATOR:
(1) 3-Methyl-1-butanol (*isopentanol, isoamyl alcohol, isobutylcarbinol*); $C_5H_{12}O$; [123-51-3] (2) Water; H_2O; [7732-18-5]	A. Maczynski, Institute of Physical Chemistry of the Polish Academy of Sciences, Warsaw, Poland; and A.F.M. Barton, Murdoch University, Perth, Western Australia. November 1982

CRITICAL EVALUATION (continued)

T/K	g(1)/100g sln literature values	graphically determined values
273	3.95 (ref 3)	
283	2.8 (ref 15)	3.2
288	3.04 (ref 4)	3.0
293	2.82(ref 4), 2.85(ref 6), 2.66(ref 8), 6.20(ref 12)	2.8
298	2.67(ref 4), 2.67(ref 6), 2.7(ref 10), 2.48(ref 13) 2.4 (ref 15)	2.7
303	2.56(ref 4), 2.53(ref 6), 2.57(ref 14), 2.41(ref 17)	2.6
313	2.2 (ref 15), 2.52 (ref 18)	2.4
323	4 (ref 12) , 2.53 (ref 13)	2.3
333	2.20 (ref 18)	2.2

Between 293 and 303 K the values of the solubility of (1) in (2) of ref 4, 6, 10, 14 are in good agreement, and were selected to estimate the recommended values. Below 293 K and above 303 K all the data are in poor agreement, and the graphically estimated values are regarded as tentative.

For the alcohol-rich phase, values of the solubility of (2) in (1) between 283 and 333 K are tablulated here.

T/K	g(1)/100g sln literature values	graphically determined values
273	91.0 (ref 3)	
283	91.0 (ref 15)	91.0
288	90.67 (ref 4)	90.7
293	90.40(ref 4), 90.53(ref 6), 85.46(ref 12)	90.4
298	90.13(ref 4), 90.39(ref 6), 90.9(ref 10), 90.25(ref 13), 90.2 (ref 15)	90.2
303	89.85 (ref 5), 90.24 (ref 6)	89.9
313	86.6 (ref 15), 89.3 (ref 18)	89.3
323	82.02 (ref 12), 88.5 (ref 13)	88.6
333	88.0 (ref 18)	88.0

At 298 K, the values of ref 4, 13 and 15 are in sufficiently good agreement to allow the values of 90.2 g(1)/100g sln to be recommended. The remaining values are tentative.

(continued next page)

COMPONENTS:	EVALUATOR:
(1) 3-Methyl-1-butanol (*isopentanol, isoamyl alcohol, isobutylcarbinol*); $C_5H_{12}O$; [123-51-3] (2) Water; H_2O; [7732-18-5]	A. Maczynski, Institute of Physical Chemistry of the Polish Academy of Sciences, Warsaw, Poland; and A.F.M. Barton, Murdoch University, Perth, Western Australia. November 1982

CRITICAL EVALUATION (continued)

Recommended and tentative values for the mutual solubilities of 3-methyl-1-butanol (1) and water (2)

T/K	Water-rich phase g(1)/100g sln	10^3x_1	Alcohol-rich phase g(2)/100g sln	x_2
283	3.2 (tentative)	6.7	9.0 (tentative)	0.33
288	3.0 (tentative)	6.3	9.3 (tentative)	0.33
293	2.8 (recommended)	5.8	9.6 (tentative)	0.34
298	2.7 (recommended)	5.6	9.8 (recommended)	0.35
303	2.6 (recommended)	5.4	10.1 (tentative)	0.36
313	2.4 (tentative)	5.0	10.7 (tentative)	0.37
333	2.2 (tentative)	4.6	12 (tentative)	0.40
459	upper critical solution temperature			

References

1. Fontein, F. *Z. phys. Chem.* 1910, *73*, 212.
2. Führer, H. *Ber. Dtsch. Chem. Ges.* 1924, *57B*, 510.
3. Brun, P. *C.R. Hebd. Seances Acad. Sci.* 1925, *180*, 1745; 1926, *183*, 207.
4. Kablukov, I.A.; Malischeva, V.T. *J. Am. Chem. Soc.* 1925, *47*, 1553.
5. Coull, J.; Hope, H.B. *J. Phys. Chem.* 1935, *39*, 967.
6. Ginnings, P.M.; Baum, R. *J. Am. Chem. Soc.* 1937, *59*, 1111.
7. Mitsui, S.; Sasaki, T. *J. Chem. Soc. Jpn.* 1942, *63*, 1766.
8. Addison, C.C. *J. Chem. Soc.* 1945, 98.
9. Booth, H.S.; Everson, H.E. *Ind. Eng. Chem.* 1948, *40*, 1491.
10. Crittenden, E.D.,Jr.; Hixon, A.N. *Ind. Eng. Chem.* 1954, *46*, 265.
11. Hyde, A.J.; Langbridge, D.M.; Lawrence, A.S.C. *Disc. Faraday Soc.* 1954, *18*, 239.
12. Krupatkin, I.L. *Zh. Obschch. Khim.* 1955, *25*, 1871; *J. Gen. Chem. USSR* 1955, *25*, 1815.
13. Weiser, R.B.; Geankopus, C.J. *Ind. Eng. Chem.* 1955, *47*, 858.
14. Hayashi, M.; Sasaki, T. *Bull. Chem. Soc. Jpn,* 1956, *29*, 857.
15. Arnold, V.W.; Washburn, E.R. *J. Phys. Chem.* 1958, *62*, 1088.
16. Ito, K. *Sci. Papers Inst. Phys. Chem. Res. (Tokyo)* 1961, *55*, 189.
17. Ratouis, M.; Dodé, M. *Bull. Soc. Chim. Fr.* 1965, 3318.
18. Lavrova, O.A.; Lesteva, T.M. *Zh. Fiz. Khim.* 1976, *50*, 1617; *Dep. Doc. VINITI* 3813-75.

COMPONENTS:

(1) 3-Methyl-1-butanol *(isopentanol, isoamyl alcohol, isobutylcarbinol)*; $C_5H_{12}O$; [123-51-3]

(2) Water; H_2O; [7732-18-5]

ORIGINAL MEASUREMENTS:

Brun, P.

C.R. Hebd. Seances Acad. Sci. 1925, *180*, 1745-7; 1926, *183*, 207-10.

VARIABLES:

One temperature: 0°C

PREPARED BY:

A.F.M. Barton

EXPERIMENTAL VALUES:

The mass percentage of 3-methyl-1-butanol (1) at equilibrium in the water-rich phase at 0°C was reported as 3.95 g(1)/100g sln, the corresponding mole fraction solubility, calculated by the compiler, is x_1 = 0.00833.

The mass percentage of 3-methyl-1-butanol in the alcohol-rich phase at equilibrium at 0°C was reported as 91.0 g(1)/100g sln; the corresponding mole fraction solubility calculated by the compiler is x_1 = 0.674.

Graphical results were reported in the 1925 paper for 40, 80, and 120°C.

AUXILIARY INFORMATION

METHOD/APPARATUS/PROCEDURE:

The synthetic method was used, in which the turbidity temperature is determined for mixtures of known composition.

SOURCE AND PURITY OF MATERIALS:

Not specified.

ESTIMATED ERROR:

Not specified.

REFERENCES:

COMPONENTS:

(1) 3-Methyl-1-butanol (*isopentanol, isoamyl alcohol isobutylcarbinol*); $C_5H_{12}O$; [123-51-3]

(2) Water; H_2O; [7732-18-5]

ORIGINAL MEASUREMENTS:

Kablukov, I.A.; Malischeva, V.T.

J. Am. Chem. Soc. 1925, *47*, 1553-61.

VARIABLES:

Temperature: 15-30°C

PREPARED BY:

S.H. Yalkowsky; S.C. Valvani; A.F.M. Barton

EXPERIMENTAL VALUES:

Mutual solubility of 3-methyl-1-butanol (1) and water (2)

t/°C	Water-rich phase				Alcohol-rich phase			
	c_1	g(1)/100g sln	x_1 (compiler)	d/g cm^{-3}	c_2	g(2)/100g sln	x_1 (compiler)	d/g cm^{-3}
15	0.0306	3.081	-	0.9934	0.0773	9.310	-	0.8308
15	0.0396	3.086	-	0.9933	0.0776	9.346	-	0.8306
15	0.0294	2.965	-	0.9921	0.0776	9.347	-	0.8306
15(mean)	-	3.04	0.00636	0.9929	-	9.33	0.665	0.8306
20	0.0283	2.854	-	0.9927	0.0793	9.578	-	0.8276
20	0.0284	2.857	-	0.9935	0.0795	9.613	-	0.8275
20	0.0273	2.752	-	0.9911	0.0796	9.617	-	0.8275
20(mean)	-	2.82	0.00589	0.9924	-	9.60	0.665	0.8275
25	0.0268	2.703	-	0.9921	0.0812	9.853	-	0.8245
25	0.0269	2.711	-	0.9916	0.0815	9.887	-	0.8241
25	0.258	2.610	-	0.9906	0.0815	9.885	-	0.8241
25(mean)	-	2.67	0.00558	0.9914	-	9.87	0.651	0.8242

(continued next page)

AUXILIARY INFORMATION

METHOD/APPARATUS/PROCEDURE:

The method is based on the measurement of the quantity of each liquid taken for mixing and the volumes of the layers at equilibrium proposed by the present authors but published first by Hill (ref 1). Alcohol was poured into the measuring tube (calibrated within ± 0.01 cm^3). Water was added from the buret (± 0.01 cm^3) and the liquids mixed by shaking or reversing the tube. After complete separation into layers, the volumes and temperature were recorded, and the refractive index and density determined.

SOURCE AND PURITY OF MATERIALS:

(1) dried over anhydrous cupric sulfate, distilled twice from calcium.

b.p. 130.6-131.1°C/760 mm Hg

(2) distilled with potassium permanganate, then with barium hydroxide, distillate redistilled, middle portion selected.

ESTIMATED ERROR:

Temperature: tenth-degree standardized thermometer used.
Solubility: not stated; deviation from mean apparent from experimental values above

REFERENCES:

1. Hill, A.E. *J. Am. Chem. Soc.* 1923, *45*, 1143.

COMPONENTS:	ORIGINAL MEASUREMENTS:
(1) 3-Methyl-1-butanol *(isopentanol, isoamyl alcohol isobutylcarbinol)*; $C_5H_{12}O$; [123-51-3]	Kablukov, I.A., Malischeva, V.T.
(2) Water; H_2O; [7732-18-5]	*J. Am. Chem. Soc.* 1925, *47*, 1553-61.

EXPERIMENTAL VALUES: (continued)

Mutual solubility of 3-methyl-1-butanol (1) and water (2)

$t/^{\circ}C$	Water-rich phase				Alcohol-rich phase			
	c_1	g(1)/100g sln	x_1 (compiler)	d/g cm^{-3}	c_2	g(2)/100g sln	x_1 (compiler)	d/g cm^{-3}
30	0.0257	2.594	-	0.9906	0.0832	10.13	-	0.8212
30	0.0257	2.599	-	0.9909	0.0834	10.16	-	0.8208
30	0.0246	2.491	-	0.9899	0.0834	10.16	-	0.8209
30(mean)	-	2.56	0.00534	0.9904	-	10.15	0.644	0.8209

COMPONENTS:

(1) 3-Methyl-1-butanol *(isopentanol, isoamyl alcohol, isobutylcarbinol)*; $C_5H_{12}O$; [123-51-3]

(2) Water; H_2O; [7732-18-5]

ORIGINAL MEASUREMENTS:

Coull, J.; Hope, H.B.

J. Phys. Chem. 1935, *39*, 967-71.

VARIABLES:

One temperature: 25°C

PREPARED BY:

S.H. Yalkowsky; S.C. Valvani; A.F.M. Barton

EXPERIMENTAL VALUES:

The proportions by volume at equilibrium at 25°C of the alcohol and water were reported as 3.10 mL(1)/100mL sln in the water-rich phase and 93.32 mL(1)/100mL sln in the alcohol-rich phase.

Properties of homogeneous mixtures of (1) and (2)

mL(1)/100mL sln	mL(2)/100mL sln	n^{25}
100.0	0.0	1.4058
95.0	5.0	1.4033
93.0	7.0	1.4018
2.0	98.0	1.3346

AUXILIARY INFORMATION

METHOD/APPARATUS/PROCEDURE:

The results from part of the investigation of the ternary system including propanol, (1), and (2). The alcohol sample (200 mL) was thoroughly agitated in a thermostat while water was added until the first appearance of turbidity. The agitation was continued for 6-8h to ensure equilibrium. In some cases (1) was added to water and propanol.

SOURCE AND PURITY OF MATERIALS:

(1) Baker A.C.S. reagent grade; fractionally distilled;

b.p. 130.5°C

(2) not stated

ESTIMATED ERROR:

Temperature: ± 0.1°C

Density: ± 0.0001 g cm^{-3}

Refractive index: read to 0.00005, rounded to 0.0001

Compositions: ± 0.1%

REFERENCES:

COMPONENTS:

(1) 3-Methyl-1-butanol (*isopentanol, isoamyl alcohol, isobutylcarbinol*); $C_5H_{12}O$; [123-51-3]

(2) Water; H_2O; [7732-18-5]

ORIGINAL MEASUREMENTS:

Ginnings, P.M.; Baum R.

J. Am. Chem. Soc. 1937, *59*, 1111-3.

VARIABLES:

Temperature: 20-30°C

PREPARED BY:

A. Maczynski; Z. Maczynski

EXPERIMENTAL VALUES:

Mutual solubility of 3-methyl-1-butanol(1) and water(2)

t/°C	g(1)/100g sln		x_1 (compiler)	
	(2)-rich phase	(1)-rich phase	(2)-rich phase	(1)-rich phase
20	2.85	90.53	0.00596	0.6614
25	2.67	90.39	0.00557	0.6577
30	2.53	90.24	0.00527	0.6539

Relative density, d_4

t/°C	Water-rich phase	Alcohol-rich phase
20	0.9941	0.8286
25	0.9932	0.8257
30	0.9921	0.8188

AUXILIARY INFORMATION

METHOD/APPARATUS/PROCEDURE:

The volumetric method was used.

Both components were introduced in known amounts into a two-bulb graduated and calibrated flask and shaken mechanically in a water-bath at constant temperature. After sufficient time the liquids were allowed to separate and the total volume was measured. Upon centrifugation, the phase separation line was read, and phase volumes were calculated. From the total weights of the components, the total volume, individual phase volumes, and component concentrations in either phase were evaluated.

SOURCE AND PURITY OF MATERIALS:

(1) Eastman best grade; extracted with water, dried with anhydrous potassium carbonate and distilled from metallic calcium; b.p. range 131.5-131.7°C, d_4^{25} 0.8071.

(2) not specified.

ESTIMATED ERROR:

Temperature: ± 0.1°C

Solubility: better than 0.1 wt % (type of error not specified)

REFERENCES:

COMPONENTS:

(1) 3-Methyl-1-butanol (*isopentanol, isoamyl alcohol, isobutylcarbinol*); $C_5H_{12}O$; [123-51-3]

(2) Water; H_2O; [7732-18-5]

ORIGINAL MEASUREMENTS:

Mitsui, S; Sasaki, T.;

J. Chem. Soc. Jpn, 1942, *63*, 1766-71.

VARIABLES:

One temperature: 19°C

PREPARED BY:

N. Tsuchida

EXPERIMENTAL VALUES:

The solubility of (1) in (2) was estimated to be 3.00 ± 0.01 g(1)/100 cm^3 sln at 19.14°C.

AUXILIARY INFORMATION

METHOD/APPARATUS/PROCEDURE:

The solubility of butanol in the (1)-(2) system was determined experimentally. By means of the relation thus obtained, the solubility of (1) in (2) was determined by interpolation.

The measurements were carried out by adding butanol to aqueous (1) containing ethanol until one drop produced a permanent turbidity; the ethanol increased the butanol solubility and permitted measurement over the saturation value.

SOURCE AND PURITY OF MATERIALS:

(1) Katayama, A.R.; boiled with aqueous KOH, washed with dilute phosphoric acid, dried with anhydrous K_2CO_3 and $CaSO_4$, distilled at 131°C.

(2) not specified

ESTIMATED ERROR:

Temperature: ± 0.02°C

Solubility: ± 0.01 g(1)/100g sln

REFERENCES:

COMPONENTS:

(1) 3-Methyl-1-butanol *(isopentanol, isoamyl alcohol isobutylcarbinol)*; $C_5H_{12}O$; [123-51-3]

(2) Water; H_2O; [7732-18-5]

ORIGINAL MEASUREMENTS:

Addison, C.C.;

J. Chem. Soc. 1945, 98-106.

VARIABLES:

One temperature: 20°C

PREPARED BY:

S.H. Yalkowsky; S.C. Valvani; A.F.M. Barton

EXPERIMENTAL VALUES:

The proportion of 3-methyl-1-butanol (1) in the water-rich phase at equilibrium at 20°C was reported to be 2.66 g(1)/100g sln.

The corresponding mole fraction solubility calculated by the compiler is x_1 = 0.00555.

AUXILIARY INFORMATION

METHOD/APPARATUS/PROCEDURE:

A surface tension method was used. Sufficient excess of (1) was added to 100 mL of (2) in a stoppered flask to form a separate lens on the surface. The mixture was swirled gently, too vigorous an agitation giving a semi-permanent emulsion and incorrect readings. After settling, a small sample of the clear aqueous sln was withdrawn into a drop weight pipe and the surface tension determined. The swirling was continued until a constant value was obtained. The surface tension-concentration curve was known, and only a slight extrapolation (logarithmic scale) was necessary to find the concentration corresponding to the equilibrium value.

SOURCE AND PURITY OF MATERIALS:

(1) impure alcohols were purified by fractional distillation, the middle fraction from a distillation being redistilled;

b.p. 131.5°C d_4^{20} 0.8127

n_D^{20} 1.4075

(2) not stated

ESTIMATED ERROR:

Solubility: ±0.5%

REFERENCES:

AWW-F*

COMPONENTS:

(1) 3-Methyl-1-butanol (*isopentanol, isoamyl alcohol*); $C_5H_{12}O$; [123-51-3]

(2) Water; H_2O; [7732-18-5]

ORIGINAL MEASUREMENTS:

Booth, H.S.; Everson, H.E.

Ind. Eng. Chem. 1948, *40*, 1491-3.

VARIABLES:

One temperature: 25°C
Sodium xylene sulfonate

PREPARED BY:

S.H. Yalkowsky; S.C.Valvani; A.F.M.Barton

EXPERIMENTAL VALUES:

It was reported that the solubility of 3-methyl-1-butanol(1) in water(2) was 3.5 mL (1)/100 mL (2) at 25.0°C.

The corresponding value in 40% sodium xylene sulfonate solution as solvent was > 400 mL(1)/100 mL solvent.

AUXILIARY INFORMATION

METHOD/APPARATUS/PROCEDURE:

A known volume of solvent (usually 50 mL) in a tightly stoppered calibrated Babcock tube was thermostatted. Successive measured quantities of solute were added and equilibrated until a slight excess of solute remained. The solution was centrifuged, returned to the thermostat bath for 10 min, and the volume of excess solute measured directly. This was a modification of the method described in ref 1.

SOURCE AND PURITY OF MATERIALS:

(1) not specified ("CP or highest grade commercial")

(2) "distilled"

ESTIMATED ERROR:

Solubility: within 0.1 mL(1)/100 mL(2)

REFERENCES:

1. Hanslick, R.S. *Dissertation*, Columbia University, 1935.

COMPONENTS:

(1) 3-Methyl-1-butanol (*isopentanol, isoamyl alcohol, isobutylcarbinol*); $C_5H_{12}O$; [123-51-3]

(2) Water; H_2O; [7732-18-5]

ORIGINAL MEASUREMENTS:

Crittenden, E.D., Jr.; Hixon, A.N.

Ind. Eng. Chem. 1954, *46*, 265-8.

VARIABLES:

One temperature: 25°C

PREPARED BY:

A. Maczynski

EXPERIMENTAL VALUES:

The solubility of 3-methyl-1-butanol in water at 25°C was reported to be 2.7g(1)/100g sln.

The corresponding mole fraction, x_1, calculated by the compiler is 0.0056.

The solubility of water in 3-methyl-1-butanol at 25°C was reported to be 9.1 g(2)/100 g sln.

The corresponding mole fraction, x_2, calculated by the compiler is 0.33.

AUXILIARY INFORMATION

METHOD/APPARATUS/PROCEDURE:

Presumably the titration method described for ternary systems containing HCl was used. In this method the solubility was determined by bringing 100-ml samples of (1) or (2) to a temperature 25° ± 0.10°C and the second component was then added from a calibrated buret, with vigorous stirring, until the solution became permanently cloudy.

SOURCE AND PURITY OF MATERIALS:

(1) source not specified; purified; purity not specified.

(2) not specified.

ESTIMATED ERROR:

Solubility: 2% (alcohol-rich)-10% (water-rich).

Temperature: ± 0.10°C

REFERENCES:

COMPONENTS:

(1) 3-Methyl-1-butanol (*isopentanol, isoamyl alcohol, isobutylcarbinol*); $C_5H_{12}O$; [123-51-3]

(2) Water; H_2O; [7732-18-5]

ORIGINAL MEASUREMENTS:

Hyde, A.J.; Langbridge, D.M.; Lawrence, A.S.C.

Disc. Faraday Soc. 1954, *18*, 239-58.

VARIABLES:

One temperature: 185.5°C

PREPARED BY:

A. Maczynski; A.F.M. Barton

EXPERIMENTAL VALUES:

The upper critical solution temperature was reported to be 185.5°C at 42.0 g(1)/100g sln (x_1 = 0.13, compiler).

The solubility of (1) in (2) was reported as 2.5%.

AUXILIARY INFORMATION

METHOD/APPARATUS/PROCEDURE:

Not specified

SOURCE AND PURITY OF MATERIALS:

(1) not specified

(2) not specified

ESTIMATED ERROR:

Not specified

REFERENCES:

COMPONENTS:

(1) 3-Methyl-1-butanol *(isopentanol, isoamyl alcohol, isobutylcarbinol)*; $C_5H_{12}O$; [123-51-3]

(2) Water; H_2O; [7732-18-5]

ORIGINAL MEASUREMENTS:

Krupatkin, I.L.

Zh. Obshch. Khim. 1955, *25*, 1871-6;

**J. Gen. Chem. USSR* 1955, *25*, 1815-9.

VARIABLES:

Temperature: 20°C and 50°C

PREPARED BY:

S.H. Yalkowsky; S.C. Valvani; A.F.M. Barton

EXPERIMENTAL VALUES:

Mutual solubility of 3-methyl-1-butanol (1) and water (2)

t/°C	Water-rich phase		Alcohol-rich phase	
	g(1)/100g sln	x_1 (compilers)	g(1)/100g sln	x_1 (compilers)
20	6.20	0.0133	85.46	0.546
50	4[a]	0.009	82.02	0.483

[a]There is an inconsistency in the text, which reads: "With temperature rise from 20 to 50° the solubility of isoamyl alcohol *increases* to 4%" (Italics inserted by compilers).

AUXILIARY INFORMATION

METHOD/APPARATUS/PROCEDURE:

The data above formed part of the results of a study of the ternary system including 2-methyl-2-butanol. The investigation used the isothermal method, in ampoules with ground-glass stoppers in a water thermostat. Samples of one component at the specified constant temperature were titrated with the other component until turbidity developed.

SOURCE AND PURITY OF MATERIALS:

(1) distilled;
b.p. 114°C

(2) distilled twice

ESTIMATED ERROR:

Not stated

REFERENCES:

COMPONENTS:

(1) 3-Methyl-1-butanol *(isopentanol; isoamyl alcohol, isobutylcarbinol)*; $C_5H_{12}O$; [123-51-3]

(2) Water; H_2O; [7732-18-5]

ORIGINAL MEASUREMENTS:

Weiser, R.B.; Geankopolis, C.J.

Ind. Eng. Chem. 1955, *47*, 858-63.

VARIABLES:

Temperature: 25 and 50 °C

PREPARED BY:

A. Maczynski

EXPERIMENTAL VALUES:

Mutual solubility of 3-methyl-1-butanol (1) and water (2)

t/°C	g(1)/100g sln		x_1 (compiler)	
	(2)-rich phase	(1)-rich phase	(2)-rich phase	(1)-rich phase
25	2.48	90.25	0.00517	0.6541
49.5	2.53	88.5	0.00527	0.611

AUXILIARY INFORMATION

METHOD/APPARATUS/PROCEDURE:

Probably the titration method was used.

No details were reported in the paper.

SOURCE AND PURITY OF MATERIALS:

(1) CP reagent, source not specified; distilled; b.p. range 131.5-132.0 °C, n^{15} 1.4085.

(2) not specified.

ESTIMATED ERROR:

Not specified.

REFERENCES:

COMPONENTS:	ORIGINAL MEASUREMENTS:
(1) 3-Methyl-1-butanol *(isopentanol, isoamyl alcohol, isobutylcarbinol)*; $C_5H_{12}O$; [123-51-3] (2) Water; H_2O; [7732-18-5]	Hayashi, M.; Sasaki, T. *Bull. Chem. Soc. Jpn.* 1956, *29*, 857-9

VARIABLES:	PREPARED BY:
One temperature: 30°C	S.H. Yalkowsky; S.C. Valvani; A.F.M. Barton

EXPERIMENTAL VALUES:

The proportion of 3-methyl-1-butanol (1) in the water-rich phase at equilibrium at 30.1°C was reported to be 2.57 g(1)/100g sln. (This was the mean of three determinations: 2.587, 2.565, and 2.577 g(1)/100g sln.)

The corresponding mole fraction solubility calculated by the compilers is x_1 = 0.00536.

The solubility of the alcohol in dilute solutions of Tween 80, obtained by extrapolating to zero turbidity the linear relation between turbidity and solute concentration in the surfactant solution, was less than that in pure water but increased with increasing concentration of Tween 80.

AUXILIARY INFORMATION

METHOD/APPARATUS/PROCEDURE:

The mixture was well shaken at a temperature below 30.1°C and then stood in the thermostat for 24h. After the excess solute particles cleared, a transparent saturated solution was obtained which was taken from the bottom of the vessel by a siphon. A known amount of this solution (about 20g) was titrated with Tween 80 solution. Concentration was determined by comparison of turbidity with that of standard samples. (The relation between the turbidity and the quantity of solute in the surfactant solution is linear, and the solubility limits are lower than the solubilities in pure water in the dilute region of the surfactant.)

SOURCE AND PURITY OF MATERIALS:

(1) boiled with concentrated sodium hydroxide sln, washed with water, dried with anhydrous potassium carbonate, distilled over calcium oxide, redistilled over calcium metal;

b.p. 131°C

(2) not stated.

ESTIMATED ERROR:

Solubility: "possible error" 0.4%

REFERENCES:

COMPONENTS:

(1) 3-Methyl-1-butanol *(isopentanol, isoamyl alcohol, isobutylcarbinol)*; $C_5H_{12}O$; [123-51-3]

(2) Water: H_2O; [7732-18-5]

ORIGINAL MEASUREMENTS:

Arnold, V.W.; Washburn, E.R.

J. Phys. Chem. 1958, *62*, 1088-90.

VARIABLES:

Temperature: 10-40°C

PREPARED BY:

A. Maczynski

EXPERIMENTAL VALUES:

Mutual solubility of 3-methyl-1-butanol (1) and water (2)

$t/^{\circ}C$	g(1)/100g sln		x_1 (compiler)	
	(2)-rich phase	(1)-rich phase	(2)-rich phase	(1)-rich phase
10	2.8	91.0	0.0059	0.674
25	2.4	90.2	0.0050	0.653
40	2.2	86.6	0.0046	0.569

AUXILIARY INFORMATION

METHOD/APPARATUS/PROCEDURE:

Two methods were used.

The solubility of (1) in (2) was determined by Alexejeff's method, ref 1, using sealed tubes.

The solubility of (2) and (1) was determined by analysing saturated solutions at each of the temperatures with Karl Fischer reagent.

SOURCE AND PURITY OF MATERIALS:

(1) prepared from isobutyl bromide; b.p. 131.9°C/760 mm Hg; n_4^{25} 1.4048, d_4^{25} 0.8051

(2) not specified.

ESTIMATED ERROR:

Not specified.

REFERENCES:

1. Alexejeff, M.W. *Bull. soc. chim.* 1882, *38*, 145.

COMPONENTS:

(1) 3-Methyl-1-butanol (*isopentanol, isoamyl alcohol isobutylcarbinol)* $C_5H_{12}O$, [123-51-3]

(2) Water; H_2O; [7732-18-5]

ORIGINAL MEASUREMENTS:

Ratouis, M.; Dodé, M.

Bull. Soc. Chim. Fr. 1965, 3318-22.

VARIABLES:

One temperature: 30°C

Ringer solution also studied

PREPARED BY:

S.C. Valvani; S.H. Yalkowsky; A.F.M. Barton.

EXPERIMENTAL VALUES:

The proportion of 3-methyl-1-butanol (1) in the water-rich phase at equilibrium at 30°C was reported to be 2.41 g(1)/100g sln.

The corresponding mole fraction solubility, calculated by the compiler, is x_1 = 0.00502.

The proportion of (1) in the water-rich phase of a mixture with Ringer solution at equilibrium at 30°C was reported to be 2.38 g(1)/100g sln.

AUXILIARY INFORMATION

METHOD/APPARATUS/PROCEDURE:

In a round bottomed flask, 50 mL of water and a sufficient quantity of alcohol were introduced until two separate layers were formed. The flask assembly was equilibrated by agitation for a least 3 h in a constant temperature bath. Equilibrium solubility was attained by first supersaturating at a slightly lower temperature (solubility of alcohols in water is inversely proportional to temperature) and then equilibrating at the desired temperature. The aqueous layer was separated after an overnight storage in a bath. The alcohol content was determined by reacting the aqueous solution with potassium dichromate and titrating the excess dichromate with ferrous sulfate solution in the presence of phosphoric acid and diphenylamine barium sulfonate as an indicator.

SOURCE AND PURITY OF MATERIALS:

(1) Prolabo, Paris;
redistilled with 10:1 reflux ratio;

b.p. 131.0-131.1°C/746 mm Hg

n_D^{25} 1.40608

(2) twice distilled from silica apparatus or ion-exchanged with Sagei A20.

ESTIMATED ERROR:

Solubility: relative error of 2 determinations less than 1%.

Temperature: ±0.05°C

REFERENCES:

COMPONENTS:

(1) 3-Methyl-1-butanol *(isopentanol, isoamyl alcohol, isobutylcarbinol)*; $C_5H_{12}O$; [123-51-3]

(2) Water; H_2O; [7732-18-5]

ORIGINAL MEASUREMENTS:

Lavrova, O.A.; Lesteva, T.M.

Zh. Fiz. Khim. 1976, *50*, 1617; *Dep. Doc. VINITI* 3813-75.

VARIABLES:

Temperature: 40 and 60°C

PREPARED BY:

A. Maczynski

EXPERIMENTAL VALUES:

Mutual solubility of 3-methyl-1-butanol (1) and water (2)

$t/^{\circ}C$	g(1)/100g sln		x_1 (compiler)	
	(2)-rich phase	(1)-rich phase	(2)-rich phase	(1)-rich phase
40	2.52	89.3	0.00525	0.630
60	2.20	88.0	0.00457	0.600

AUXILIARY INFORMATION

METHOD/APPARATUS/PROCEDURE:

The titration method was used.

No details were reported in the paper.

SOURCE AND PURITY OF MATERIALS:

(1) source not specified; distilled with heptane; purity 99.93 wt %; 0.07 wt % of water;
n_D^{20} 1.4070
d_4^{20} 0.8112
b.p. 130.9°C.

(2) not specified.

ESTIMATED ERROR:

Not specified.

REFERENCES:

COMPONENTS:	EVALUATOR:
(1) 3-Methyl-2-butanol *(methylisopropylcarbinol)*; $C_5H_{12}O$; [598-75-4] (2) Water; H_2O; [7732-18-5]	Z. Maczynska, Institute of Physical Chemistry of the Polish Academy of Sciences, Warsaw, Poland. November 1982

CRITICAL EVALUATION:

Solubilities in the system comprising 3-methyl-2-butanol (1) and water (2) have been reported in two publications. Ginnings and Baum (ref 1) carried out measurements of the mutual solubilities of the two components at 293, 298 and 303 K by the volumetric method (Figure 1). Ratouis and Dodé (ref 2) determined the solubility of (1) in the water-rich phase at 303 K by an analytical method. Their value of 4.85 g(1)/100g sln is in reasonable agreement with the value 5.10 ± 0.1 g(1)/100g sln at 303 K of ref 1. The data are regarded as tentative, since comparison can be made at only one temperature and the other five points are derived from a single source.

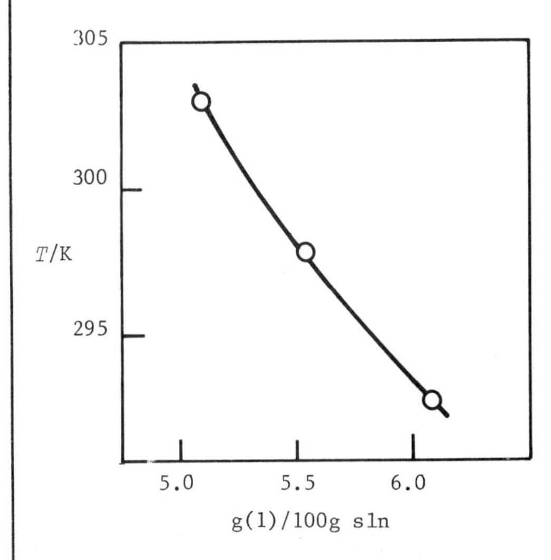

Fig. 1. Water-rich phase (ref 1)

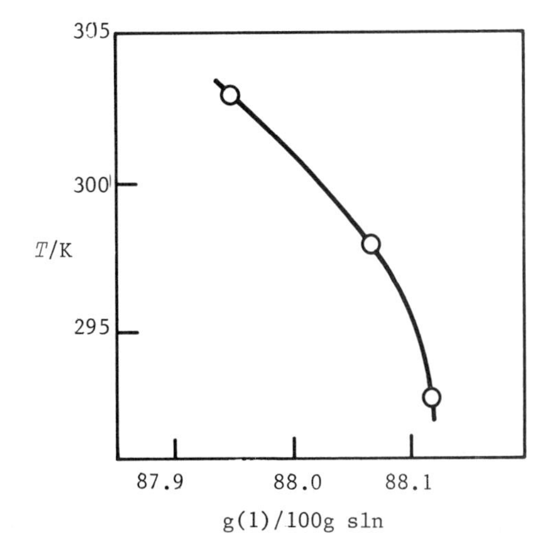

Fig. 2. Alcohol-rich phase (ref 1)

Tentative values for the mutual solubilities of 3-methyl-2-butanol (1) and water (2)

T/K	Water-rich phase		Alcohol-rich phase	
	g(1)/100g sln	$10^3 x_1$	g(2)/100g sln	x_2
293	6.1	13.0	11.9	0.397
298	5.6	11.9	11.9	0.399
303	5.1	10.9	12.0	0.401

References

1. Ginnings, P.M.; Baum, R. *J. Am. Chem. Soc.* 1937, *59*, 1111.
2. Ratouis, M.; Dodé, M. *Bull. Soc. Chim. Fr.* 1965, 3318.

COMPONENTS:

(1) 3-Methyl-2-butanol *(methylisopropylcarbinol)*; $C_5H_{12}O$, [598-75-4]

(2) Water; H_2O; [7732-18-5]

ORIGINAL MEASUREMENTS:

Ginnings, P.M.; Baum, R.

J. Am. Chem. Soc. 1937, *59*, 1111-3.

VARIABLES:

Temperature: 20-30°C

PREPARED BY:

A. Maczynski

EXPERIMENTAL VALUES:

Mutual solubility of 3-methyl-2-butanol(1) and water(2)

$t/°C$	g(1)/100g sln		x_1 (compiler)	
	(2)-rich phase	(1)-rich phase	(2)-rich phase	(1)-rich phase
20	6.07	88.12	0.01303	0.6025
25	5.55	88.07	0.01186	0.6013
30	5.10	87.95	0.01086	0.5986

Relative density d_4

$t/°C$	Water-rich phase	Alcohol-rich phase
20	0.9909	0.8390
25	0.9902	0.8352
30	0.9879	0.8348

AUXILIARY INFORMATION

METHOD/APPARATUS/PROCEDURE:

Hill's volumetric method was adopted.

Both components were introduced in known amounts into a two-bulb graduated and calibrated flask and shaken mechanically in a water-bath at constant temperature. After sufficient time the liquids were allowed to separate and the total volume was measured. Upon centrifugation, the phase separation line was read, and phase volumes were calculated. From the total weights of the components, the total volume, individual phase volumes, and component concentrations in either phase were evaluated.

SOURCE AND PURITY OF MATERIALS:

(1) Eastman best grade; distilled from metallic calcium; b.p. range 111.1-111.9°C, d_4^{25} 0.8134.

(2) not specified.

ESTIMATED ERROR:

Temperature: ± 0.1°C
Solubility : better than 0.1 wt %
(type of error not specified)

REFERENCES:

COMPONENTS:

(1) 3-Methyl-2-butanol *(methylisopropylcarbinol)* $C_5H_{12}O$; [598-75-4]

(2) Water; H_2O; [7732-18-5]

ORIGINAL MEASUREMENTS:

Ratouis, M., Dodé, M.

Bull. Soc. Chim. Fr. 1965, 3318-22.

VARIABLES:

One temperature: 30°C

Ringer solution also studied

PREPARED BY:

S.C. Valvani; S.H. Yalkowsky; A.F.M. Barton

EXPERIMENTAL VALUES:

The proportion of 3-methyl-2-butanol (1) in the water-rich phase at equilibrium at 30°C was reported to be 4.85 g(1)/100g sln.

The corresponding mole fraction solubility calculated by the compiler is x_1 = 0.0103.

The proportion of (1) in the water-rich phase of a mixture with Ringer solution at equilibrium at 30°C was reported to be 4.75 g(1)/100g sln.

AUXILIARY INFORMATION

METHOD/APPARATUS/PROCEDURE:

In a round bottomed flask, 50mL of water and a sufficient quantity of alcohol were introduced until two separate layers were formed. The flask assembly was equilibrated by agitation for a least 3 h in a constant temp bath. Equilibrium solubility was attained by first supersaturating at a slightly lower temperature (solubility of alcohols in water is inversely proportional to temperature) and then equilibrating at the desired temperature. The aqueous layer was separated after an overnight storage in a bath. The alcohol content was determined by reacting the aqueous solution with potassium dichromate and titrating the excess dichromate with ferrous sulfate solution in the presence of phosphoric acid and diphenylamine barium sulfonate as an indicator.

SOURCE AND PURITY OF MATERIALS:

(1) Fluka A.G., Buchs S.G., redistilled with 10:1 reflux ratio, b.p. 112.4 - 112.5°C/766.7 mm Hg n_D^{25} 1.40758.

(2) twice distilled from silica apparatus or ion-exchanged with Sagei A20.

ESTIMATED ERROR:

Solubility: relative error of 2 determinations less than 1%.

Temperature: ±0.05°C

REFERENCES:

COMPONENTS:	EVALUATOR:
(1) 1-Pentanol *(n-pentyl alcohol, n-amyl alcohol, n-butylcarbinol)*; $C_5H_{12}O$; [71-41-0] (2) Water; H_2O; [7732-18-5]	G.T. Hefter and A.F.M. Barton, Murdoch University, Perth, Western Australia. July, 1983.

CRITICAL EVALUATION:

Solubilities in the system comprising 1-pentanol (1) and water (2) have been reported in the following publications:

Reference	T/K	Phase	Method
Verschaffelt (ref 1)	279-309	alcohol-rich	synthetic
Herz (ref 2)	295	mutual	densimetric
Timmermans (ref 3)	Not compiled:	insufficient information	
Fontein (ref 4)	Not compiled:	mixture of pentyl alochols	
Butler *et al.* (ref 5)	298	water-rich	interferometric
Ginnings and Baum (ref 6)	293-303	mutual	volumetric
Jasper *et al.* (ref 7)	368	mutual	refractometric
Addison (ref 8)	293	water-rich	surface tension
Booth and Everson (ref 9)	298	water-rich	titration
Laddha and Smith (ref 10)	293	mutual	titration
Hansen *et al.* (ref 11)	298	water-rich	interferometric
Donahue and Bartell (ref 12)	298	mutual	analytical
Erichsen (ref 13)	273-455	mutual	synthetic
Erichsen (ref 14)	273-323	water-rich	synthetic
Crittenden and Hixon (ref 15)	298	mutual	titration
Hyde *et al.* (ref 16)	457.5	u.c.s.t.	
Kinoshita *et al.* (ref 17)	298	water-rich	surface tension
Ratouis and Dodé (ref 18)	303	water-rich	analytical
Ionin and Shanina (ref 19)	Not compiled:	unspecified pentyl alcohol with wide b.p. range	
Hanssens (ref 20)	298	mutual	interferometric
Krasnov and Gartseva (ref 21)	285,313	alcohol-rich	analytical
Zhuravleva and Zhuravlev (ref 22)	285-360	mutual	synthetic
Mullens *et al* (ref 23)	298	water-rich	interferometric
Vochten and Petre (ref 24)	288	water-rich	surface tension
Korenman *et al.* (ref 25)	298	mutual	analytical
Lavrova and Lesteva (ref 26)	313,333	mutual	titration
Charykov *et al.* (ref 27)	293	alcohol-rich	analytical
Evans *et al.* (ref 28)	310	water-rich	analytical
Nishino and Nakamura (ref 29)	Not compiled:	graphical information only	
Singh and Haque (ref 30)	303	mutual	titration
Tokunaga *et al.* (ref 31)	288-308	alcohol-rich	analytical

With the exception of those reports noted above (ref 3, 4, 19, 29) all original data are compiled in the data sheets immediately following this Critical Evaluation.

The data of Herz (ref 2), Booth and Everson (ref 9), Hanssens (ref 20), Mullens *et al.* (ref 23), Vochten and Petre (ref 24), Korenman (ref 25) and Evans *et al.* (ref 28) are given in weight/volume fractions without densities and so have been excluded from consideration in this Critical Evaluation. (continued next page)

COMPONENTS:	EVALUATOR:
(1) 1-Pentanol (*n-pentyl alcohol, n-amyl alcohol, n-butylcarbinol*); $C_5H_{12}O$: [71-41-0] (2) Water; H_2O; [7732-18-5]	G.T. Hefter and A.F.M. Barton, Murdoch University, Perth, Western Australia. July, 1983.

CRITICAL EVALUATION (continued)

In the water-rich phase the data of Jasper *et al.* (ref 7), Laddha and Smith (ref 10), Hansen *et al.* (ref 11), Zhuravleva and Shuravlev (ref 22), and Singh and Haque (ref 30) disagree markedly with all other studies and are rejected.

In the alcohol-rich phase the datum of Jasper *et al.* (ref 7) appears to be inconsistent with other studies (ref 13,22) and is also rejected. The data of Erichsen (ref 14) are considered to refer to the same primary data as ref 13.

All other data are included in the Tables below. Values obtained by the Evaluators by graphical interpolation or extrapolation from the data sheets are indicated by an asterisk.* "Best" values have been obtained by simple averaging. The uncertainty limits (σ_n) attached to the "best" values do not have statistical significance and should be regarded only as a convenient representation of the spread of reported values and not as error limits. The letter (*R*) indicates "Recommended" data. Data are "Recommended" if two or more apparently reliable studies are in reasonable agreement (< ± 5% relative).

For convenience, further discussion of the two phases is given separately.

The solubility of 1-pentanol (1) in water (2)

Surprisingly little information is available except over the temperature range 293-303. In this range agreement is excellent and the average values can be recommended. Outside this range only the data of Erichsen (ref 13) are available and hence must be considered only as tentative.

Recommended (*R*) and tentative solubilities of 1-pentanol (1) in water (2)

T/K	Solubility, g(1)/100g sln: Reported values	"Best" value (± σ_n)
273	3.05 (ref 13)	3.1
283	2.70 (ref 13)	2.7
293	2.36 (ref 6), 2.21 (ref 8), 2.35 (ref 13)	2.31 ± 0.07 (*R*)
298	2.208(ref 5), 2.19(ref 6), 2.21(ref 12), 2.22*(ref 13), 2.2 (ref 15), 2.2 (ref 17)	2.20 ± 0.01 (*R*)
303	2.03 (ref 6), 2.10 (ref 13), 2.00 (ref 18)	2.04 ± 0.04 (*R*)
313	1.90 (ref 13), 2.1 (ref 26)	2.0 ± 0.1
323	1.80 (ref 13)	1.8
333	1.80 (ref 13)	1.8
343	1.85 (ref 13)	1.9
353	1.90 (ref 13)	1.9
363	2.00 (ref 13)	2.0
373	2.25 (ref 13)	2.3
383	2.60 (ref 13)	2.6
393	3.00 (ref 13)	3.0

(continued next page)

COMPONENTS:	EVALUATOR:
(1) 1-Pentanol (*n-pentyl alcohol, n-amyl alcohol, n-butylcarbinol*); $C_5H_{12}O$: [71-41-0] (2) Water; H_2O; [7732-18-5]	G.T. Hefter and A.F.M. Barton, Murdoch University, Perth, Western Australia. July, 1983.

CRITICAL EVALUATION (continued)

T/K	Solubility, g(1)/100g sln Reported values	"Best" value (± σ_n)
403	3.55 (ref 13)	3.6
413	4.30 (ref 13)	4.3
423	5.35 (ref 13)	5.4
433	6.90 (ref 13)	6.9
443	9.55 (ref 13)	9.6
453	17.50 (ref 13)	17.5

The solubility of water (2) in 1-pentanol (1).

Agreement between the numerous studies is extremely poor at almost all temperatures and the "best" values should be regarded as very tentative, including those where agreement appears reasonable. Further studies of this system are clearly required. It is possible that the isomeric purity of the 1-pentanol is a significant factor.

Tentative solubilities of water (2) in 1-pentanol (1)

T/K	Solubility, g(2)/100g sln Reported values	"Best" value (± σ_n)
273	6.30 (ref 13)	6.3
283	8.92* (ref 1), 6.80 (ref 13), 10.2* (ref 21)	8.6 ± 1.4
293	9.42* (ref 1), 7.48(ref 6), 8.9(ref 10), 7.45(ref 13), 9.1(ref 27), 10.5(ref 31)	8.8 ± 1.1
298	9.67* (ref 1), 7.46(ref 6), 10.2(ref 12), 7.78* (ref 13), 7.5(ref 15), 9.8* (ref 22), 10.6(ref 31)	9.0 ± 1.3
303	9.91* (ref 1), 7.65(ref 6), 8.10(ref 13), 10.0* (ref 22), 9.3 (ref 30), 10.7 (ref 31)	9.3 ± 1.1
313	8.90 (ref 13), 10.48(ref 21), 10.6* (ref 22), 10.5(ref 26)	10.1 ± 0.7
323	9.75 (ref 13), 11.2* (ref 22)	10.5 ± 0.7
333	10.65 (ref 13), 11.8* (ref 22), 12.05 (ref 26)	11.5 ± 0.6
343	11.75 (ref 13), 12.7* (ref 22)	12.2 ± 0.5
353	12.95 (ref 13), 13.7* (ref 22)	13.3 ± 0.4
363	14.35 (ref 13), 14.8* (ref 22)	14.6 ± 0.2
373	15.85 (ref 13)	15.9
383	17.60 (ref 13)	17.6
393	19.60 (ref 13)	19.6
403	21.90 (ref 13)	21.9
413	24.65 (ref 13)	24.7
423	27.95 (ref 13)	28.0
433	32.00 (ref 13)	32.0
443	37.55 (ref 13)	37.6
453	49.70 (ref 13)	49.7

(continued next page)

COMPONENTS:

(1) 1-Pentanol (*n-pentyl alcohol, n-amyl alcohol, n-butylcarbinol*); $C_5H_{12}O$; [71-41-0]

(2) Water; H_2O; [7732-18-5]

EVALUATOR:

G.T. Hefter and A.F.M. Barton, Murdoch University, Perth, Western Australia, July, 1983.

CRITICAL EVALUATION (continued)

The upper critical solution temperature

The UCST has been reported as 457.6 K (184.4°C) by Hyde *et al.* (ref 16) which is in reasonable agreement with Erichsen's smoothed data (ref 13,14).

The dependence of the mutual solubility of 1-pentanol and water is illustrated in Figure 1. For ease of presentation only the data of Erichsen are plotted.

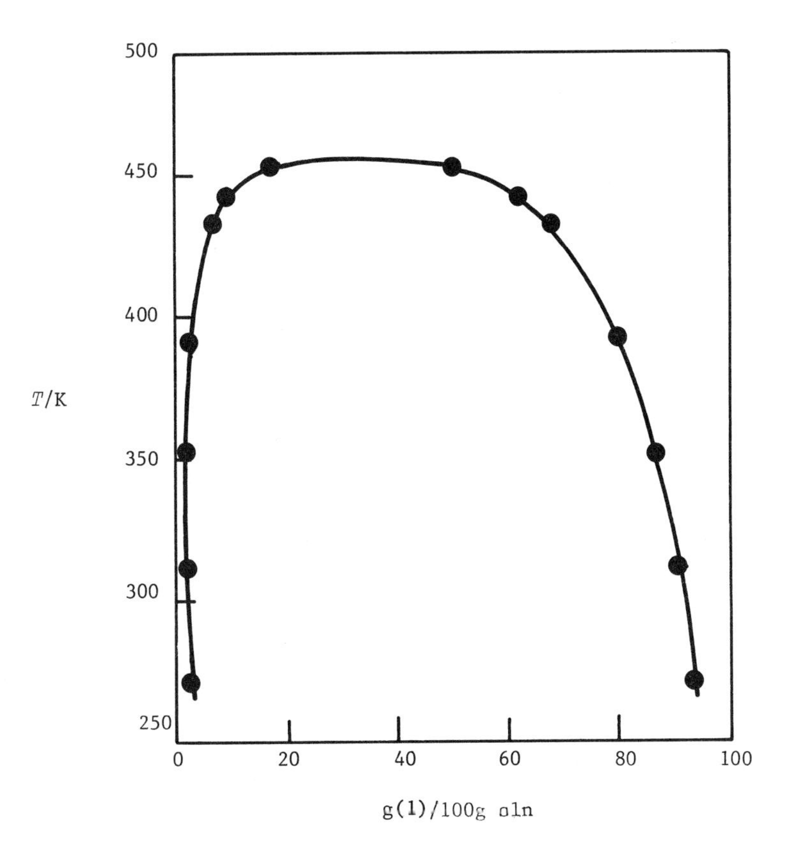

Fig. 1. Mutual solubility of (1) and (2) (data points from ref 13).

(continued next page)

COMPONENTS:	EVALUATOR:
(1) 1-Pentanol *(n-pentyl alcohol, n-amyl alcohol, n-butylcarbinol)*; $C_5H_{12}O$; [71-41-0] (2) Water; H_2O; [7732-18-5]	G.T. Hefter and A.F.M. Barton, Murdoch University, Perth, Western Australia. July, 1983.

CRITICAL EVALUATION (continued)

References

1. Verschaffelt, J. *Z. Phys. Chem.* 1884, *15*, 437.
2. Herz, W. *Ber.* 1898, *31*, 2669; *Boll. Chim. Farm.* 1915, *54*, 37.
3. Timmermans, J. *Z. Phys. Chem.* 1907, *58*, 129.
4. Fontein, F. *Z. Phys. Chem.* 1910, *73*, 212.
5. Butler, J.A.V.; Thomson, D.W.; Maclennan, W.H. *J. Chem. Soc.* 1933, 674.
6. Ginnings, P.M.; Baum, R. *J. Am. Chem. Soc.* 1937, *59*, 1111.
7. Jasper, J.J.; Farrell, L.G.; Madoff, M. *J. Chem. Educ.* 1944, *21*, 536.
8. Addison, C.C. *J. Chem. Soc.* 1945, 98.
9. Booth, H.S.; Everson, H.E. *Ind. Eng. Chem.* 1948, *40*, 1491.
10. Laddha, G.S.; Smith, J.M. *Ind. Eng. Chem.* 1948, *40*, 494.
11. Hansen, R.S.; Fu, Y.; Bartell, F.E. *J. Phys. Chem.* 1949, *53*, 769.
12. Donahue, D.J.; Bartell, F.E. *J. Phys. Chem.* 1952,
13. Erichsen, L. von. *Brennst. Chem.* 1952, *33*, 166.
14. Erichsen, L. von. *Naturwissenschaften* 1952, *39*, 41.
15. Crittenden, E.D.,Jr.; Hixon, A.N. *Ind. Eng. Chem.* 1954, *46*, 265.
16. Hyde, A.J.; Langbridge, D.M.; Lawrence, A.S.C. *Disc. Faraday Soc.* 1954, *18*, 239.
17. Kinoshita, K.; Ishikawa, H.; Shinoda, K. *Bull. Chem. Soc. Jpn.* 1958, *31*, 1081.
18. Ratouis, M.; Dodé, M. *Bull. Soc. Chem. Fr.* 1965, 3318.
19. Ionin, M.V.; Shanina, P.I. *Zh. Obshch. Khim.* 1967, *37*, 749; *J. Gen. Chem. USSR* 1975, *37*, 703.
20. Hanssens, I. *Associatie van normale alcoholen en hun affiniteit voor water en organische solventen*, Doctoraatsproefschrift, Leuven, 1969.
21. Krasnov, K.S.; Gartseva, L.A. *Izv. Vyssh. Uchebn. Zaved, Khim. Khim. Tekhnol.* 1970, *13*, 952.
22. Zhuravleva, I.K.; Zhuravlev, E.F. *Izv. Vyssh. Uchebn. Zaved., Khim. Khim. Tekhnol.* 1970, *13*, 480.
23. Mullens, J. *Alcoholassociaten*, Doctoraatsproefschift, Leuven, 1971; Huyskens, P.; Mullens, J.; Gomez, A.; Tack, J. *Bull. Soc. Chim. Belg.* 1975, *84*, 253.
24. Vochten, R.; Petre, G. *J. Colloid Interface Sci.* 1973, *42*, 320.
25. Korenman, I.M.; Gorokhov, A.A.; Polozenko, G.N. *Zh. Fiz. Khim.* 1974, *48*, 1810; 1975, *49*, 1490; *Russ. J. Phys. Chem.* 1974, *48*, 1065; 1975, *49*, 877.

(continued next page)

COMPONENTS:	EVALUATOR:
(1) 1-Pentanol (*n-pentyl alcohol, n-amyl alcohol, n-butylcarbinol*); $C_5H_{12}O$; [71-41-0] (2) Water; H_2O; [7732-18-5]	G.T. Hefter and A.F.M. Barton, Murdoch University, Perth, Western Australia. July, 1983

CRITICAL EVALUATION (continued)

26. Lavrova, O.A.; Lesteva, T.M. *Zh. Fiz. Khim.* 1976, *50*, 1617; *Dep. Doc. VINITI* 3813-75.

27. Charykov, A.K.; Tikhomirov, V.I.; Potapova, T.M. *Zh. Obshch. Khim.* 1978, *48*, 1916.

28. Evans, B.K.; James, K.C.; Luscombe, D.K. *J. Pharm. Sci.* 1978, *67*, 277.

29. Nishino, N.; Nakamura, M. *Bull. Chem. Soc. Jpn.* 1978, *51*, 1617; 1981, *54*, 545.

30. Singh, R.P.; Haque, M.M. *Indian J. Chem.* 1979, *17A*, 449.

31. Tokunaga, S.; Manabe, M.; Koda, M. *Niihama Kogyo Koto Semmon Gakko Kiyo, Rikogaku Hen, (Memoirs Niihama Technical College, Sci. and Eng.)* 1980, *16*, 96.

COMPONENTS:

(1) 1-Pentanol *(n-amyl alcohol, n-butylcarbinol)*; $C_5H_{12}O$; [71-41-0]

(2) Water; H_2O; [7732-18-5]

ORIGINAL MEASUREMENTS:

Verschaffelt, J.

Z. Phys. Chem. 1884, *15*, 437-57.

VARIABLES:

Temperature: 6-36 °C

PREPARED BY:

A. Maczynski; Z. Maczynska; A Szafranski

EXPERIMENTAL VALUES:

Solubility of water (2) in 1-pentanol (1)

$t/^{\circ}C$	g(2)/100g sln	x_2 (compiler)
6	8.72	0.319
15	9.14	0.330
25	9.67	0.344
32	10.00	0.352
36	10.20	0.357

$$g(2)/100g\ sln = 0.05(168 + t/^{\circ}C)$$

AUXILIARY INFORMATION

METHOD/APPARATUS/PROCEDURE:

The synthetic method was used.

Weighed amounts of (1) and (2) were placed in an Eykmann's freezing point apparatus immersed in a water bath and equipped with a calibrated thermometer. Cloud points were measured.

SOURCE AND PURITY OF MATERIALS:

(1) not specified.

(2) not specified.

ESTIMATED ERROR:

Temperature: ± 0.5 °C
Solubility : ± 0.025 g(2)/100g sln (maximum error)

REFERENCES:

COMPONENTS:

(1) 1-Pentanol *(n-amyl alcohol, n-butylcarbinol)*; $C_5H_{12}O$; [71-41-0]

(2) Water; H_2O; [7732-18-5]

ORIGINAL MEASUREMENTS:

Herz, W.

Ber. 1898, *31*, 2669-72.

Boll. chim. farm. 1915, *54*, 37.

VARIABLES:

One temperature: 22°C

PREPARED BY:

A. Maczynski, Z. Maczynska; A. Szafranski; A.F.M. Barton

EXPERIMENTAL VALUES:

The solubility of 1-pentanol in water at 22°C was reported to be 3.284 ml(1)/100 ml(2).

The 1915 reference reported (without details) that the solubility of water in 1-pentanol at 22°C was 2.214 ml(2)/100 ml(1).

AUXILIARY INFORMATION

METHOD/APPARATUS/PROCEDURE:

The densimetric method was used.

The composition of the saturated solution was evaluated by extrapolation of calibration density measurements (carried out on a series of synthetic solutions) to the measured density of the saturated solution. The maximum difference between the actual and the synthetic densities was a few in the third decimal place.

SOURCE AND PURITY OF MATERIALS:

(1) not specified.

(2) d^{22} 0.9980 g/cm^3.

ESTIMATED ERROR:

Not specified.

REFERENCES:

COMPONENTS:

(1) 1-Pentanol (*n-amyl alcohol, n-butylcarbinol*); $C_5H_{12}O$; [71-41-0]

(2) Water; H_2O; [7732-18-5]

ORIGINAL MEASUREMENTS:

Butler, J.A.V.; Thomson, D.W.; Maclennan, W.H.

J. Chem. Soc. 1933, 674-80.

VARIABLES:

One temperature: 25°C

PREPARED BY:

S.H. Yalkowsky; S.C. Valvani; A.F.M. Barton

EXPERIMENTAL VALUES:

The proportion of 1-pentanol (1) in the water-rich phase at equilibrium at 25°C was reported to be 2.208 g(1)/100g sln, the mean of five determinations (2.209, 2.203, 2.207, 2.212, 2.211 g(1)/100g sln). The corresponding mole fraction was reported as x_1 = 0.00460.

An approximate determination of the mole fraction solubility in the alcohol-rich phase gave x_1 = 0.71.

AUXILIARY INFORMATION

METHOD/APPARATUS/PROCEDURE:

An analytical method was used, with a U-tube apparatus having two internal stoppers. Suitable quantities of (1) and (2) were placed in one of the connected vessels and shaken in the thermostat for some hours. The liquid was allowed to separate into two layers, the heavier aqueous layer being separated by raising the stoppers and allowing part of the liquid to run into the connected vessel. A weighed portion of the separated sln was diluted with about an equal quantity of (2) and the resulting sln compared with calibration slns in an interferometer. To avoid the possibility of reading the position of the wrong fringe 2 cells (1 cm and 5 cm) were used. The method was unsuitable for analysis of alcohol-rich slns as no stoppered interferometer cell was available.

SOURCE AND PURITY OF MATERIALS:

(1) B.D.H.; repeatedly fractionated under 30 cm Hempel column in all-glass apparatus, middle fraction dried with Ca and fractionated;

b.p. 137.60 - 137.70°C (corr.)

d_4^{25} 0.81146 n_D^{20} 1.41043

(2) not stated

ESTIMATED ERROR:

Solubility: the result is the mean of five determinations.

Temperature: not stated (but in related experiments it was ± 0.03°C).

REFERENCES:

COMPONENTS:

(1) 1-Pentanol *(n-amyl alcohol, n-butylcarbinol)*; $C_5H_{12}O$; [71-41-0]

(2) Water; H_2O; [7732-18-5]

ORIGINAL MEASUREMENTS:

Ginnings, P.M.; Baum, R.

J. Am. Chem. Soc. 1937, *59*, 1111-3.

VARIABLES:

Temperature: 20-30°C

PREPARED BY:

A. Maczynski and Z. Maczynska

EXPERIMENTAL VALUES:

Mutual solubility of 1-pentanol(1) and water(2)

$t/^{\circ}C$	g(1)/100g sln		x_1(compiler)	
	(2)-rich phase	(1)-rich phase	(2)-rich phase	(1)-rich phase
20	2.36	92.52	0.00491	0.7165
25	2.19	92.54	0.00455	0.7171
30	2.03	92.35	0.00421	0.7115

Relative density, d_4

$t/^{\circ}C$	Water-rich phase	Alcohol-rich phase
20	0.9939	0.8317
25	0.9930	0.8287
30	0.9919	0.8253

AUXILIARY INFORMATION

METHOD/APPARATUS/PROCEDURE:

Hill's volumetric method was adopted.

Both components were introduced in known amounts into a two-bulb graduated and calibrated flask and shaken mechanically in a water-bath at constant temperature. After sufficient time the liquids were allowed to separate and the total volume was measured. Upon centrifugation, the phase separation line was read, and phase volumes were calculated. From the total weights of the components, the total volume, individual phase volumes, and component concentrations in either phase were evaluated.

SOURCE AND PURITY OF MATERIALS:

(1) Eastman best grade; distilled from metallic calcium; b.p. range 137.6-138.3°C, d_4^{25} 0.8110.

(2) not specified.

ESTIMATED ERROR:

Temperature: ± 0.1°C
Solubility: better than 0.1 wt% (type of error not specified)

REFERENCES:

COMPONENTS:

(1) 1-Pentanol *(n-amyl alcohol, n-butyl carbinol)*; $C_5H_{12}O$; [71-41-0]

(2) Water; H_2O; [7732-18-5]

ORIGINAL MEASUREMENTS:

Jasper, J.J.; Farrell, L.G.; Madoff, M. *J. Chem. Educ.* 1944, *21*, 536-8.

VARIABLES:

One temperature: 95.3°C

PREPARED BY:

A.F.M. Barton

EXPERIMENTAL VALUES:

The mole fractions of 1-pentanol (1) at equilibrium at the 1 atm boiling point (95.3°C) of the binary mixture with water (2) were x_1 = 0.024 in the water-rich phase and x_1 = 0.481 in the alcohol-rich phase.

The corresponding mass percentage solubilities, calculated by the compilers, are 10.7 g(1)/100g sln and 81.9 g(1)/100g sln, respectively.

(Actual temperatures and pressures observed were 94.7°C at 744.6 mm Hg and 94.9°C at 750.3 mm Hg respectively.)

AUXILIARY INFORMATION

METHOD/APPARATUS/PROCEDURE:

The vapor-liquid temperature-composition diagram was determined by the method previously described in ref 1. Samples of distillate and residue from distillation with both excess water and excess 1-pentanol were analyzed by refractive index. Ethanol was added to the samples to ensure homogeneity during analysis, and compositions were determined from a calibration curve. The results reported were obtained while two layers remained in the distillation flask.

SOURCE AND PURITY OF MATERIALS:

Not stated.

Not stated.

ESTIMATED ERROR:

Not stated.

REFERENCES:

(1) Jasper, J.J.; Campbell, C.J.; Marshall, D.E. *J. Chem. Educ.* 1941, *18*, 540-2.

COMPONENTS:

(1) 1-Pentanol *(n-amyl alcohol, n-butylcarbinol)*; $C_5H_{12}O$; [71-41-0]

(2) Water; H_2O; [7732-18-5]

ORIGINAL MEASUREMENTS:

Addison, C.C.

J. Chem. Soc. 1945, 98-106.

VARIABLES:

One temperature: 20°C

PREPARED BY:

S.H. Yalkowsky; S.C. Valvani; A.F.M. Barton

EXPERIMENTAL VALUES:

The proportion of 1-pentanol (1) in the water-rich phase at equilibrium at 20°C was reported to be 2.21 g(1)/100g sln.

The corresponding mole fraction solubility calculated by the compilers is x_1 = 0.00460.

AUXILIARY INFORMATION

METHOD/APPARATUS/PROCEDURE:

A surface tension method was used. Sufficient excess of (1) was added to 100 mL of (2) in a stoppered flask to form a separate lens on the surface. The mixture was swirled gently, too vigorous an agitation giving a semi-permanent emulsion and incorrect readings. After settling, a small sample of the clear aqueous sln was withdrawn into a drop weight pipet and the surface tension determined. The swirling was continued until a constant value was obtained. The surface tension-concentration curve was known, and only a slight extrapolation (logarithmic scale) was necessary to find the concentration corresponding to the equilibrium value.

SOURCE AND PURITY OF MATERIALS:

(1) impure alcohols were purified by fractional distillation, the middle fraction from a distillation being redistilled;

b.p. 138.0°C d_4^{20} 0.8154

n_D^{20} 1.4102

(2) not stated

ESTIMATED ERROR:

Solubility: ± 0.5%

REFERENCES:

AWW-G

COMPONENTS:

(1) 1-Pentanol (*n-amyl alcohol; n-butylcarbinol*); $C_5H_{12}O$; [71-41-0]

(2) Water; H_2O; [7732-18-5]

ORIGINAL MEASUREMENTS:

Booth, H.S.; Everson, H.E.

Ind. Eng. Chem. 1948, *40*, 1491-3.

VARIABLES:

One temperature: 25°C

Sodium xylene sulfonate

PREPARED BY:

S.H. Yalkowsky; S.C. Valvani; A.F.M. Barton

EXPERIMENTAL VALUES:

It was reported that the solubility of 1-pentanol (1) in water (2) was 3.4 mL(1)/100mL (2) at 25.0°C.

The corresponding value in 40% sodium xylene sulfonate solution as solvent was >400 mL/(1)/100mL solvent.

AUXILIARY INFORMATION

METHOD/APPARATUS/PROCEDURE:

A known volume of (2) or aqueous solvent (usually 50mL) in a tightly stoppered calibrated Babcock tube was thermostatted. Successive measured quantities (1) were added and equilibrated until a slight excess of solute remained. The solution was centrifuged, returned to the thermostat bath for 10 min. and the volume of excess solute measured directly. This was a modification of the method described in ref 1.

SOURCE AND PURITY OF MATERIALS:

(1) "CP or highest grade commercial".

(2) "distilled"

ESTIMATED ERROR:

Solubility: within 0.1 mL(1)/100mL (2)

REFERENCES:

1. Henslick, R.S. *Dissertation*, Columbia University, 1935.

COMPONENTS:

(1) 1-Pentanol *(n-amyl alcohol, n-butylcarbinol)*; $C_5H_{12}O$; [71-41-0]

(2) Water; H_2O; [7732-18-5]

ORIGINAL MEASUREMENTS:

Laddha, G.S.; Smith, J.M.

Ind. Eng. Chem., 1948, *40*, 494-6.

VARIABLES:

One temperature: 20°C

PREPARED BY:

A. Maczynski

EXPERIMENTAL VALUES:

The solubility of 1-pentanol in water at 20°C was reported to be 1.5 g(1)/100g sln.

The corresponding mole fraction, x_1, calculated by compiler is 0.0031.

The solubility of water in 1-pentanol at 20°C was reported to be 8.9 g(2)/100g sln.

The corresponding mole fraction, x_2, calculated by compiler is 0.32.

AUXILIARY INFORMATION

METHOD/APPARATUS/PROCEDURE:

The titration method was used.

One component was placed in a 20°C constant-temperature bath for 1 h . Then titration was carried out in several steps, in order that the mixture could be frequently returned to the constant-temperature bath to ensure maintenance of the 20°C temperature. The end point was taken when turbidity appeared over the entire solution.

SOURCE AND PURITY OF MATERIALS:

(1) Mallinckrodt Chemical Co., reagent grade; b.p. range 134-137°C, but nearly all distilled at 137°C; d^{20} 0.817.

(2) distilled.

ESTIMATED ERROR:

Not specified.

REFERENCES:

COMPONENTS:

(1) 1-Pentanol (*n-amyl alcohol, n-butylcarbinol*); $C_5H_{12}O$; [71-41-0]

(2) Water; H_2O; [7732-18-5]

ORIGINAL MEASUREMENTS:

Hansen, R.S.; Fu, Y.; Bartell, F.E.

J. Phys. Chem. 1949, *53*, 769-85

VARIABLES:

One temperature: 25°C

PREPARED BY:

S.H. Yalkowsky; S.C. Valavani; A.F.M. Barton

EXPERIMENTAL VALUES:

The proportion of 1-pentanol in the water-rich phase at equilibrium at 25°C was reported to be 2.54 g(1)/100g sln, a concentration of 0.287 mol(1)/L sln. The corresponding mole fraction solubility, calculated by the compilers, is x_1 = 0.00529.

AUXILIARY INFORMATION

METHOD/APPARATUS/PROCEDURE:

An excess of the alcohol was added to water in a mercury-sealed flask which was shaken mechanically for 48h in an air chamber thermostat. The flask was then allowed to stand for 3h in the airbath, after which a portion of the water-rich phase was removed by means of a hypodermic syringe and was compared interferometrically with the most concentrated alcohol solution which could be prepared conveniently. The solubility determination was associated with a study of multimolecular adsorption from binary liquid solution.

SOURCE AND PURITY OF MATERIALS:

(1) CP;
extracted into large volume of water and steam distilled to remove diamyl ether contaminant, the procedure repeated, then dried over anhydrous magnesium sulfate, distilled with 60 cm glass-packed reflux column;

b.p. 136°C/735 mm Hg

(2) distilled laboratory water, redistilled from alkaline permanganate solution.

ESTIMATED ERROR:

Temperature: ± 0.1°C

Solubility: deviation from mean of three determinations ± 0.03 wt %

REFERENCES:

COMPONENTS:

(1) 1-Pentanol (*n-amyl alcohol, n-butylcarbinol*); $C_5H_{12}O$; [71-41-0]

(2) Water; H_2O; [7732-18-5]

ORIGINAL MEASUREMENTS:

Donahue, D.J.; Bartell, F.E.

J. Phys. Chem. 1952, *56*, 480-4.

VARIABLES:

One temperature: 25°C

PREPARED BY:

A.F.M. Barton.

EXPERIMENTAL VALUES:

	Density g mL^{-1}	x_1	Mutual solubilities g(1)/100g sln (compiler)
Alcohol-rich phase	0.8325	0.643	89.8
Water-rich phase	0.9935	0.00459[a]	2.21

[a] From ref 1 and 2

AUXILIARY INFORMATION

METHOD/APPARATUS/PROCEDURE:

Mixtures were placed in glass stoppered flasks and were shaken intermittently for at least 3 days in a water bath. The organic phase was analyzed for water content by the Karl Fischer method and the aqueous phase was analyzed interferometrically. The solubility measurements formed part of a study of water-organic liquid interfacial tensions.

SOURCE AND PURITY OF MATERIALS:

(1) "best reagent grade", fractional distillation.

(2) "purified"

ESTIMATED ERROR:

Temperature: ± 0.1°C

REFERENCES:

1. Butler, J.A.V.; Thomson, D.W.; Maclennan, W.H. *J. Chem. Soc.* 1933, 674.
2. Hansen, R.S.; Fu, Y.; Bartell, F.E. *J. Phys. Chem.* 1949, *53*, 769.

COMPONENTS:

(1) 1-Pentanol *(n-amyl alcohol, n-butylcarbinol)*; $C_5H_{12}O$; [71-41-0]

(2) Water; H_2O; [7732-18-5]

ORIGINAL MEASUREMENTS:

Erichsen, L. von

Brennst. Chem. 1952, *33*, 166-72.

VARIABLES:

Temperature: 0-180° C

PREPARED BY:

S.H. Yalkowsky and Z. Maczynska

EXPERIMENTAL VALUES:

Mutual solubility of 1-pentanol and water

$T/^{\circ}C$	(2)-rich phase		(1)-rich phase	
	g(1)/100g sln	x_1	g(1)/100g sln	x_1
0	3.05	0.0064	93.70	0.7526
10	2.70	0.0055	93.20	0.7370
20	2.35	0.0049	92.55	0.7175
30	2.10	0.0044	91.90	0.6988
40	1.90	0.0039	91.10	0.6767
50	1.80	0.0036	90.25	0.6543
60	1.80	0.0036	89.35	0.6317
70	1.85	0.0038	88.25	0.6057
80	1.90	0.0039	87.05	0.5788
90	2.00	0.0041	85.65	0.5497
100	2.25	0.0047	84.15	0.5205
110	2.60	0.0054	82.40	0.4890
120	3.00	0.0063	80.40	0.4561
130	3.55	0.0075	78.10	0.4217
140	4.30	0.0092	75.35	0.3846
150	5.35	0.0114	72.05	0.3451
160	6.90	0.0149	68.00	0.3029
170	9.55	0.0211	62.45	0.2548
180	17.50	0.0416	50.30	0.1714

AUXILIARY INFORMATION

METHOD/APPARATUS/PROCEDURE:

The synthetic method was used.

The measurements were carried out in 2 mL glass ampules placed in an aluminum block equipped with two glass windows. Cloud points were measured with a thermocouple wound around the ampule. Each measurement was repeated twice.

SOURCE AND PURITY OF MATERIALS:

(1) Merck, or Ciba, or industrial products; distilled and chemically free from isomers;
b.p. 137.8-137.9°C (757 mm Hg)
n_D^{20} 1.4098.

(2) not specified.

ESTIMATED ERROR:

Not specified.

REFERENCES:

COMPONENTS:

(1) 1-Pentanol *(n-amyl alcohol, n-butylcarbinol)*; $C_5H_{12}O$; [71-41-0]

(2) Water; H_2O; [7732-18-5]

ORIGINAL MEASUREMENTS:

Erichsen, L. von

Naturwissenschaften 1952, *39*, 41-2.

VARIABLES:

Temperature: 0-50°C

PREPARED BY:

A. Maczynski; Z Maczynska

EXPERIMENTAL VALUES:

Solubility of 1-pentanol in water

$t/^{\circ}C$	x_1	g(1)/100g sln (compiler)
0	0.0064	3.1
10	0.0055	2.6
20	0.0049	2.4
30	0.0044	2.1
40	0.0039	1.9
50	0.0036	1.7

AUXILIARY INFORMATION

METHOD/APPARATUS/PROCEDURE:

The synthetic method was used.

No details were reported in the paper.

SOURCE AND PURITY OF MATERIALS:

(1) not specified.

(2) not specified.

ESTIMATED ERROR:

Not specified.

REFERENCES:

COMPONENTS:

(1) 1-Pentanol *(n-amyl alcohol, n-butylcarbinol)*; $C_5H_{10}O$; [71-41-0]

(2) Water; H_2O; [7732-18-5]

ORIGINAL MEASUREMENTS:

Crittenden, E.D.,Jr.; Hixon, A.N.

Ind. Eng. Chem. 1954, *46*, 265-8.

VARIABLES:

One temperature: 25°C

PREPARED BY:

A. Maczynski

EXPERIMENTAL VALUES:

The solubility of 1-pentanol in water at 25°C was reported to be 2.2g(1)/100g sln.

The corresponding mole fraction, x_1, calculated by the compiler is 0.0046.

The solubility of water in 1-pentanol at 25°C was reported to be 7.5g(2)/100g sln.

The corresponding mole fraction, x_2, calculated by the compiler is 0.28.

AUXILIARY INFORMATION

METHOD/APPARATUS/PROCEDURE:

Presumably the titration method described for ternary systems containing HCl was used. In this method the solubility was determined by bringing 100-ml samples of (1) or (2) to a temperature 25 ± 0.10°C and the second component was then added from a calibrated buret, with vigorous stirring, until the solution became permanently cloudy.

SOURCE AND PURITY OF MATERIALS:

(1) source not specified; purified; purity not specified.

(2) not specified.

ESTIMATED ERROR:

Solubility: 2% (alcohol-rich)- 10% (water-rich)

Temperature: ± 0.10°C.

REFERENCES:

COMPONENTS:

(1) 1-Pentanol *(n-amyl alcohol, n-butylcarbinol)*; $C_5H_{12}O$; [71-41-0]

(2) Water; H_2O; [7732-18-5]

ORIGINAL MEASUREMENTS:

Hyde, A.J.; Langbridge, D.M.; Lawrence, A.S.C.

Disc. Faraday Soc., 1954, *18*, 239-58.

VARIABLES:

One temperature: 184.4°C

PREPARED BY:

A. Maczynski; A.F.M. Barton

EXPERIMENTAL VALUES:

The upper critical solution temperature was reported to be 184.4°C at 43.0 g(1)/100g sln (x_1 = 0.134, compiler).

The room temperature solubility of (1) and (2) was reported as 2.0%

AUXILIARY INFORMATION

METHOD/APPARATUS/PROCEDURE:

Not specified.

SOURCE AND PURITY OF MATERIALS:

(1) not specified.

(2) not specified.

ESTIMATED ERROR:

Not specified.

REFERENCES:

COMPONENTS:

(1) 1-Pentanol *(n-amyl alcohol; n-butylcarbinol)*; $C_5H_{12}O$; [71-41-0]

(2) Water; H_2O; [7732-18-5]

ORIGINAL MEASUREMENTS:

Kinoshita, K; Ishikawa, H; Shinoda, K; *Bull. Chem. Soc. Jpn.* 1958, *31*, 1081-4.

VARIABLES:

One temperature: 25°C

PREPARED BY:

S.H. Yalkowsky; S.C. Valvani; A.F.M. Barton

EXPERIMENTAL VALUES:

The concentration of 1-pentanol (1) in the water-rich phase at equilibrium at 25.0°C was reported to be 0.25 mol(1) L^{-1}. The weight percentage solubility was reported as 2.2 g(1)/100g sln, and the corresponding mole fraction solubility, calculated by the compilers, is x_1 = 0.0046.

AUXILIARY INFORMATION

METHOD/APPARATUS/PROCEDURE:

The surface tension in aqueous solutions of alcohols monotonically decreases up to their saturation concentration and remains constant in the heterogeneous region of (ref 1-4). Surface tension was measured by the drop weight method, using a tip 6 mm in diameter, the measurements being carried out in a water thermostat. From the (surface tension) - (logarithm of concentration) curves the saturation points were determined as the intersections of the curves with the horizontal straight lines passing through the lowest experimental points.

SOURCE AND PURITY OF MATERIALS:

(1) purified by vacuum distillation through 50-100 cm column; b.p. 138°C

(2) not stated

ESTIMATED ERROR:

Temperature: 0.05°C

Solubility: within 4%

REFERENCES:

1. Motylewski, S. *Z.Anorg.Chem.* 1904, *38*, 410.
2. Taubamann, A. *Z. Physik. Chem* 1932, *A161*, 141.
3. Zimmerman, H.K.,Jr. *Chem. Rev.* 1952, *51*, 25.
4. Shinodo, K.; Yamanaka, T.; Kinoshita, K. *J.Phys. Chem.* 1959, *63*, 648.

COMPONENTS:

(1) 1-Pentanol (*n-amyl alcohol, n-butylcarbinol*); $C_5H_{12}O$; [71-41-0]

(2) Water; H_2O; [7732-18-5]

ORIGINAL MEASUREMENTS:

Ratouis, M.; Dodé, M.;

Bull. Soc. Chim. Fr. 1965, 3318-22.

VARIABLES:

One temperature: 30°C

Ringer solution also studied

PREPARED BY:

S.C. Valvani; S.H. Yalkowsky; A.F.M. Barton

EXPERIMENTAL VALUES:

The proportion of 1-pentanol (1) in the water-rich phase at equilibrium at 30°C was reported to be 2.00 g(1)/100g sln.

The corresponding mole fraction solubility, calculated by the compilers, is x_1 = 0.00415.

The proportion of (1) in the water-rich phase of a mixture with Ringer solution at equilibrium at 30°C was reported to be 1.92 g(1)/100g sln.

AUXILIARY INFORMATION

METHOD/APPARATUS/PROCEDURE:

In a round bottomed flask, 50 mL of water and a sufficient quantity of alcohol were introduced until two separate layers were formed. The flask assembly was equilibrated by agitation for at least 3 h in a constant temperature bath. Equilibrium solubility was attained by first supersaturating at a slightly lower temperature (solubility of alcohols in water decreases with increasing temperature) and the equilibrating at the desired temperature. The aqueous layer was separated after an overnight storage in a bath. The alcohol content was determined by reacting the aqueous solution with potassium dichromate and titrating the excess dichromate with ferrous sulfate solution in the presence of phosphoric acid and diphenylamine barium sulfonate as an indicator.

SOURCE AND PURITY OF MATERIALS:

(1) Prolabo, Paris; redistilled with 10:1 reflux ratio;

b.p. 138.6-138.7/763 mm Hg

n_D^{25} = 1.40800

(2) twice distilled from silica apparatus or ion-exchanged with Sagei A20

ESTIMATED ERROR:

Solubility: relative error of 2 determinations less than 1%

Temperature: ± 0.05°C

REFERENCES:

COMPONENTS:

(1) 1-Pentanol (*n-amyl alcohol, n-butylcarbinol*); $C_5H_{12}O$; [71-41-0]

(2) Water: H_2O; [7732-18-5]

ORIGINAL MEASUREMENTS:

Hanssens, I.

Associatie van normalle alcoholen en hun affiniteit voor water en organische solventen

Doctoraatsproefschrift, Leuven, 1969.

VARIABLES:

One temperature: 298K

PREPARED BY:

M.C. Haulait-Pirson; A.F.M. Barton

EXPERIMENTAL VALUES:

The concentration of 1-pentanol (1) in the water-rich phase was reported as 0.2405 mol(1)/L sln, and the concentration of water (2) in the alcohol-rich phase was reported as 8.290 mol(2)/L sln.

The corresponding solubilities on a mass/volume basis, calculated by the compilers, are 21.2 g(1)/L sln, and 149.4 g(2)/L sln respectively.

(The temperature was unspecified in the Thesis, but reported as 298 K for related investigations in ref 2.)

AUXILIARY INFORMATION

METHOD/APPARATUS/PROCEDURE:

(1) and (2) were equilibrated using a cell described in ref 1. The Rayleigh M75 interference refractometer with the cell M160 for liquids was used for the determination of the concentrations. Cell thicknesses were 1, 3 and 10 cm depending on the concentration range. Standard solutions covering the whole range of concentrations investigated were used for the calibration.

SOURCE AND PURITY OF MATERIALS:

(1) Merck p.a.

(2) distilled

ESTIMATED ERROR:

Solubility: ± 0.00036 - 0.05 mol(1) depending on concentration.

REFERENCES:

(1) Meeussen, E.; Huyskens, P. *J.Chim.Phys.* 1966, *63*, 845.

(2) Huyskens, P; Mullens, J.; Gomez, A.; Tack, J. *Bull. Soc. Chim. Belg.* 1975, *84*, 253-62.

COMPONENTS:

(1) 1-Pentanol *(n-amyl alcohol, n-butylcarbinol)*; $C_5H_{12}O$; [71-41-0]

(2) Water; H_2O; [7732-18-5]

ORIGINAL MEASUREMENTS:

Krasnov, K.S.; Gartseva, L.A.

Izv. Vyssh. Ucheb. Zaved. Khim. Khim. Tekhnol. 1970, *13(7)*, 952-6.

VARIABLES:

Temperature: 12 and 40°C

PREPARED BY:

A. Maczynski and Z. Maczynska

EXPERIMENTAL VALUES:

Solubility of water (2) in 1-pentanol (1)

$t/^{\circ}C$	g(2)/100g sln	x_2 (compiler)
12	10.20	0.357
40	10.48	0.364

AUXILIARY INFORMATION

METHOD/APPARATUS/PROCEDURE:

The analytical method was used.

A saturated mixture of (1) and (2) was placed in a thermostat and the phases allowed to separate. Then (2) was determined in the organic layer by the Karl Fischer analysis.

SOURCE AND PURITY OF MATERIALS:

(1) source not specified; CP reagent; distilled; no isomers by GLC; d_4^{25} 0.8106.

(2) not specified.

ESTIMATED ERROR:

Temperature: ± 0.05°C
Solubility: ± 0.05 wt % (not specified)

REFERENCES:

COMPONENTS:

(1) 1-Pentanol *(n-amyl alcohol, n-butylcarbinol)*; $C_5H_{12}O$; [71-41-0]

(2) Water; H_2O; [7732-18-5]

ORIGINAL MEASUREMENTS:

Zhuravleva, I.K.; Zhuravlev, E.F.

Izv. Vyssh. Ucheb. Zaved. Khim. Khim. Tekhnol. 1970, *13*, 480-5.

VARIABLES:

Temperature: 12-94°C

PREPARED BY:

Z. Maczynska

EXPERIMENTAL VALUES:

Solubility of 1-pentanol (1) in water (2)

$t/^{\circ}C$	g(1)/100g sln	x_1(compiler)
12.0	2.0	0.0042
63.0	1.6	0.0033
94.0	2.0	0.0042

Solubility of water (2) in 1-pentanol (1)

$t/^{\circ}C$	g(2)/100g sln	x_2(compiler)
29.0	10.0	0.352
41.5	10.7	0.370
56.0	11.5	0.389
66.0	12.4	0.409
73.5	13.0	0.422
86.5	14.4	0.452

AUXILIARY INFORMATION

METHOD/APPARATUS/PROCEDURE:

Alekseev's method (ref 1) was used.

No details were reported in the paper.

SOURCE AND PURITY OF MATERIALS:

(1) source not specified; freshly distilled; purity not specified.

(2) twice distilled.

ESTIMATED ERROR:

Not specified.

REFERENCES:

1. Alekseev, V.F. *Zh. russk. Khim., o-va,* 1876, *8*, 249.

COMPONENTS:

(1) 1-Pentanol (*n-amyl alcohol; n-butylcarbinol*); $C_5H_{12}O$; [71-41-0]

(2) Water; H_2O; [7732-18-5]

ORIGINAL MEASUREMENTS:

Mullens, J. *Alcoholassociation;* Doctoraatsporoefschrift, Leuven, 1971.

Huyskens, P.; Mullens, J.; Gomez, A.; Tack, J.; *Bull. Soc. Chim. Belg.* 1975, *84*, 253-62

VARIABLES:

One temperature: 25°C

PREPARED BY:

M.C. Haulait-Pirson; A.F.M. Barton

EXPERIMENTAL VALUES:

At equilibrium at 25°C the concentration of 1-pentanol (1) in the water-rich phase was reported as 0.241 mol(1)/L sln, and the concentration of water (2) in the alcohol-rich phase was reported as 8.290 mol(2)/L sln.

The corresponding solubilities on a mass/volume basis, calculated by the compilers, are 21.24 g(1)/L sln, and 149.4 g(2)/L sln, respectively.

AUXILIARY INFORMATION

METHOD/APPARATUS/PROCEDURE:

The partition of the two components was made using a cell described in ref 1. The Rayleigh Interference Refractometer M154 was used for the determination of the concentrations.

Standard solutions covering the whole range of concentration investigated were used for the calibration.

SOURCE AND PURITY OF MATERIALS:

(1) Merck (p.a.).

(2) distilled

ESTIMATED ERROR:

Solubility: ± 0.001 mol(1)/L sln.

REFERENCES:

1. Meeussen, E.; Huyskens, P. *J. Chim. Phys.* 1966, *63*, 845

COMPONENTS:

(1) 1-Pentanol (*n-amyl alcohol, n-butylcarbinol*); $C_5H_{12}O$; [71-41-0]

(2) Water; H_2O; [7732-18-5]

ORIGINAL MEASUREMENTS:

Vochten, R.; Petre, G.;
J. Colloid Interface Sci. 1973, *42*, 320-7.

VARIABLES:

One temperature: 15°C

PREPARED BY:

S.H. Yalkowsky; S.C. Valvani; A.F.M. Barton

EXPERIMENTAL VALUES:

The concentration of 1-pentanol (1) in the water-rich phase at equilibrium at 15°C was reported to be 0.26 mol(1)/L sln.

The corresponding mass/volume solubility, calculated by the compilers, is 22.9 g(1)/L sln.

AUXILIARY INFORMATION

METHOD/APPARATUS/PROCEDURE:

The solubility was obtained from the surface tension of saturated solvents measured by the staticmethod of Wilhelmy (platinum plate). The apparatus consisted of an electrobalance (R.G. Cahn) connected with a high impedance null detector (FLUKE type 845 AR). An all-Pyrex vessel was used.

SOURCE AND PURITY OF MATERIALS:

(1) purified by distillation and preparative gas chromatography;
b.p. 127.8°C/760 mm Hg

(2) triply distilled from permanganate solution.

ESTIMATED ERROR:

Temperature: ± 0.1°C

Solubility: ± 0.01 mol(1)/L sln. (probably standard deviation)

REFERENCES:

COMPONENTS:

(1) 1-Pentanol *(n-amyl alcohol, n-butylcarbinol)*; $C_5H_{12}O$; [71-41-0]

(2) Water; H_2O; [7732-18-5]

ORIGINAL MEASUREMENTS:

Korenman, I.M.; Gorokhov, A.A.; Polozenko, G.N.
Zhur. Fiz. Khim. 1974, *48*, 1810-2;
**Russ. J. Phys. Chem.* 1974, *48*, 1065-7;
Zhur. Fiz. Khim. 1975, *49*, 1490-3;
**Russ. J. Phys. Chem.* 1975, *49*, 877-8.

VARIABLES:

One temperature: 25°C

PREPARED BY:

A.F.M. Barton

EXPERIMENTAL VALUES:

The equilibrium concentration of 1-pentanol (1) in the water-rich phase at 25.0°C was reported to be 0.27 mol (1)/L sln, and the concentration of water (2) in the alcohol-rich phase was reported to be 4.37 mol (2)/L sln.

The corresponding solubilities on a mass/volume basis, calculated by the compiler, are 23.8 g(1)/L sln, and 78.7 g(2)/L sln respectively.

AUXILIARY INFORMATION

METHOD/APPARATUS/PROCEDURE:

The two liquids were shaken in a closed vessel at 25.0 ± 0.1°C until equilibrium was established. The soly of the alcohol in the aqueous phase was determined on a Tsvet-1 chromatograph with a flame-ionization detector. The sorbent was a polyethylene glycol adipate deposited on Polychrom-1 (10% of the mass of the carrier). The column had an internal diameter 4 mm, its temp. was 140°C, and the flow of the carrier gas (nitrogen) was 50 mL min^{-1}, The soly of water in the alcohol was determined on a UKh-2 universal chromatograph under isothermal conditions (150°C) with a heat-conductivity detector. The 1 m by 6 mm column was filled with Polysorb. The carrier gas was helium (50 mL min^{-1}). The study formed part of an investigation of salting-out by alkali halides of higher alcohol-water systems.

SOURCE AND PURITY OF MATERIALS:

Not stated.

ESTIMATED ERROR:

Temperature: ± 0.1°C

Solubility: not stated; the results reported are the arithmetic means from four sets of experiments.

REFERENCES:

COMPONENTS:

(1) 1-Pentanol *(n-amyl alcohol, n-butylcarbinol)*; $C_5H_{12}O$; [71-41-0]

(2) Water; H_2O; [7732-18-5]

ORIGINAL MEASUREMENTS:

Lavrova, O.A.; Lesteva, T.M.

Zh. Fiz. Khim., 1976, *50*, 1617; *Dep. Doc. VINITI* 3813-75.

VARIABLES:

Temperature: 40 and 60°C

PREPARED BY:

A. Maczynski

EXPERIMENTAL VALUES:

Mutual solubility of 1-pentanol and water

$t/°C$	g(1)/100g sln		x_1 (compiler)	
	(2)-rich phase	(1)-rich phase	(2)-rich phase	(1)-rich phase
40	2.1	89.5	0.0044	0.635
60	2.0	87.95	0.0041	0.5986

AUXILIARY INFORMATION

METHOD/APPARATUS/PROCEDURE:

The titration method was used.

No details were reported in the paper.

SOURCE AND PURITY OF MATERIALS:

(1) source not specified; distilled with heptane; purity 99.93 wt % with 0.07 wt % of water; n_D^{20} 1.4100, d_4^{20} 0.8146; b.p. 138.0°C.

(2) not specified.

ESTIMATED ERROR:

Not specified.

REFERENCES:

COMPONENTS:

(1) 1-Pentanol *(n-amyl alcohol, n-butylcarbinol)*; $C_5H_{12}O$: [71-41-0]

(2) Water; H_2O; [7732-18-5]

ORIGINAL MEASUREMENTS:

Charykov, A.K.; Tikhomirov, V.I.; Potapova, T.M.

Zh. Obshch. Khim. 1978, *48*, 1916-21.

VARIABLES:

One temperature: 20°C

PREPARED BY:

A. Maczynski

EXPERIMENTAL VALUES:

The solubility of water in 1-propanol at 20°C was reported to be x_2 = 0.33.

The corresponding mass per cent value calculated by the compiler is 9.1 g(2)/100 sln.

AUXILIARY INFORMATION

METHOD/APPARATUS/PROCEDURE:

The analytical method was used.

The solubility of (2) in (1) was determined by Karl Fischer reagent method. Three determinations were made.

SOURCE AND PURITY OF MATERIALS:

(1) not specified.

(2) not specified.

ESTIMATED ERROR:

Not specified.

REFERENCES:

COMPONENTS:

(1) 1-Pentanol *(n-amyl alcohol, n-butylcarbinol)*; $C_5H_{12}O$; [71-41-0]

(2) Water; H_2O; [7732-18-5]

ORIGINAL MEASUREMENTS:

Evans, B.K.; James, K.C.; Luscombe, D.K.
J. Pharm. Sci. 1978, *67*, 277-8.

VARIABLES:

One temperature: 37°C

PREPARED BY:

S.H. Yalkowsky; S.C. Valvani; A.F.M. Barton

EXPERIMENTAL VALUES:

The concentration of 1-pentanol (1) in the water-rich phase at equilibrium at 37°C was reported to be 0.213 mol(1) L^{-1} sln.

The corresponding solubility on a mass/volume basis, calculated by the compilers, is 18.8 g(1) L^{-1} sln.

AUXILIARY INFORMATION

METHOD/APPARATUS/PROCEDURE:

This determination is one of a large number in the paper reported only briefly; the following is the procedure assumed by the compiler to have been used. A moderate excess of solute was stirred continuously with water for up to 10h in a sealed conical flask immersed in a water thermostat bath, and allowed to stand overnight in the bath. After separation, analysis was by g.l.c. using a hydrogen flame-ionization detection system. Two columns were used for the range of solutes, one consisting of 10% Apiezon L on 80-100 mesh Chromosorb W, and the other of Poropak Q polymer beads. The columns were conditioned before use for 48h at 210-220°C with a nitrogen flow rate of 60 mL min^{-1}.

SOURCE AND PURITY OF MATERIALS:

(1) "purest product commercially available" no further purification

(2) not stated.

ESTIMATED ERROR:

not stated

REFERENCES:

COMPONENTS:	ORIGINAL MEASUREMENTS:
(1) 1-Pentanol (*n-amyl alcohol, n-butylcarbinol*); $C_5H_{12}O$; [71-41-0] (2) Water; H_2O; [7732-18-5]	Nishino, N.; Nakamura, M. *Bull. Chem. Soc. Japan* 1978, *51*, 1617-20; 1981, *54*, 545-8.
VARIABLES:	PREPARED BY:
Temperature: 275-360 K	G.T. Hefter

EXPERIMENTAL VALUES:

The mutual solubility of (1) and (2) in mole fractions are reported over the temperature range in graphical form. Graphical data are also presented for the heat of evaporation of (1).

AUXILIARY INFORMATION

METHOD/APPARATUS/PROCEDURE:

The turbidimetric method was used. Twenty to thirty glass ampoules containing aqueous solutions of *ca.* 5 cm^3 of various concentrations near the solubility at room temperature were immersed in a water thermostat. The distinction between clear and turbid ampoules was made after equilibrium was established (*ca.* 2h). The smooth curve drawn to separate the clear and turbid regions was regarded as the solubility curve.

SOURCE AND PURITY OF MATERIALS:

(1) G.R. grade (various commercial sources given); dried over calcium oxide; kept in ampoules over magnesium powder.

(2) Deionized, refluxed for 15h with potassium permanganate then distilled.

ESTIMATED ERROR:

Not stated.

REFERENCES:

COMPONENTS:

(1) 1-Pentanol *(n-amyl alcohol, n-butylcarbinol)*; $C_5H_{12}O$; [71-41-0]

(2) Water; H_2O; [7732-18-5]

ORIGINAL MEASUREMENTS:

Singh, R.P.; Haque, M.M.
Indian J. Chem. 1979, *17A*, 449-51.

VARIABLES:

One temperature: 30°C

PREPARED BY:

A.F.M. Barton

EXPERIMENTAL VALUES:

Mutual solubility of 1-pentanol (1) and water (2)

	mol(2)/mol(1)	x_1	g(1)/100g sln (compiler)
Alcohol-rich phase	0.5014	0.6662	90.7
Water-rich phase	291.30	0.00342	1.65

AUXILIARY INFORMATION

METHOD/APPARATUS/PROCEDURE:

Titrations of one component with the other (ref 1) were carried out in well-stoppered volumetric flasks. The shaking after each addition was done ultrasonically for at least 30 minutes. These results form part of a study of the ternary system 1-pentanol/methanol/water.

SOURCE AND PURITY OF MATERIALS:

(1) BDH AR;
purified;
density and refractive index checked

(2) conductivity water from all-glass still

ESTIMATED ERROR:

Temperature: ± 0.1°C
Solubility: each titration was repeated at least three times

REFERENCES:

(1) Simonsen, D.R.; Washburn, E.R.
J. Am. Chem. Soc. 1946, *68*, 235.

COMPONENTS:

(1) 1-Pentanol (*n-amyl alcohol, n-butylcarbinol*); $C_5H_{12}O$; [71-41-0]

(2) Water; H_2O; [7732-18-5]

ORIGINAL MEASUREMENTS:

Tokunaga, S.; Manabe, M.; Koda, M.; *Niihama Kogyo Koto Semmon Gakko Kiyo, Rikogaku Hen (Memoirs Niihama Technical College, Sci. and Eng.)* 1980, *16*, 96-101.

VARIABLES:

Temperature: 15 - 35°C

PREPARED BY:

A.F.M. Barton

EXPERIMENTAL VALUES:

Solubility of water (2) in the alcohol-rich phase

t/°C	g(2)/100g sln	x_2	mol (1)/mol (2)
15	10.4	0.362	1.77
20	10.5	0.365	1.74
25	10.6	0.367	1.72
30	10.7	0.369	1.69
35	11.0	0.377	1.67

AUXILIARY INFORMATION

METHOD/APPARATUS/PROCEDURE:

The mixtures of 1-pentanol (~5 mL) and water (~10 mL) were stirred magnetically in a stoppered vessel and allowed to stand for 10-12 h in a water thermostat. The alcohol phase was analyzed for water by Karl Fischer titration.

SOURCE AND PURITY OF MATERIALS:

(1) distilled; no impurities detectable by gas chromatography.

(2) deionized; distilled prior to use.

ESTIMATED ERROR:

Temperature: ± 0.1°C

Solubility: each result is the mean of three determinations.

REFERENCES:

COMPONENTS:	EVALUATOR:
(1) 2-Pentanol (*methyl-n-propylcarbinol*); $C_5H_{12}O$; [6032-29-7] (2) Water; H_2O; [7732-18-5]	A. Maczynski, Institute of Physical Chemistry of the Polish Academy of Sciences, Warsaw, Poland. November 1982

CRITICAL EVALUATION:

Solubilities in the system comprising 2-pentanol (1) and water (2) have been reported in six publications. Clough and Johns (ref 1) determined the mutual solubilities of the two components at 293 K, but neither the method nor the estimated reliability was reported. Ginnings and Baum (ref 2) measured mutual solubilities at 293, 298 and 303 K by the volumetric method (Figures 1 and 2). Ratouis and Dodé (ref 3) determined the solubility of (1) in (2) at one temperature (303 K) by an analytical method. Mullens (ref 4) also determined the solubility in the water-rich phase at one temperature (298 K). Nishino and Nakamura provided graphical information only of the solubility of (1) in (2) (ref 5) and of (2) in (1) (ref 6).

The value 4.09 g(1)/100g sln at 303 K of Ratouis and Dode (ref 3) is in excellent agreement with that of 4.13 ± 0.1 g(1)/100g sln of ref 2. The data given in ref 1 and 4 are both lower than these, but the values from ref 2 and 3 are considered more reliable, only these sources providing an estimated error. Since the comparison is possible for only one point, and since the other five points are from a single report, the data are regarded as tentative.

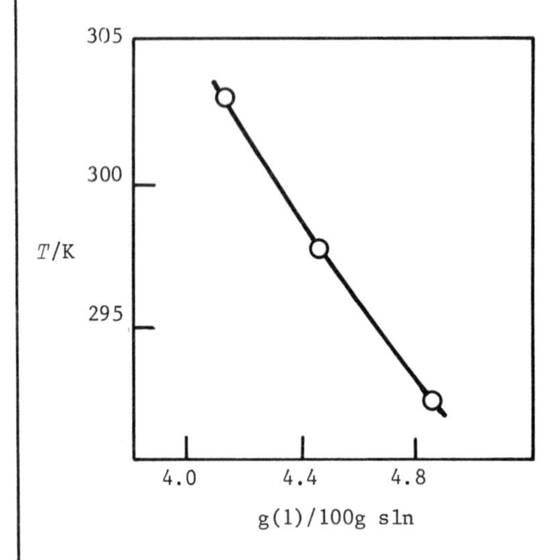

Fig. 1. Water-rich phase (ref 2)

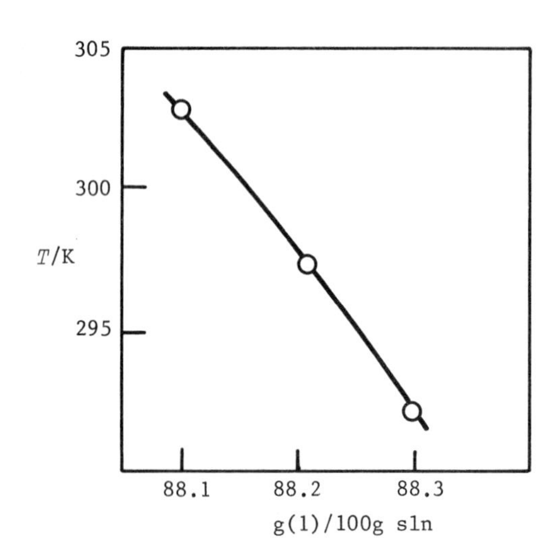

Fig. 2. Alcohol-rich phase (ref 2)

Tentative values for the mutual solubilities of 2-pentanol (1) and water (2)

T/K	Water-rich phase g(1)/100g sln	10^3x_1	Alcohol-rich phase g(2)/100g sln	x_2
293	4.9	10.3	11.7	0.393
298	4.5	9.4	11.8	0.395
303	4.1	8.7	11.9	0.398

(continued next page)

COMPONENTS:	EVALUATOR:
(1) 2-Pentanol (*methyl-n-propylcarbinol*); $C_5H_{10}O$; [6032-29-7] (2) Water; H_2O; [7732-18-5]	A. Maczynski, Institute of Physical Chemistry of the Polish Academy of Sciences, Warsaw, Poland. November 1982

CRITICAL EVALUATION (continued)

References

1. Clough, W.W.; Johns, C.O. *Ind. Eng. Chem.* 1923, *15*, 1030.
2. Ginnings, P.M.; Baum, R. *J. Am. Chem. Soc.* 1937, *59*, 1111.
3. Ratouis, M.; Dodé, M. *Bull. Soc. Chim. Fr.* 1965, 3318.
4. Mullens, J. *Alcoholassociaten*, doctoraatsproefschrift, Leuven, 1971; Huyskens, P.; Mullens, J.; Gomez, A.; Tack, J. *Bull. Soc. Chim. Belg.* 1975, *84*, 253.
5. Nishino, N.; Nakamura, M. *Bull. Chem. Soc. Japan* 1978, *51*, 1617.
6. Nishino, N.; Nakamura, M. *Bull. Chem. Soc. Japan* 1981, *54*, 545.

COMPONENTS:

(1) 2-Pentanol *(methyl-n-propylcarbinol)*; $C_5H_{12}O$; [6032-29-7]

(2) Water; H_2O; [7732-18-5]

ORIGINAL MEASUREMENTS:

Clough, W.W.; Johns, C.O.

Ind. Eng. Chem. 1923, *15*, 1030-2.

VARIABLES:

One temperature: 20°C

PREPARED BY:

S.H. Yalkowsky; S.C. Valvani; A.F.M. Barton

EXPERIMENTAL VALUES:

Mutual solubility of 2-pentanol (1) and water (2) at 20°C

	g(1)/100g(2)	g(2)/100g(1)	g(1)/100g sln (compiler)	x_1 (compiler)
Alcohol-rich phase	4.2	-	4.0	0.0085
Water-rich phase	-	11.2	89.9	0.644

AUXILIARY INFORMATION

METHOD/APPARATUS/PROCEDURE:

Not stated

SOURCE AND PURITY OF MATERIALS:

(1) fractionated;

b.p. 119.2°C/760 mm Hg

d_4^{20} 0.8088

ESTIMATED ERROR:

Not stated

REFERENCES:

COMPONENTS:

(1) 2-Pentanol *(methyl-n-propylcarbinol)*; $C_5H_{12}O$; [6032-19-7]

(2) Water; H_2O; [7732-18-5]

ORIGINAL MEASUREMENTS:

Ginnings, P.M.; Baum, R.

J. Am. Chem. Soc. 1937, *59*, 1111-3.

VARIABLES:

Temperature: 20-30°C

PREPARED BY:

A. Maczynski and Z. Maczynska

EXPERIMENTAL VALUES:

Mutual solubility of 2-pentanol(1) and water(2)

$t/°C$	g(1)/100g sln		x_1 (compiler)	
	(2)-rich phase	(1)-rich phase	(2)-rich phase	(1)-rich phase
20	4.86	88.30	0.01032	0.6066
25	4.46	88.21	0.00944	0.6045
30	4.13	88.10	0.00872	0.6020

Relative density, d_4

$t/°C$	Water-rich phase	Alcohol-rich phase
20	0.9914	0.8317
25	0.9909	0.8280
30	0.9898	0.8243

AUXILIARY INFORMATION

METHOD/APPARATUS/PROCEDURE:

The volumetric method was used.

Both components were introduced in known amounts into a two-bulb graduated and calibrated flask and shaken mechanically in a water-bath at constant temperature. After sufficient time the liquids were allowed to separate and the total volume was measured. Upon centrifugation, the phase separation line was read, and phase volumes were calculated. From the total weights of the components, the total volume, individual phase volumes, and component concentrations in either phase were evaluated.

SOURCE AND PURITY OF MATERIALS:

(1) Eastman best grade; distilled from metallic calcium; b.p. range 119.2-2.119.7°C, d_4^{25} 0.8056.

(2) not specified.

ESTIMATED ERROR:

Temperature: ± 0.1°C
Solubility: better than 0.1 wt % (type of error not specified.)

REFERENCES:

COMPONENTS:

(1) 2-Pentanol (*methyl-n-propylcarbinol*); $C_5H_{12}O$; [6032-29-7]

(2) Water; H_2O; [7732-18-5]

ORIGINAL MEASUREMENTS:

Ratouis, M.; Dodé, M.;

Bull. Soc. Chim. Fr. 1965, 3318-22.

VARIABLES:

One temperature: 30°C

Ringer solution also studied

PREPARED BY:

S.C. Valvani; S.H. Yalkowsky; A.F.M. Barton

EXPERIMENTAL VALUES:

The proportion of 2-pentanol (1) in the water-rich phase at equilibrium at 30°C was reported to be 4.09 g(1)/100g sln.

The corresponding mole fraction solubility, calculated by the compilers, is x_1 = 0.00864.

The proportion of (1) in the water-rich phase of a mixture with Ringer solution at equilibrium at 30°C was reported to be 3.93 g(1)/100g sln.

AUXILIARY INFORMATION

METHOD/APPARATUS/PROCEDURE:

In a round bottomed flask, 50 mL of water and a sufficient quantity of alcohol were introduced until two separate layers were formed. The flask assembly was equilibrated by agitation for at least 3 h in a constant temperature bath. Equilibrium solubility was attained by first supersaturating at a slightly lower temperature (solubility of alcohols in water decreases with increasing temperature) and then equilibrating at the desired temperature. The aqueous layer was separated after an overnight storage in a bath. The alcohol content was determined by reacting the aqueous solution with potassium dichromate and titrating the excess dichromate with ferrous sulfate solution in the presence of phosphoric acid and diphenylamine barium sulfonate as an indicator.

SOURCE AND PURITY OF MATERIALS:

(1) Fluke A.G. Buchs S.G.; redistilled with 10:1 reflux ratio; b.p. 119°C/760 mm Hg n_D^{25} = 1.40454

(2) twice distilled from silica apparatus or ion-exchanged with Sagei A20

ESTIMATED ERROR:

Solubility: relative error of 2 determinations less than 1%.

Temperature: ± 0.05°C

REFERENCES:

COMPONENTS:

(1) 2-Pentanol *(methyl-n-propylcarbinol)*; $C_5H_{12}O$; [6032-29-7]

(2) Water; H_2O; [7732-18-5]

ORIGINAL MEASUREMENTS:

*Mullens, J.
Alcoholassociaten, doctoraatsproefschrift, Leuven, 1971

Huyskens, P.; Mullens, J.; Gomez, A.; Tack, J.; *Bull. Soc. Chim. Belg.* 1975, *84*, 253-62.

VARIABLES:

One temperature: 25°C

PREPARED BY:

M.C. Haulait-Pirson; A.F.M. Barton

EXPERIMENTAL VALUES:

At equilibrium at 25°C the concentration of 2-pentanol (2) in the water-rich phase was reported as 0.478 mol(1) L^{-1} sln, and the concentration of water (2) in the alcohol-rich phase was reported as 8.369 mol(2) L^{-1} sln.

The corresponding solubilities on a mass/volume basis calculated by the compilers are 42.1 g(1) L^{-1} sln and 150.8 g(2) L^{-1} sln, respectively.

AUXILIARY INFORMATION

METHOD/APPARATUS/PROCEDURE:

The partition of the two components was made using a cell described in ref 1. The Rayleigh Interference Refractometer M154 was used for the determination of the concentrations. Standard solutions covering the whole range of concentration investigated were used for the calibration.

SOURCE AND PURITY OF MATERIALS:

(1) Merck (p.a.)

(2) distilled

ESTIMATED ERROR:

Solubility: ± 0.001 mol(1) L^{-1} sln

REFERENCES:

(1) Meeussen, E.; Huyskens, P. *J. Chim. Phys.* 1966, *63*, 845.

COMPONENTS:	ORIGINAL MEASUREMENTS:
(1) 2-Pentanol (*methyl-n-propylcarbinol*); $C_5H_{12}O$; [6032-29-7] (2) Water; H_2O; [7732-18-5]	Nishino, N.; Nakamura, M. *Bull. Chem. Soc. Jpn.* 1978, *51*, 1617-20; 1981, *54*, 545-8.

VARIABLES:	PREPARED BY:
Temperature; 275-360 K	G.T. Hefter

EXPERIMENTAL VALUES:

The mutual solubility of (1) and (2) in mole fractions are reported over the temperature range in graphical form. Graphical data are also presented for the heat of evaporation of (1).

AUXILIARY INFORMATION

METHOD/APPARATUS/PROCEDURE:

The turbidimetric method was used. Twenty to thirty glass ampoules containing aqueous solutions of *ca.* 5 cm^3 of various concentrations near the solubility at room temperature were immersed in a water thermostat. The distinction between clear and turbid ampoules was made after equilibrium was established (*ca.* 2h). The smooth curve drawn to separate the clear and turbid regions was regarded as the solubility curve.

SOURCE AND PURITY OF MATERIALS:

(1) G.R. grade (various commercial sources given); dried over calcium oxide; kept in ampoules over magnesium powder.

(2) Deionized, refluxed for 15h with potassium permanganate then distilled.

ESTIMATED ERROR:

Not stated.

REFERENCES:

COMPONENTS:	EVALUATOR:
(1) 3-Pentanol (*diethylcarbinol*); $C_5H_{12}O$; [584-02-1] (2) Water; H_2O; [7732-18-5]	A. Maczynski, Institute of Physical Chemistry of the Polish Academy of Sciences, Warsaw, Poland. November 1982

CRITICAL EVALUATION:

Solubilities in the system comprising 3-pentanol (1) and water (2) have been reported in six publications. Ginnings and Baum (ref 1) carried out measurements of the mutual solubilities of the two components at 293, 298 and 303 K by the volumetric method. Crittenden and Hixon determined mutual solubilities at one temperature (298 K) presumably by titration method. Hyde *et al.* (ref 3) reported the upper critical point and a "room temperature" solubility and Ratouis and Dodé (ref 4) measured the solubility of (1) in (2) at one temperature (303 K) by an analytical method. Nishino and Nakamura provided graphical information only of the solubility of (1) in (2) (ref 5) and of (2) in (1) (ref 6).

For the water-rich phase, values 5.1 g(1)/100g sln at 298 K (ref 2), and 4.65 g(1)/100g sln at 303 K (ref 4) are in good agreement with the values of ref 1 (5.15 and 4.75 g(1)/100g sln, respectively) (Figure 1). Accordingly, the water-rich phase solubility data of ref 1 are recommended. For the alcohol-rich phase, the value 91.7 g(1)/100g sln of ref 2 at 298 K is in very good agreement with the value 91.68 g(1)/100g sln of ref 1, (Figure 2) so these data of ref 1 are also recommended. The upper critical solution temperature given in ref 3 is regarded as tentative, as there are no supporting data.

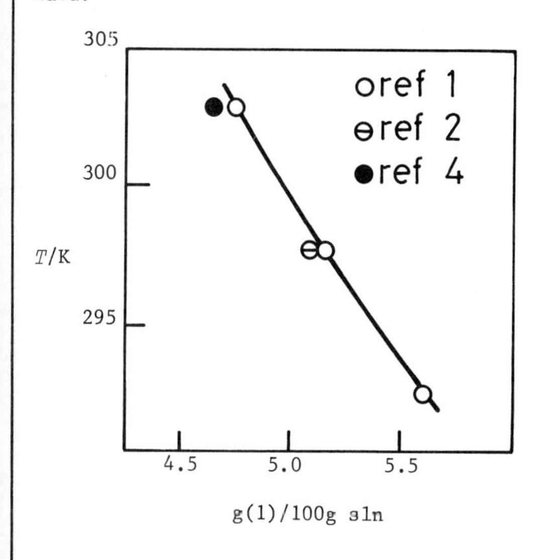

Fig. 1. Water-rich phase

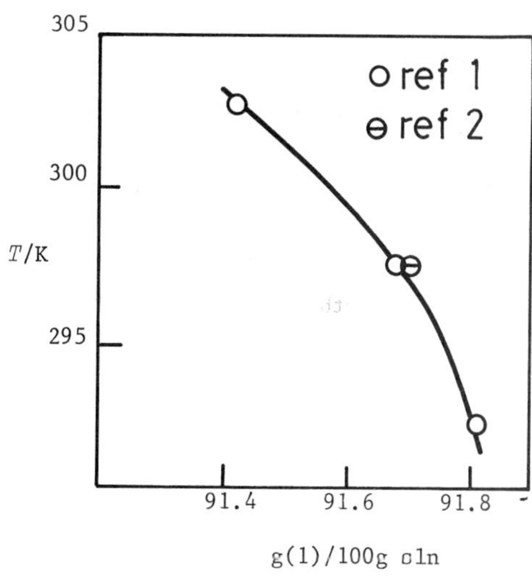

Fig. 2. Alcohol-rich phase

(continued next page)

COMPONENTS:	EVALUATOR:
(1) 3-Pentanol (*diethylcarbinol*); $C_5H_{12}O$; [584-02-1] (2) Water; H_2O; [7732-18-5]	A. Maczynski, Institute of Physical Chemistry of the Polish Academy of Sciences, Warsaw, Poland. November 1982

CRITICAL EVALUATION: (continued)

Recommended and tentative values for the mutual solubilities of 3-pentanol (1) and water (2)

T/K	Water-rich phase			Alcohol-rich phase		
	g(1)/100g sln		x_1	g(2)/100g sln		x_2
293	5.6	(tentative)	0.0120	8.2	(tentative)	0.304
298	5.2	(recommended)	0.0110	8.3	(recommended)	0.308
303	4.8	(recommended)	0.0101	8.6	(tentative)	0.315
489.4	Upper critical solution temperature (tentative)					

References

1. Ginnings, P.M.; Baum, R.J. *J. Am. Chem. Soc.* 1937, *59*, 1111.
2. Crittenden, E.D.,Jr.; Hixon, A.N. *Ind. Eng. Chem.* 1954, *46*, 265.
3. Hyde, A.J.; Langbridge, D.M.; Lawrence, A.S.C. *Disc. Faraday Soc.* 1954, *18*, 239.
4. Ratouis, M.; Dodé, M. *Bull. Soc. Chim. Fr.* 1965, 3318.
5. Nishino, N.; Nakamura, M. *Bull. Chem. Soc. Jpn.* 1978, *51*, 1617.
6. Nishino, N.; Nakamura, M. *Bull. Chem. Soc. Jpn.* 1981, *54*, 545.

COMPONENTS:

(1) 3-Pentanol *(diethylcarbinol)*; $C_5H_{12}O$; [584-02-1]

(2) Water; H_2O; [7732-18-5]

ORIGINAL MEASUREMENTS:

Ginnings, P.M.; Baum, R.

J. Am. Chem. Soc. 1937, *59*, 1111-3.

VARIABLES:

Temperature: 20-30°C

PREPARED BY:

A. Maczynski and Z. Maczynska

EXPERIMENTAL VALUES:

Mutual solubility of 3-pentanol(1) and water(2)

$t/^{\circ}C$	g(1)/100g sln		x_1 (compiler)	
	(2)-rich phase	(1)-rich phase	(2)-rich phase	(1)-rich phase
20	5.61	91.81	0.01200	0.6961
25	5.15	91.68	0.01097	0.6924
30	4.75	91.42	0.01008	0.6852

Relative density, d_4

$t/^{\circ}C$	Water-rich phase	Alcohol-rich phase
20	0.9920	0.8368
25	0.9914	0.8330
30	0.9903	0.8294

AUXILIARY INFORMATION

METHOD/APPARATUS/PROCEDURE:

Hill's volumetric method was adopted.

Both components were introduced in known amounts into a two-bulb graduated and calibrated flask and shaken mechanically in a water-bath at constant temperature. After sufficient time the liquids were allowed to separate and the total volume was measured. Upon centrifugation, the phase separation line was read, and phase volumes were calculated. From the total weights of the components, the total volume, individual phase volumes, and component concentrations in either phase were evaluated.

SOURCE AND PURITY OF MATERIALS:

(1) Eastman best grade; distilled from metallic calcium; b.p. range 115.4-115.9°C, d_4^{25} 0.8195.

(2) not specified.

ESTIMATED ERROR:

Temperature: ± 0.1°C

Solubility: better than 0.1 wt % (type of error not specified)

REFERENCES:

AWW-H

COMPONENTS:	ORIGINAL MEASUREMENTS:
(1) 3-Pentanol *(diethylcarbinol)*; $C_5H_{12}O$; [584-02-1] (2) Water; H_2O; [7732-18-5]	Crittenden, E.D., Jr.; Hixon, A.N.; *Ind. Eng. Chem.*, 1954, *46*, 265-8.
VARIABLES:	**PREPARED BY:**
One temperature: 25°C	A. Maczynski

EXPERIMENTAL VALUES:

The solubility of 3-pentanol in water at 25°C was reported to be 5.1 g(1)/100g sln.

The corresponding mole fraction, x_1, calculated by the compiler is 0.011.

The solubility of water in 3-pentanol at 25°C was reported to be 8.3 g(2)/100g sln-

The corresponding mole fraction, x_2, calculated by the compiler is 0.31.

AUXILIARY INFORMATION

METHOD/APPARATUS/PROCEDURE:

Presumably the titration method described for ternary systems containing HCl was used. In this method the solubility was determined by bringing 100-mL samples of (1) or (2) to a temperature 25.0 ± 0.1°C and the second component was then added from a calibrated buret, with vigorous stirring, until the solution became permanently cloudy.

SOURCE AND PURITY OF MATERIALS:

(1) source not specified;
purified;
purity not specified.

(2) not specified.

ESTIMATED ERROR:

Solubility: 2% (alcohol-rich) - 10% water-rich)

Temperature: ± 0.10°C.

REFERENCES:

COMPONENTS:

(1) 3-Pentanol (*diethylcarbinol*); $C_5H_{12}O$; [584-01-1]

(2) Water; H_2O; [7732-18-5]

ORIGINAL MEASUREMENTS:

Hyde, A.J.; Langbridge, D.M.; Lawrence, A.S.C.

Disc. Faraday Soc. 1954, *18*, 239-58.

VARIABLES:

One temperature: 216°C

PREPARED BY:

A. Maczynski, and A.F.M. Barton

EXPERIMENTAL VALUES:

The upper critical solution temperature was reported to be 216.3°C at 42.8 g(1)/100g sln (x_1 = 0.133, compiler).

The solubility of (1) in (2) was reported as 4.8%.

AUXILIARY INFORMATION

METHOD/APPARATUS/PROCEDURE:

Not specified

SOURCE AND PURITY OF MATERIALS:

(1) not specified.

(2) not specified.

ESTIMATED ERROR:

Not specified.

REFERENCES:

COMPONENTS:	ORIGINAL MEASUREMENTS:
(1) 3-Pentanol (*diethylcarbinol*); $C_5H_{12}O$; [584-02-1] (2) Water; H_2O; [7732-18-5]	Ratouis, M.; Dodé, M.; *Bull. Soc. Chim. Fr.* 1965; 3318-22.
VARIABLES:	**PREPARED BY:**
One temperature: 30°C Ringer solution also studied	S.C. Valvani; S.H. Yalkowsky; A.F.M. Barton

EXPERIMENTAL VALUES:

The proportion of 3-pentanol (1) in the water-rich phase at equilibrium at 30°C was reported to be 4.65 g(1)/100g sln.

The corresponding mole fraction solubility, calculated by the compilers, is x_1 = 0.00987.

The proportion of (1) in the water-rich phase of a mixture with Ringer solution at equilibrium at 30°C was reported to be 4.36 g(1)/100g sln.

AUXILIARY INFORMATION

METHOD/APPARATUS/PROCEDURE:

In a round bottomed flask, 50 mL of water and a sufficient quantity of alcohol were introduced until two separate layers were formed. The flask assembly was equilibrated by agitation for at least 3 h in a constant temperature bath. Equilibrium solubility was attained by first supersaturating at a slightly lower temperature (solubility of alcohols in water decreases with increasing temperature) and then equilibrating at the desired temperature. The aqueous layer was separated after an overnight storage in a bath. The alcohol content was determined by reacting the aqueous solution with potassium dichromate and titrating the excess dichromate with ferrous sulfate solution in the presence of phosphoric acid and diphenylamine barium sulfonate as an indicator.

SOURCE AND PURITY OF MATERIALS:

(1) Fluka A.G. Buchs S.G.; redistilled with 10:1 reflux ratio

b.p. 115.7°C/759.6 mm Hg

n_D^{25} = 1.40842

(2) twice distilled from silica apparatus or ion-exchanged with Sagei A20

ESTIMATED ERROR:

Solubility: relative error of 2 determinations less than 1%.

Temperature: ± 0.05°C

REFERENCES:

COMPONENTS:

(1) 3-Pentanol (*diethylcarbinol*); $C_5H_{12}O$; [584-02-1]

(2) Water; H_2O; [7732-18-5]

ORIGINAL MEASUREMENTS:

Nishino, N.; Nakamura, M.

Bull. Chem. Soc. Jpn. 1978, *51*, 1617-20; 1981, *54*, 545-8.

VARIABLES:

Temperature: 275-360 K

PREPARED BY:

G.T. Hefter

EXPERIMENTAL VALUES:

The mutual solubility of (1) and (2) in mole fractions are reported over the temperature range in graphical form. Graphical data are also presented for the heat of evaporation of (1).

AUXILIARY INFORMATION

METHOD/APPARATUS/PROCEDURE:

The turbidimetric method was used. Twenty to thirty glass ampoules containing aqueous solutions of *ca.* 5 cm^3 of various concentrations near the solubility at room temperature were immersed in a water thermostat. The distinction between clear and turbid ampoules was made after equilibrium was established (*ca.* 2h). The smooth curve drawn to separate the clear and turbid regions was regarded as the solubility curve.

SOURCE AND PURITY OF MATERIALS:

(1) G.R. grade (various commercial sources given); dried over calcium oxide; kept in ampoules over magnesium powder.

(2) Deionized, refluxed for 15h with potassium permanganate then distilled.

ESTIMATED ERROR:

Not stated.

REFERENCES:

COMPONENTS:	EVALUATOR:
(1) Cyclohexanol; $C_6H_{12}O$; [108-93-0] (2) Water; H_2O; [7732-18-5]	G.T. Hefter and A.F.M. Barton, Murdoch University, Perth, Western Australia. June, 1983.

CRITICAL EVALUATION:

Solubilities in the system comprising cyclohexanol (1) and water (2) have been reported in the following publications:

Reference	*T*/K	Phase	Method
Forcrand (ref 1)	284	mutual	analytical
Sidgwick and Sutton (ref 2)	258-459	mutual	synthetic
Booth and Everson (ref 3,4)	298,333	(1) in (2)	titration
Hansen *et al.* (ref 5)	298	(1) in (2)	interferometric
Zil'berman (ref 6)	273-457	mutual	synthetic
Skrzec and Murphy (ref 7)	300	mutual	titration
Tettamanti *et al.* (ref 8)	293	mutual	titration
Lavrova and Lesteva (ref 9)	313, 333	mutual	titration

The original data are compiled in the data sheets immediately following this Critical Evaluation.
In preparing this Critical Evaluation the data of Booth and Everson (ref 3,4) in volume/volume fractions were excluded as no density data were included in the original references.

In both the water-rich and alcohol-rich phases the data of Forcrand (ref 1), Skrzec and Murphy (ref 7) and Tettamanti *et al.* (ref 8) disagree markedly with other studies (ref 2,5,6) and are rejected. All other data are included in the tables below. Values obtained by the Evaluators by graphical interpolation or extrapolation from the data sheets are indicated by an asterisk (*). "Best" values have been obtained by simple averaging. The uncertainty limits (σ_n) attached to the "best" values do not have statistical significance and should be regarded only as a convenient representation of the spread of reported values and not as error limits. The letter (*R*) designates "Recommended" data. Data are "Recommended" if two or more apparently reliable studies are in reasonable agreement ($\leq \pm 5\%$).

For convenience further discussion for the two phases will be given separately.

Solubility of cyclohexanol (1) in water (2)

Excluding the data already rejected (see above) the reported solubilities for the water-rich phase are in good agreement except at very high temperatures (> 423 K).

(continued next page)

COMPONENTS:	EVALUATOR:
(1) Cyclohexanol; $C_6H_{12}O$; [108-93-0] (2) Water; H_2O; [7732-18-5]	G.T. Hefter and A.F.M. Barton, Murdoch University, Perth, Western Australia. June, 1983.

CRITICAL EVALUATION (continued)

Tentative and Recommended (*R*) values for the solubility of cyclohexanol (1) in water (2).

T/K	Solubility, g(1)/100g sln Reported values	"Best" value ($\pm\sigma_n$)
273	5.34 (ref 6)	5.3
283	4.66* (ref 2), 4.57 (ref 6)	4.62 ± 0.05 (*R*)
293	3.95* (ref 2), 4.00 (ref 6)	3.97 ± 0.03 (*R*)
298	3.70* (ref 2), 3.92 (ref 5)	3.8 ± 0.1 (*R*)
303	3.50* (ref 2), 3.60 (ref 6)	3.55 ± 0.05 (*R*)
313	3.23* (ref 2), 3.33 (ref 6), 3.33 (ref 9)	3.30 ± 0.05 (*R*)
323	3.14 (ref 6)	3.1
333	3.1* (ref 6)	3.1
343	3.19 (ref 6)	3.2
353	3.41 (ref 6)	3.4
363	3.65 (ref 6)	3.7
373	3.93 (ref 6)	3.9
383	4.28 (ref 6)	4.3
393	5.0* (ref 2), 4.7 (ref 6)	4.9 ± 0.2 (*R*)
403	5.8* (ref 2), 5.3 (ref 6)	5.6 ± 0.3 (*R*)
413	6.6* (ref 2), 6.1 (ref 6)	6.4 ± 0.3 (*R*)
423	8.2* (ref 2), 7.2 (ref 6)	7.7 ± 0.5
433	9.8* (ref 2), 8.8 (ref 6)	9.3 ± 0.5
443	13.0* (ref 2), 11.5 (ref 6)	12.3 ± 0.8
453	20.0* (ref 2), 17.8 (ref 6)	19 ± 1

Solubility of water (2) in cyclohexanol (1)

The existence of two extensive sets of data in reasonable (although not fully satisfactory) agreement enables values to be recommended over a wide range of temperatures. Insufficient data are available below 313 K to enable a realistic assessment of their reliability to be made.

(continued next page)

COMPONENTS:	EVALUATOR:
(1) Cyclohexanol; $C_6H_{12}O$; [108-93-0] (2) Water; H_2O; [7732-18-5]	G.T. Hefter and A.F.M. Barton, Murdoch University, Perth, Western Australia. June, 1983

CRITICAL EVALUATION (continued)

Tentative and Recommended (*R*) values for the solubility of water (2) in cyclohexanol (1)

T/K	Solubility, g(1)/100g sln Reported Values	"Best" value ($\pm\sigma_n$)
273	10.13 (ref 6)	10.1
283	10.53 (ref 6)	10.5
293	11.07 (ref 6)	11.1
298	11.35* (ref 6)	11.4
303	11.63 (ref 6)	11.6
313	12.23 (ref 6), 12.21 (ref 9)	12.2 (*R*)
323	12.00* (ref 2), 12.89 (ref 6)	12.4 ± 0.5 (*R*)
333	12.7* (ref 2), 13.6 (ref 6), 13.52 (ref 9)	13.3 ± 0.4 (*R*)
343	13.3* (ref 2), 14.34 (ref 6)	13.8 ± 0.5 (*R*)
353	14.0* (ref 2), 15.17 (ref 6)	14.6 ± 0.6 (*R*)
363	14.5* (ref 2), 16.07 (ref 6)	15.3 ± 0.8
373	15.4* (ref 2), 17.04 (ref 6)	16.2 ± 0.8
383	16.5* (ref 2), 18.1 (ref 6)	17.3 ± 0.8 (*R*)
393	18.0* (ref 2), 19.3 (ref 6)	18.7 ± 0.7 (*R*)
403	19.5* (ref 2), 20.7 (ref 6)	20.1 ± 0.6 (*R*)
413	22.0* (ref 2), 23.6 (ref 6)	22.8 ± 0.8 (*R*)
423	25.0* (ref 2), 25.3 (ref 6)	25.2 ± 0.2 (*R*)
433	28.5* (ref 2), 29.5 (ref 6)	29.0 ± 0.5 (*R*)
443	32.5* (ref 2), 36.2 (ref 6)	34 ± 2
453	41.0* (ref 2), 47.7 (ref 6)	44 ± 3

Upper critical solution temperature

The UCST has been reported as 457 K (184 °C) at 33.9 g(1)/100g sln (ref 6).

Representative data for the mutual solubilities of cyclohexanol and water are plotted in Figure 1.

(continued next page)

COMPONENTS:	EVALUATOR:
(1) Cyclohexanol; $C_6H_{12}O$; [108-93-0] (2) Water; H_2O; [7732-18-5]	G.T. Hefter and A.F.M. Barton, Murdoch University, Perth, Western Australia. June, 1983.

CRITICAL EVALUATION (continued)

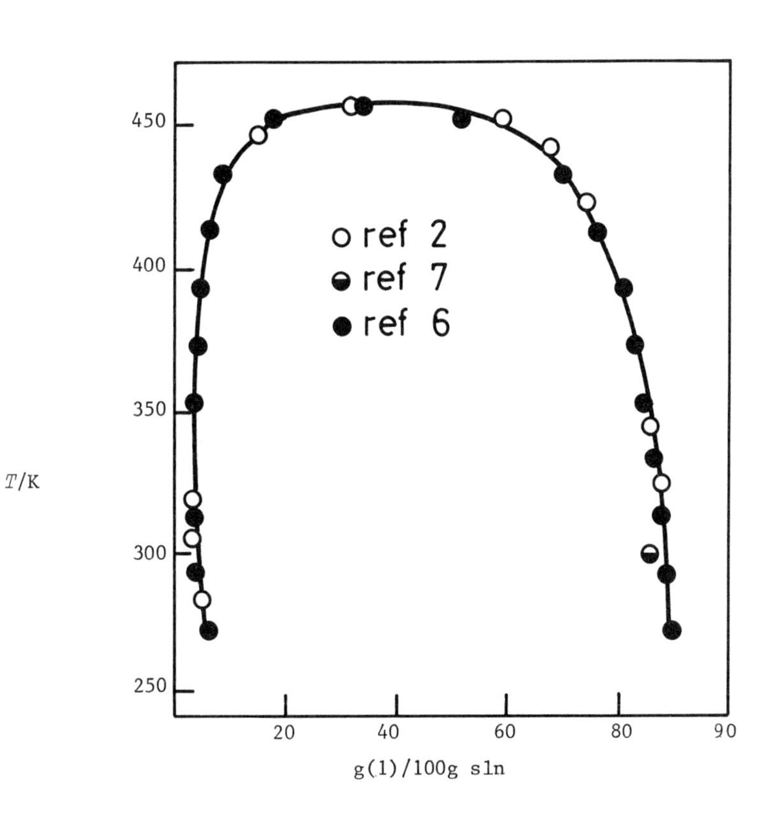

Fig. 1. Mutual solubility of (1) and (2)

References

1. de Forcrand, M. *C.R. Hebd. Seances Acad. Sci.* 1912, *154*, 1327.
2. Sidgwick, N.V.; Sutton, L.E. *J. Chem. Soc.* 1930, 1323.
3. Booth, H.S.; Everson, H.E. *Ind. Eng. Chem.* 1948, *40*, 1491.
4. Booth, H.S.; Everson, H.E. *Ind. Eng. Chem.* 1949, *41*, 2627; 1950, *42*, 1536.
5. Hansen, R.S.; Fu, Y.; Bartell, F.E. *J. Phys. Chem.* 1949, *53*, 769.
6. Zil'berman, E.N. *Zh. Fiz. Khim.* 1951, *24*, 776.
7. Skrzec, A.E.; Murdphy, N.F. *Ind. Eng. Chem.* 1954, *46*, 2245.
8. Tettamanti, K.; Nogradi, M.; Sawinsky, J. *Periodica Politechnica* 1960, *4*, 201.
9. Lavrova, O.A.; Lesteva, T.M. *Zh. Fiz. Khim.* 1976, *50*, 1617; *Dep. Doc. VINITI* 3813-75.

COMPONENTS:	ORIGINAL MEASUREMENTS:
(1) Cyclohexanol; $C_6H_{12}O$; [108-93-0] (2) Water; H_2O; [7732-18-5]	de Forcrand, M. *C.R. Hebd. Seances Acad. Sci.* 1912, *154*, 1327-30.
VARIABLES:	PREPARED BY:
Temperature: 11 °C	A. Maczynski

EXPERIMENTAL VALUES:

The solubility of cyclohexanol in water at 11 °C was reported to be 5.67 g(1)/100g (2). The corresponding mass percent and mole fraction, x_1, calculated by the compiler are 5.37 g(1)/100g sln and 0.0101.

The solubility of water in cyclohexanol at 11 °C was reported to be 11.27 g(2)/100g (1). The corresponding mass percent and mole fraction, x_2, calculated by the compiler are 10.13 g(2)/100g sln and 0.3853.

AUXILIARY INFORMATION

METHOD/APPARATUS/PROCEDURE:	SOURCE AND PURITY OF MATERIALS:
An analytical method was used. No details were reported in the paper.	(1) Laboratoire de catalyse de Toulouse; recrystallized; b.p. 160.9 °C (766 mm Hg) m.p. 22.45 °C, $d_4^{22.5}$ 0.9471. (2) not specified.
	ESTIMATED ERROR:
	Not specified.
	REFERENCES:

COMPONENTS:

(1) Cyclohexanol, $C_6H_{12}O$; [108-93-0]

(2) Water, H_2O [7732-18-5]

ORIGINAL MEASUREMENTS:

Sidgwick, N.V.; Sutton, L.E.

J. Chem. Soc. 1930, 1323-6.

VARIABLES:

Temperature: (-15) - 185°C

PREPARED BY:

A. Maczynski and Z. Maczynska

EXPERIMENTAL VALUES:

The Ice Line

t/°C	g(1)/100g sln	x_1 (compiler)
-15.4	93.0	0.705
-10.2	92.3	0.683
- 7.6	91.2	0.652
- 4.9	90.45	0.630
- 4.1	90.08	0.620
- 2.0	89.0	0.593
- 1.2	88.3	0.579
- 1.1	88.45	0.579
- 0.9	5.00	0.0093
- 0.6	3.33	0.00616
- 0.32	1.67	0.00304
- 0.3	1.67	0.00304

(continued next page)

AUXILIARY INFORMATION

METHOD/APPARATUS/PROCEDURE:

The usual synthetic method was used.

At the higher temperatures sealed glass tubes were used. At large proportions, the turbidity appearance temperature of (1) rose with successive heatings (decomposition ?) and the first reading was taken. Upward of 87.9% of (1) the curve was too steep and turbidity too faint to take reliable temperature measurements.

SOURCE AND PURITY OF MATERIALS:

(1) British Drug Houses Ltd.; fractionally distilled at about 10 mm Hg; 99.85 wt % purity (if 23.6°C is taken as the true m.p.)

(2) not specified.

ESTIMATED ERROR:

Not specified.

REFERENCES:

COMPONENTS:	ORIGINAL MEASUREMENTS:
(1) Cyclohexanol; $C_6H_{12}O$; [108-93-0] (2) Water; H_2O; [7732-18-5]	Sidgwick, N.V.; Sutton, L.E. *J. Chem. Soc.* 1930, 1323-6.

EXPERIMENTAL VALUES: (continued)

The Liquid-Liquid Curve

$t/^{o}C$	g(1)/100g sln		x_1 (compiler)	
	(2)-rich phase	(1)-rich phase	(2)-rich phase	(1)-rich phase
7.2	5.00	-	0.00938	-
9.4	4.78	-	0.00894	-
9.7	4.58		0.00856	-
11.2	4.41	-	0.00823	-
12.0	4.55	-	0.00850	-
14.2	4.23	-	0.00788	-
15.2	4.29	-	0.00800	-
16.3	4.09	-	0.00762	-
20.6	3.95	-	0.00734	-
20.8	3.82	-	0.00709	-
24.6	3.75	-	0.00696	-
27.55	3.52	-	0.00652	-
28.7	3.57	-	0.00661	-
31.85	3.37	-	0.00623	-
33.6	3.41	-	0.00631	-
40.4	3.26	-	0.00602	-
40.45	3.18	-	0.00587	-
45.8	3.19	-	0.00589	-
51.55	-	87.9	-	0.5664
71.5	-	86.75	-	0.5407
72.75	-	85.95	-	0.5219
93.63	-	85.3	-	0.5106
121.95	5.14	-	0.00965	-
130.9	-	80.2	-	0.4214
150.35	-	74.6	-	0.3456
156.9	9.22	-	0.01793	-
163.03	-	70.1	-	0.2965
168.64	-	68.5	-	0.2811
169.7	-	68.0	-	0.2764
174.3	15.00	-	0.0307	-
179.4	19.2	-	0.0410	-
180.1	-	59.4	-	0.2083
183.66	-	52.3	-	0.1647
184.72	32.4	-	0.0793	-

COMPONENTS:

(1) Cyclohexanol; $C_6H_{12}O$; [108-93-0]

(2) Water; H_2O; [7732-18-5]

ORIGINAL MEASUREMENTS:

Booth, H.S.; Everson, H.E.

Ind. Eng. Chem. 1948, *40*, 1491-3.

VARIABLES:

One temperature: 25°C

Sodium xylene sulfonate

PREPARED BY:

S.H. Yalkowsky; S.C. Valvani; A.F.M. Barton

EXPERIMENTAL VALUES:

It was reported that the solubility of cyclohexanol (1) in water (2) was 3.4 mL(1)/100mL (2) at 25.0°C.

The corresponding value in 40% sodium xylene sulfonate solution as solvent was >400 mL(1)/ 100mL solvent.

AUXILIARY INFORMATION

METHOD/APPARATUS/PROCEDURE:

A known volume of solvent (usually 50mL) in a tightly stoppered Babcock tube was thermostatted. Successive measured quantities of solute were added and equilibrated until a slight excess of solute remained. The solution was centrifuged, returned to the thermostat bath for 10 min, and the volume of excess solute measured directly. This was a modification of the method described in ref 1.

SOURCE AND PURITY OF MATERIALS:

(1) "CP or highest grade commercial".

(2) distilled.

ESTIMATED ERROR:

Solubility: within 0.1 mL(1)/100mL (2)

REFERENCES:

(1) Hanslick, R.S. *Dissertation*, Columbia University, 1935.

COMPONENTS:

(1) Cyclohexanol; $C_6H_{12}O$; [108-93-0]

(2) Water; H_2O; [7732-18-5]

ORIGINAL MEASUREMENTS:

Booth, H.S.; Everson, H.E.

Ind. Eng. Chem. 1949, *41*, 2627-8; 1950, *42*, 1536-7.

VARIABLES:

Temperature: 25°C and 60°C

Hydrotropic solutes (sodium arylsulfonates)

PREPARED BY:

S.H. Yalkowsky; S.C. Valvani; A.F.M. Barton

EXPERIMENTAL VALUES:

It was reported that the solubilities of cyclohexanol (1) in water (2) at 25.0°C and 60°C were 3.40 mL(1)/100mL(2) and 3.38 mL(1)/100mL(2) respectively. (The units quoted for the table of data in the 1949 paper were mL(1)/mL(2) but those reported in 1950 agreed with ref 1 and with information elsewhere in the paper. Solubilities in aqueous solutions of hydrotropic salts were also reported:

Arylsulfonate		Solubility, mL(1)/100mL solvent, for various solvent compositions (%)								
		5.0	8.0	10.0	15.0	20.0	22.0	25.0	34.6	40.0
Sodium benzenesulfonate	25°C	4.08	-	4.43	-	14.1	28.8	>400	-	-
	60°C	3.71	-	4.05	-	13.8	27.0	>400	-	-
Sodium *p*-cymenesulfonate	25°C	13.6	30.1	-	54.2	76.7	-	-	145	-
	60°C	11.0	25.9	43.2	61.6	70.6	-	-	119.4	-
Sodium toluenesulfonate	25°C	4.42	-	8.68	54.1	>400	-	-	-	>400
	60°C	4.13	-	8.62	61.8	365	-	-	-	>400
Sodium xylenesulfonate	25°C	4.0	-	17.3	131	>400	-	-	-	>400
	60°C	5.03	-	20.2	127	>400	-	-	-	>400
Sodium *o*-xylenesulfonate	25°C	5.18	-	17.5	-	210	-	-	-	-
	60°C	4.57	-	15.9	-	200	-	-	-	-

(continued next page)

AUXILIARY INFORMATION

METHOD/APPARATUS/PROCEDURE:

A known volume of solvent (usually 50 mL) in a tightly stoppered vessel was thermostatted. Successive measured quantities of solute were added and equilibrated until a slight excess of solute remained. Unlike the previous study (ref 1), the mixtures were not centrifuged, but separated by gentle rotation of the solubility tubes while they were immersed in the thermostat bath, in order to maintain the samples at constant temperature. The volume of excess solute was measured directly.

SOURCE AND PURITY OF MATERIALS:

(1) "CP or highest grate commercial"

(2) distilled.

Sodium arylsulfonates: Wyandotte Chemicals Corp., Michigan.

Sodium *p*-xylenesulfonate: prepared from *p*-xylenesulfonic acid (Eastman Kodak)

ESTIMATED ERROR:

Not stated.

REFERENCES:

1. Booth, H.S.; Everson, H.E. *Ind. Eng. Chem.* 1948, *40*, 1491.

COMPONENTS:

(1) Cyclohexanol; $C_6H_{12}O$; [108-93-0]

(2) Water; H_2O; [7732-18-5]

ORIGINAL MEASUREMENTS:

Booth, H.S.; Everson, H.E.

Ind. Eng. Chem. 1949, *41*, 2627-8; 1950, *42*, 1536-7.

EXPERIMENTAL VALUES (continued)

Solubility, mL(1)/100mL solvent, for various solvent compositions (%)

		5.0	8.0	10.0	15.0	20.0	22.0	25.0	34.6	40.0
Sodium *m*-xylenesulfonate	25°C	5.02	-	19.5	-	196	-	-	-	>400
	60°C	5.00	-	17.8	-	197	-	-	-	>400
Sodium *p*-xylenesulfonate	25°C	5.02	-	19.0	89.6	-	-	-	-	-
	60°C	4.57	-	17.0	74.2	-	-	-	-	-

COMPONENTS:

(1) Cyclohexanol; $C_6H_{12}O$; [108-93-0]

(2) Water; H_2O; [7732-18-5]

ORIGINAL MEASUREMENTS:

Hansen, R.S.,; Fu, Y.; Bartell, F.E.

J. Phys. Chem. 1949, *53*, 769-85.

VARIABLES:

One temperature: 25°C

PREPARED BY:

A. Maczynski and Z. Maczynska

EXPERIMENTAL VALUES:

The solubility of cyclohexanol in water at 25°C was reported to be 3.92 g(1)/100g sln.

The corresponding mole fraction, x_1, calculated by the compilers is 7.28×10^{-3}.

AUXILIARY INFORMATION

METHOD/APPARATUS/PROCEDURE:

The interferometer method was used.

An excess of (1) was added to (2) in a mercury-sealed flask which was shaken mechanically for 48 h in an air chamber thermostatted at 25.0 ± 1°. The flask was then allowed to stand for 3 h in an air bath, after which a portion of the water-rich phase was removed by means of a hypodermic syringe and compared interferometrically with the most concentrated solution of (2) that could be prepared.

SOURCE AND PURITY OF MATERIALS:

(1) source not specified; reagent grade; distilled and redistilled; b.p. 158°C (730 mm Hg).

(2) distilled from alkaline permanganate solution.

ESTIMATED ERROR:

Temperature: ± 0.1°C

Solubility: ± 0.01 wt % (mean from three determinations)

REFERENCES:

COMPONENTS:

(1) Cyclohexanol; $C_6H_{12}O$ [108-93-0]

(2) Water, H_2O; [7732-18-5]

ORIGINAL MEASUREMENTS:

Zil'berman, E.N.

Zh. Fiz. Khim. 1951, *24*, 776-8.

VARIABLES:

Temperature: 0-184°C

PREPARED BY:

A. Maczynski and Z. Maczynska

EXPERIMENTAL VALUES:

$t/^{\circ}C$	g(1)/100g sln		x_1 (compiler)	
0	5.34	89.87	0.01006	0.6147
10	4.57	89.47	0.00853	0.6041
20	4.00	88.93	0.00744	0.5909
30	3.60	88.37	0.00667	0.5754
40	3.33	87.77	0.00616	0.5634
50	3.14	87.11	0.00580	0.5486
54	3.1	-	0.00572	-
60	-	86.42	-	0.5336
62	3.1	-	0.00572	-
70	3.19	85.66	0.00589	0.5175
80	3.41	84.83	0.00631	0.5014
90	3.65	83.93	0.00677	0.4843
100	3.93	82.96	0.00730	0.4668
110	4.28	81.9	0.00796	0.449
120	4.7	80.7	0.00879	0.429
130	5.3	79.3	0.00996	0.408
140	6.1	76.4	0.0115	0.368
150	7.2	74.7	0.0137	0.347
160	8.8	70.5	0.0171	0.301
170	11.5	63.8	0.0228	0.241
180	17.8	52.3	0.0375	0.165
184	33.9	33.9	0.0844	0.0844

UCST is 184°C at 33.9 wt % of cyclohexanol

AUXILIARY INFORMATION

METHOD/APPARATUS/PROCEDURE:

Alekseev's synthetic method (ref 1) was used.

The procedure involved heating and cooling weighed amounts of (1) and (2) in a sealed tube, 10 mm across and 80 mm long, kept in an oil bath. Turbidity disappearance and reappearance temperatures were recorded, and the mean was adopted as the solubility temperature.

SOURCE AND PURITY OF MATERIALS:

(1) commercial product; twice distilled; m.p. 24.95°C, d_4^{25} 0.9437, n_D^{25} 1.4669; 0.02% of phenol impurity.

(2) not specified.

ESTIMATED ERROR:

Temperature: > 100°C ± 0.5°C

< 100°C ± 1°C

REFERENCES:

1. Anosov, V. Ya.; Pogodin, S.A. *Osnovyne nachala fizykokhimicheskogo analiza, Izd. AN.SSSR, M.-L.* 1947, 122.

COMPONENTS:

(1) Cyclohexanol; $C_6H_{12}O$; [108-93-0]

(2) Water; H_2O; [7732-18-5]

ORIGINAL MEASUREMENTS:

Skrzec, A.E.; Murphy, N.F.

Ind. Eng. Chem. 1954, *46*, 2245-7.

VARIABLES:

One temperature: 27°C

PREPARED BY:

A. Maczynski

EXPERIMENTAL VALUES:

The solubility of cyclohexanol in water at 26.7°C was reported to be 3.60 g (1)/100g sln.

The corresponding mole fraction x_1, calculated by the compiler is 6.67×10^{-3}.

The solubility of water in cyclohexanol at 26.7°C was reported to be 14.00 g(2)/100g sln.

The corresponding mole fraction, x_2, calculated by the compiler is 0.4752.

AUXILIARY INFORMATION

METHOD/APPARATUS/PROCEDURE:

Probably the titration method was used. This method was described for the determination of mutual solubilities in ternary systems but nothing is reported on the binary system determinations.

SOURCE AND PURITY OF MATERIALS:

(1) E.I. du Pont de Nemours and Co., Inc.,; purified; 100% as alcohol.

(2) not specified.

ESTIMATED ERROR:

Temperature: ± 0.5°C

REFERENCES:

COMPONENTS:	ORIGINAL MEASUREMENTS:
(1) Cyclohexanol; $C_6H_{12}O$; [108-93-0] (2) Water; H_2O; [7732-18-5]	Tettamanti, K.; Nogradi, M; Sawinsky, J. *Periodica Politechnica*, 1960, 4, 201-18.
VARIABLES:	PREPARED BY:
One temperature: 20°C	A. Maczynski

EXPERIMENTAL VALUES:

The solubility of cyclohexanol in water at 20°C was reported to be 5.61 g(1)/100g sln.

The corresponding mole fraction x_1, calculated by the compiler is 0.0106.

The solubility of water in cyclohexanol at 20°C was reported to be 5.3 g(2)/100g sln.

The corresponding mole fraction, x_2, calculated by the compiler is 0.24.

AUXILIARY INFORMATION

METHOD/APPARATUS/PROCEDURE:

The titration method was used.

Into well-stoppered flasks one component was weighed to an accuracy of 0.01 g. Its temperature was adjusted to, and kept constant at 20.0 + 0.1°C in a thermostat, then titrated with second component from a micro-buret till the appearance of turbidity.

SOURCE AND PURITY OF MATERIALS:

(1) not specified.

(2) not specified.

ESTIMATED ERROR:

Not specified.

REFERENCES:

COMPONENTS:

(1) Cyclohexanol; $C_6H_{12}O$; [108-93-0]

(2) Water; H_2O; [7732-18-5]

ORIGINAL MEASUREMENTS:

Lavrova, O.A.; Lesteva, T.M.

Zh. Fiz. Khim., 1976, *50*, 1617; *Dep. Doc. VINITI*, 3813-75.

VARIABLES:

Temperature: 40 and 60°C

PREPARED BY:

A. Maczynski

EXPERIMENTAL VALUES:

Mutual solubility of cyclohexanol (1) and water (2)

t/°C	g(1)/100g sln		x_1 (compiler)	
	(2)-rich phase	(1)-rich phase	(2)-rich phase	(1)-rich phase
40	3.33	87.79	0.00616	0.5639
60	3.11	86.48	0.00574	0.5349

AUXILIARY INFORMATION

METHOD/APPARATUS/PROCEDURE:

The titration method was used.

No details were reported in the paper.

SOURCE AND PURITY OF MATERIALS:

(1) source not specified; distilled with heptane; purity 99.93 wt % with 0.07 wt % of water, b.p. 161.3°C.

(2) not specified.

ESTIMATED ERROR:

Not specified.

REFERENCES:

COMPONENTS:	ORIGINAL MEASUREMENTS:
(1) 4-Methyl-1-penten-3-ol; $C_6H_{12}O$; [4798-45-2] (2) Water; H_2O; [7732-18-5]	Ginnings, P.M.; Herring E.; Coltrane, D. *J. Am. Chem. Soc.* 1939, *61*, 807-8.

VARIABLES:	PREPARED BY:
Temperature: 20-30 °C	A. Maczynski and Z. Maczynska

EXPERIMENTAL VALUES:

Mutual solubility of 4-methyl-1-penten-3-ol(1) and water(2)

t/°C	g(1)/100g sln		x_1(compiler)	
	(2)-rich phase	(1)-rich phase	(2)-rich phase	(1)-rich phase
20	3.29	94.16	0.00612	0.7397
25	3.06	94.10	0.00564	0.7376
30	2.89	93.97	0.00526	0.7331

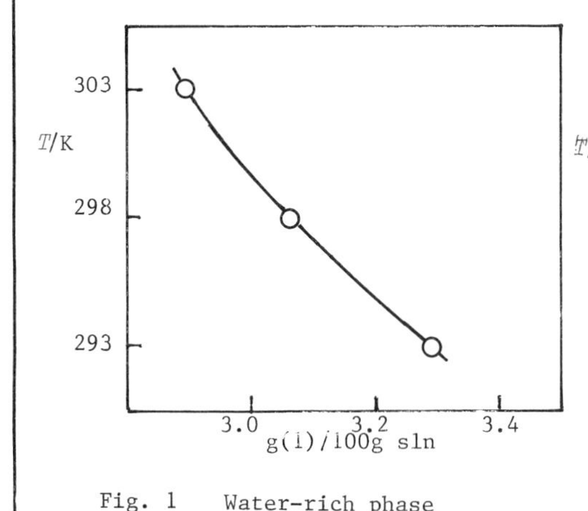

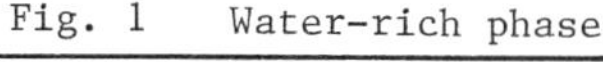

Fig. 1 Water-rich phase

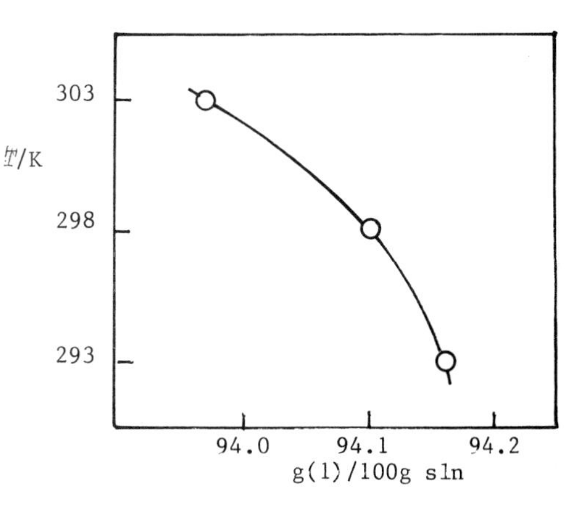

Fig. 2 Alcohol-rich phase

AUXILIARY INFORMATION

METHOD/APPARATUS/PROCEDURE:

The volumetric method was used as described in ref 1.

Both components were introduced in known amounts to a two-bulb graduated and calibrated flask and shaken mechanically in a water-bath at constant temperature. After sufficient time the liquids were allowed to separate and the total volume was measured. Upon centrifugation, the phase separation line was read, and phase volumes were calculated. From the total weights of the components, the total volume, individual phase volumes, and component concentrations in either phase were evaluated.

Relative densities of both phases were also determined.

SOURCE AND PURITY OF MATERIALS:

(1) prepared by Grignard synthesis; distilled; b.p. range 125.1-126.1 °C, d_4^{25} 0.8364; purity not specified.

(2) not specified.

ESTIMATED ERROR:

Solubility: 0.3 wt %
(type of error not specified)

REFERENCES:

1. Ginnings, P.M.; Baum, R.J. *J. Am. Chem. Soc.* 1937, *59*, 1111.

COMPONENTS:	ORIGINAL MEASUREMENTS:
(1) 1-Hexen-3-ol; $C_6H_{12}O$; [4798-44-1] (2) Water; H_2O; [7732-18-5]	Ginnings, P.M.; Herring, E.; Coltrane, D. *J. Am. Chem. Soc.* 1939, *61*, 807-8.
VARIABLES:	**PREPARED BY:**
Temperature: 20-30°C	A. Maczynski and Z. Maczynska

EXPERIMENTAL VALUES:

Mutual solubility of 1-hexen-3-ol(1) and water(2)

t/°C	g(1)/100g sln (2)-rich phase	g(1)/100g sln (1)-rich phase	x_1 (compiler) (2)-rich phase	x_1 (compiler) (1)-rich phase
20	2.72	94.12	0.00500	0.7383
25	2.52	93.92	0.00463	0.7314
30	2.36	93.90	0.00432	0.7307

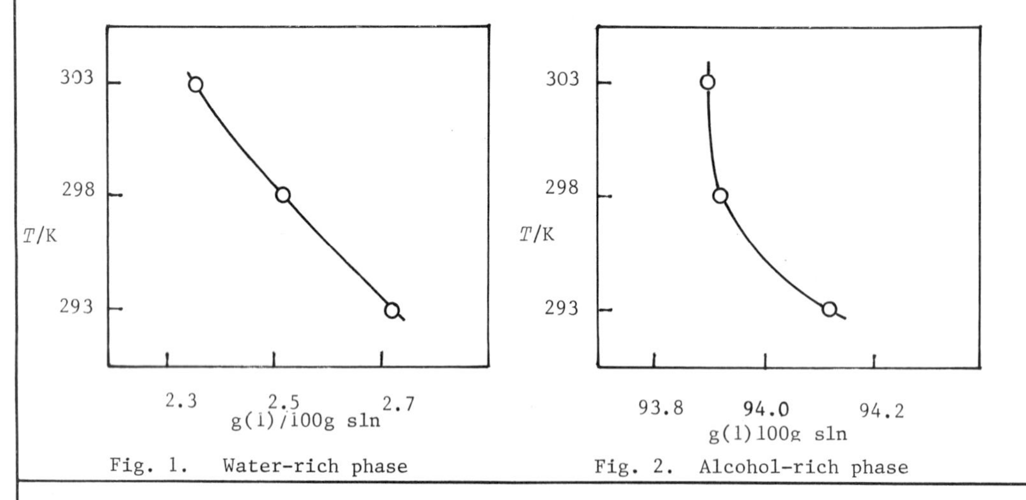

Fig. 1. Water-rich phase

Fig. 2. Alcohol-rich phase

AUXILIARY INFORMATION

METHOD/APPARATUS/PROCEDURE:

The volumetric method was used as described in ref 1.

Both components were introduced in known amounts into a two-bulb graduated and calibrated flask and shaken mechanically in a water-bath at constant temperature. After sufficient time the liquids were allowed to separate and the total volume was measured. Upon centrifugation, the phase separation line was read, and phase volumes were calcualted. From the total weights of the components, the total volume, individual phase volumes, and component concentrations in either phase were evaluated.

Relative densities of both phases were also determined.

SOURCE AND PURITY OF MATERIALS:

(1) prepared by Grignard synthesis; distilled; b.p. range 133.5-134.0°C d_4^{25} 0.8318; purity not specified.

(2) not specified.

ESTIMATED ERROR:

Solubility: 0.3 wt %
(type of error not specified)

REFERENCES:

1. Ginnings, P.M.; Baum, R.J. *J. Am. Chem. Soc.* 1937, *59*, 1111.

COMPONENTS:

(1) 4-Hexen-3-ol; $C_6H_{12}O$; [4798-58-7]

(2) Water; H_2O; [7732-18-5]

ORIGINAL MEASUREMENTS:

Ginnings, P.M.; Herring, E.; Coltrane, D.

J. Am. Chem. Soc. 1939, *61*, 807-8.

VARIABLES:

Temperature: 20-30°C

PREPARED BY:

A. Maczynski and Z. Maczynska

EXPERIMENTAL VALUES:

Mutual solubility of 4-hexen-3-ol(1) and water(2)

t/°C	g(1)/100g sln		x_1 (compiler)	
	(2)-rich phase	(1)-rich phase	(2)-rich phase	(1)-rich phase
20	4.06	96.07	0.00755	0.8116
25	3.81	95.85	0.00707	0.8028
30	3.58	95.74	0.00663	0.7984

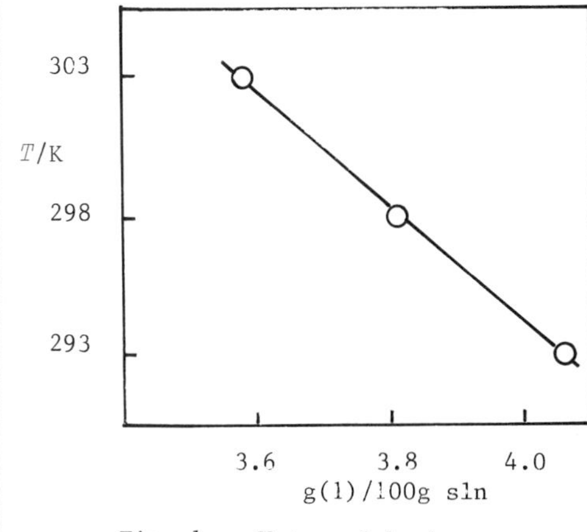

Fig. 1 Water-rich phase

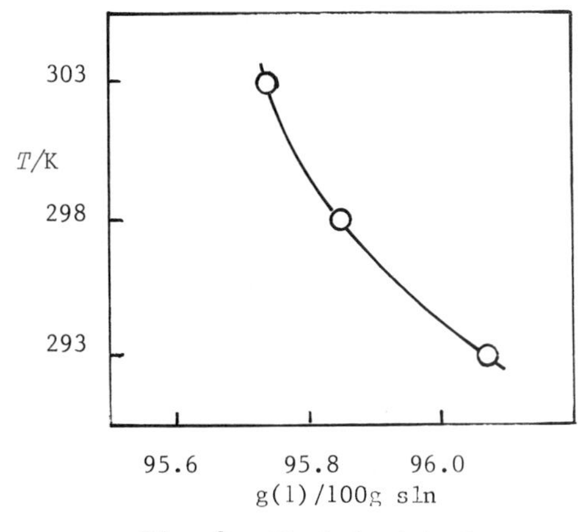

Fig. 2 Alcohol-rich phase

AUXILIARY INFORMATION

METHOD/APPARATUS/PROCEDURE:

The volumetric method was used as described in ref 1.

Both components were introduced in known amounts into a two bulb graduated and calibrated flask and shaken mechanically in a water-bath at constant temperature. After sufficient time the liquids were allowed to separate and the total volume was measured. Upon centrifugation, the phase separation line was read, and phase volumes were calculated. From the total weights of the components, the total volume, individual phase volumes, and component concentrations in either phase were evaluated.

Relative densities of both phases were also determined.

SOURCE AND PURITY OF MATERIALS:

(1) prepared by Grignard synthesis; distilled; b.p. range 133.8-136.2°C, d_4^{25} 0.8415; purity not specified.

(2) not specified.

ESTIMATED ERROR:

Solubility: 0.3 wt % (type of error not specified)

REFERENCES:

1. Ginnings, P.M.; Baum, R.J. *J. Am. Chem. Soc.* 1937, *59*, 1111.

COMPONENTS:	EVALUATOR:
(1) 2,2-Dimethyl-1-butanol (*tert-pentylcarbinol*); $C_6H_{14}O$; [1185-33-7] (2) Water; H_2O; [7732-18-5]	Z. Maczynska, Institute of Physical Chemistry of the Polish Academy of Sciences, Warsaw, Poland; and A.F.M. Harton, Murdoch University, Perth, Western Australia. November 1982

CRITICAL EVALUATION:

Solubilities in the system comprising 2,2-dimethyl-1-butanol (1) and water (2) have been reported in only two publications. Ginnings and Webb (ref 1) carried out measurements of the mutual solubilities of the two components at 293, 298 and 303 K by the volumetric method (Figures 1 and 2). Ratouis and Dodé (ref 2) determined the solubility of (1) in the water-rich phase at 298 and 303 K by an analytical method, obtaining values 50% higher than ref 1. The values below, based on ref 1, are therefore tentative.

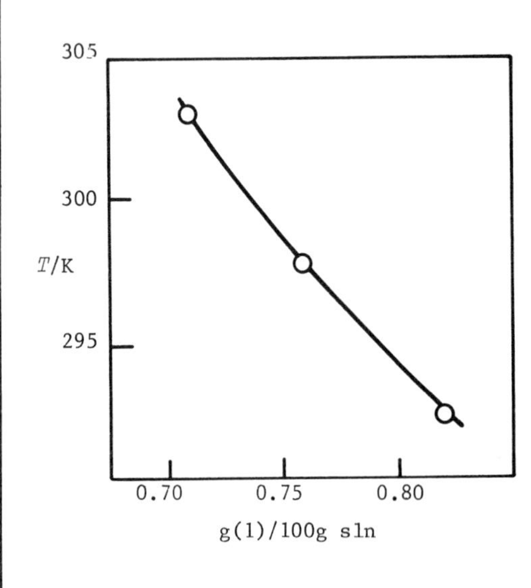

Fig. 1. Water-rich phase

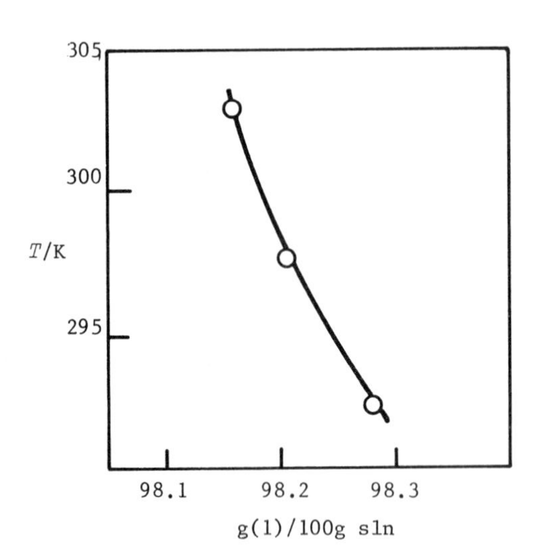

Fig. 2. Alcohol-rich phase

Tentative values for the mutual solubilities of 2,2-dimethyl-1-butanol (1) and water (2).

T/K	Water-rich phase		Alcohol-rich phase	
	g(1)/100g sln	10^3x_1	g(2)/100g sln	x_2
293	0.8	1.5	1.7	0.190
298	0.8	1.4	1.8	0.194
303	0.7	1.3	1.8	0.196

References

1. Ginnings, P.M.; Webb, R. *J. Am. Chem. Soc.* 1938, *60*, 1388.

2. Ratouis, M.; Dodé, M. *Bull. Soc. Chim. Fr.* 1965, 3318.

COMPONENTS:

(1) 2,2-Dimethyl-1-butanol (*tert-pentylcarbinol*); $C_6H_{14}O$; [1185-33-7]

(2) Water; H_2O; [7732-18-5]

ORIGINAL MEASUREMENTS:

Ginnings, P.M.; Webb, R.

J. Am. Chem. Soc. 1938, *60*, 1388-9.

VARIABLES:

Temperature: 20-30°C

PREPARED BY:

A. Maczynski and Z. Maczynska

EXPERIMENTAL VALUES:

Mutual solubility of 2,2-dimethyl-1-butanol(1) and water(2)

$t/°C$	g(1)/100g sln		x_1 (compiler)	
	(2)-rich phase	(1)-rich phase	(2)-rich phase	(1)-rich phase
20	0.82	98.28	0.00146	0.9097
25	0.76	98.21	0.00135	0.9063
30	0.71	98.16	0.00126	0.9039

Relative density d_4

$t/°C$	Water-rich phase	Alcohol-rich phase
20	0.9971	0.8598
25	0.9963	0.8551
30	0.9951	0.8511

AUXILIARY INFORMATION

METHOD/APPARATUS/PROCEDURE:

The volumetric method was used as described in ref 1.

Both components were introduced in known amounts into a two-bulb graduated and calibrated flask and shaken mechanically in a water-bath at constant temperature. After sufficient time the liquids were allowed to separate and the total volume was measured. Upon centrifugation, the phase separation line was read, and phase volumes were calculated. From the total weights of the components, the total volume, individual phase volumes, and component concentrations in either phase were evaluated.

SOURCE AND PURITY OF MATERIALS:

(1) prepared by Grignard synthesis; distilled from calcium oxide; b.p. range 136.9-137.9°C d_4^{25} 0.8498; purity not specified.

(2) not specified.

ESTIMATED ERROR:

Temperature: ± 0.1°C
Solubility: better than 0.1 wt % (type of error not specified)

REFERENCES:

1. Ginnings, P.M.; Baum, R.J. *J. Am. Chem. Soc.* 1937, *59*, 1111.

COMPONENTS:

(1) 2,2-Dimethyl-1-butanol; *(tert-pentylcarbinol)*; $C_6H_{14}O$; [1185-33-7]

(2) Water; H_2O; [7732-18-5]

ORIGINAL MEASUREMENTS:

Ratouis, M.; Dode, M.;
Bull. Soc. Chim. Fr. 1965, 3318-22.

VARIABLES:

Temperature: 25-30°C

Ringer solution also studied

PREPARED BY:

S.C. Valvani; S.H. Yalkowsky; A.F.M. Barton

EXPERIMENTAL VALUES:

Proportion of 2,2-dimethyl-1-butanol (1) in water-rich phase

$t/^{\circ}C$	g(1)/100g sln	x_1 (compilers)
25	1.235	0.00220
30	1.17	0.00208

The proportion of (1) in the water-rich phase of a mixture with Ringer solution at equilibrium at 30°C was reported to be 1.11 g(1)/100g sln.

AUXILIARY INFORMATION

METHOD/APPARATUS/PROCEDURE:

In a round bottomed flask, 50 mL of water and a sufficient quantity of alcohol were introduced until two separate layers were formed. The flask assembly was equilibrated by agitation for at least 3 h in a constant temperature bath. Equilibrium solubility was attained by first supersaturating at a slightly lower temperature (solubility of alcohols in water decrease with increasing temperature) and then equilibrating at the desired temperature. The aqueous layer was separated after an overnight storage in a bath. The alcohol content was determined by reacting the aqueous solution with potassium dichromate and titrating the excess dichromate with ferrous sulfate solution in the presence of phosphoric acid and diphenylamine barium sulfonate as an indicator.

SOURCE AND PURITY OF MATERIALS:

(1) laboratory preparation; redistilled with 10:1 reflux ratio;

b.p. 136.5-136.7°C/770 mm Hg

n_D^{25} = 1.41894

(2) twice distilled from silica apparatus or ion-exchanged with Sagei A20

ESTIMATED ERROR:

Solubility: relative error of 2 determinations less than 1%.

Temperature: ± 0.05°C

REFERENCES:

COMPONENTS:	EVALUATOR:
(1) 2,3-Dimethyl-2-butanol *(dimethylisopropylcarbinol)*; $C_6H_{14}O$; [594-60-5] (2) Water; H_2O; [7732-18-5]	Z. Maczynska, Institute of Physical Chemistry of the Polish Academy of Sciences, Warsaw, Poland. November 1982

CRITICAL EVALUATION:

Solubilities in the system comprising 2,3-dimethyl-2-butanol (1) and water (2) have been reported in only two publications (Figures 1 and 2). Ginnings and Webb (ref 1) carried out measurements of the mutual solubilities of the two components at 293, 298 and 303 K by the volumetric method. Ratouis and Dodé (ref 2) determined the solubility of (1) in the water-rich phase at one temperature (303 K) by an analytical method. Their value of 3.47 g(1)/100g sln is in reasonable agreement with the value 3.76 ± 0.1 g(1)/100g sln at this temperature of ref 1. The data are regarded as tentative, since comparison can be made only at a single temperature and the other five points are derived from a single source.

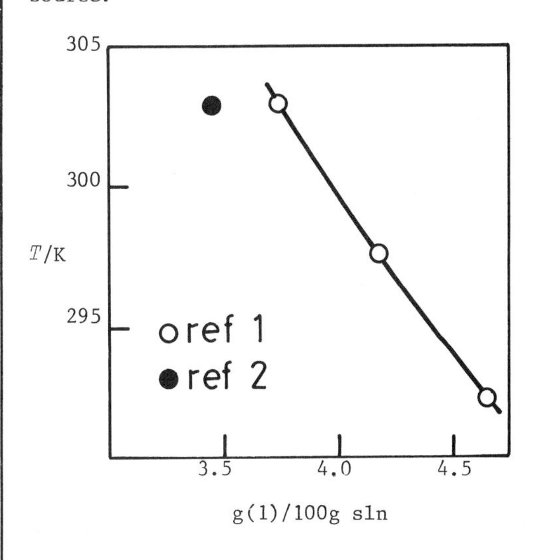

Fig. 1. Water-rich phase

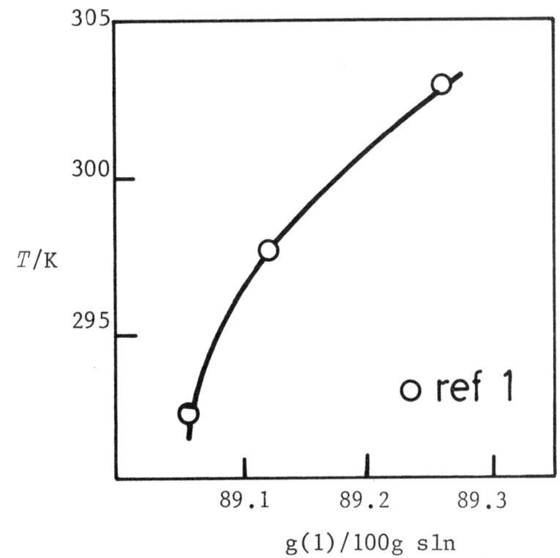

Fig. 2. Alcohol-rich phase

Tentative values for the mutual solubilties of 2,3-dimethyl-2-butanol (1) and water (2)

T/K	Water-rich phase		Alcohol-rich phase	
	g(1)/100g sln	$10^3 x_1$	g(2)/100g sln	x_2
293	4.7	8.5	10.9	0.411
298	4.2	7.6	10.9	0.409
303	3.8	6.8	10.7	0.406

References

1. Ginnings, P.M.; Webb, R. *J. Am. Chem. Soc.* 1938, *60*, 1388.
2. Ratouis, M.; Dodé, M. *Bull. Soc. Chim. Fr.* 1965, 3318.

COMPONENTS:

(1) 2,3-Dimethyl-2-butanol (*dimethylisopropylcarbinol*); $C_6H_{14}O$; [594-60-5]

(2) Water; H_2O; [7732-18-5]

ORIGINAL MEASUREMENTS:

Ginnings, P.M.; Webb, R.

J. Am. Chem. Soc. 1938, *60*, 1388-9.

VARIABLES:

Temperature: 20-30°C

PREPARED BY:

A. Maczynski and Z. Maczynska

EXPERIMENTAL VALUES:

Mutual solubility of 2,3-dimethyl-2-butanol(1) and water(2)

$t/^{\circ}C$	g(1)/100g sln		x_1 (compiler)	
	(2)-rich phase	(1)-rich phase	(2)-rich phase	(1)-rich phase
20	4.65	89.06	0.00852	0.5893
25	4.18	89.12	0.00763	0.5908
30	3.76	89.26	0.00684	0.5943

Relative density, d_4

$t/^{\circ}C$	Water-rich phase	Alcohol-rich phase
20	0.9934	0.8464
25	0.9929	0.8421
30	0.9910	0.8372

AUXILIARY INFORMATION

METHOD/APPARATUS/PROCEDURE:

The volumetric method was used as described in ref 1.

Both components were introduced in known amounts into a two-bulb graduated and calibrated flask and shaken mechanically in a water-bath at constant temperature. After sufficient time the liquids were allowed to separate and the total volume was measured. Upon centrifugation, the phase separation line was read, and phase volumes were calculated. From the total weights of the components, the total volume, individual phase volumes, and component concentrations in either phase were evaluated.

SOURCE AND PURITY OF MATERIALS:

(1) prepared by Grignard synthesis; distilled from calcium oxide; b.p. range 118.0-118.8°C d_4^{25} 0.8118; purity not specified.

(2) not specified.

ESTIMATED ERROR:

Temperature: ± 0.1°C
Solubility: better than 0.1 wt % (type of error not specified)

REFERENCES:

1. Ginnings, P.M.; Baum, R.J. *J. Am. Chem. Soc.* 1937, *59*, 1111.

COMPONENTS:

(1) 2,3-Dimethyl-2-butanol; *(dimethylisopropylcarbinol)* $C_6H_{14}O$; [594-60-5]

(2) Water; H_2O; [7732-18-5]

ORIGINAL MEASUREMENTS:

Ratouis, M.; Dodé, M.;

Bull. Soc. Chim. Fr. 1965, 3318-22.

VARIABLES:

One temperature: 30°C

Ringer solution also studied

PREPARED BY:

S.C. Valvani; S.H. Yalkowsky; A.F.M. Barton

EXPERIMENTAL VALUES:

The proportion of 2,3-dimethyl-2-butanol (1) in the water-rich phase at equilibrium at 30°C was reported to be 3.47 g(1)/100g sln.

The corresponding mole fraction solubility, calculated by the compiler, is x_1 = 0.00630.

The proportion of (1) in the water-rich phase of a mixture with Ringer solution at equilibrium at 30°C was reported to be 3.09 g(1)/100g sln.

AUXILIARY INFORMATION

METHOD/APPARATUS/PROCEDURE:

In a round bottomed flask, 50 mL of water and a sufficient quantity of alcohol were introduced until two separate layers were formed. The flask assembly was equilibrated by agitation for at least 3 h in a constant temperature bath. Equilibrium solubility was attained by first supersaturating at a slightly lower temperature (solubility of alcohols in water decreases with increasing temperature) and then equilibrating at the desired temperature. The aqueous layer was separated after an overnight storage in a bath. The alcohol content was determined by reacting the aqueous solution with potassium dichromate and titrating the excess dichromate with ferrous sulfate solution in the presence of phosphoric acid and diphenylamine barium sulfonate as an indicator.

SOURCE AND PURITY OF MATERIALS:

(1) laboratory preparation; redistilled with 10:1 reflux ratio;

b.p. 120.6°C/758.2 mm Hg

n_D^{25} = 1.40926

(2) twice distilled from silica apparatus or ion-exchanged with Sagei A20

ESTIMATED ERROR:

Solubility: relative error of 2 determinations less than 1%.

Temperature: ± 0.05°C

REFERENCES:

COMPONENTS:	EVALUATOR:
(1) 3,3-Dimethyl-2-butanol (*tert-butylmethylcarbinol*); $C_6H_{14}O$; [464-07-3] (2) Water; H_2O; [7732-18-5]	Z. Maczynska, Institute of Physical Chemistry of the Polish Academy of Sciences, Warsaw, Poland. November 1982

CRITICAL EVALUATION:

Solubilities in the system comprising 3,3-dimethyl-2-butanol (1) and water (2) have been reported in only two publications. Ginnings and Webb (ref 1) determined the mutual solubilities of the two components at 293, 298 and 303 K by the volumetric method. Ratouis and Dodé (ref 2) measured the solubility of (1) in the water-rich phase at one temperature (303 K) by an analytical method. Their value of 1.94 g(1)/100g sln is in reasonable agreement with the value 2.26 ± 0.1 g(1)/100g sln at this temperature of ref 1. The data are regarded as tentative, since comparison can be made only at a single temperature and the other five points are derived from a single source.

Tentative values for the mutual solubilities of 3,3-dimethyl-2-butanol (1) and water (2).

T/K	Water-rich phase		Alcohol-rich phase	
	g(1)/100g sln	10^3x_1	g(2)/100g sln	x_2
293	2.6	4.8	7.3	0.307
298	2.4	4.4	7.3	0.308
303	2.3	4.1	7.3	0.310

References

1. Ginnings, P.M.; Webb, R. *J. Am. Chem. Soc.* 1938, *60*, 1388.
2. Ratouis, M.; Dodé, M. *Bull. Soc. Chim. Fr.* 1965, 3318.

COMPONENTS:

(1) 3,3-Dimethyl-2-butanol *(tert-methylcarbinol)*; $C_6H_{14}O$: [464-07-3]

(2) Water; H_2O; [7732-18-5]

ORIGINAL MEASUREMENTS:

Ginnings, P.M.; Webb, R.

J. Am. Chem. Soc. 1938, *60*, 1388-9.

VARIABLES:

Temperature: 20-30°C

PREPARED BY:

A. Maczynski and Z. Maczynska

EXPERIMENTAL VALUES:

Mutual solubility of 3,3-dimethyl-2-butanol(1) and water(2)

$t/°C$	g(1)/100g sln		x_1 (compiler)	
	(2)-rich phase	(1)-rich phase	(2)-rich phase	(1)-rich phase
20	2.64	92.74	0.00476	0.6925
25	2.43	92.71	0.00437	0.6915
30	2.26	92.67	0.00406	0.6902

Relative density, d_4

$t/°C$	Water-rich phase	Alcohol-rich phase
20	0.9955	0.8356
25	0.9946	0.8313
30	0.9936	0.8272

AUXILIARY INFORMATION

METHOD/APPARATUS/PROCEDURE:

The volumetric method was used as described in ref 1.

Both components were introduced in known amounts into a two-bulb graduated and calibrated flask and shaken mechanically in a water-bath at constant tmperature. After sufficient time the liquids were allowed to separate and the total volume was measured. Upon centrifugation, the phase separation line was read, and phase volumes were calculated. From the total weights of the components, the total volume, individual phase volumes, and component concentrations in either phase were evaluated.

SOURCE AND PURITY OF MATERIALS:

(1) prepared by Grignard synthesis; distilled from calcium oxide; b.p. range 119.9-120.9°C d_4^{25} 0.8157; purity not specified.

(2) not specified.

ESTIMATED ERROR:

Temperature: ± 0.1°C
Solubility: better than 0.1 wt %
(type of error not specified)

REFERENCES:

1. Ginnings, P.M.; Baum, R.J. *J. Am. Chem. Soc.* 1937, *59*, 1111.

COMPONENTS:

(1) 3,3-Dimethyl-2-butanol (*tert-butylmethylcarbinol*); $C_6H_{14}O$; [464-07-3]

(2) Water; H_2O; [7732-18-5]

ORIGINAL MEASUREMENTS:

Ratouis, M.; Dodé, M.;

Bull. Soc. Chim. Fr. 1965, 3318-22.

VARIABLES:

One temperature: 30°C

Ringer solution also studied

PREPARED BY:

S.C. Valvani; S.H. Yalkowsky; A.F.M. Barton

EXPERIMENTAL VALUES:

The proportion of 3,3-dimethyl-2-butanol (1) in the water-rich phase at equilibrium at 30°C was reported to be 1.94 g(1)/100g sln.

The corresponding mole fraction solubility, calculated by the compilers, is x_1 = 0.00348.

The proportion of (1) in the water-rich phase of a mixture with Ringer solution at equilibrium at 30°C was reported to be 1.85 g(1)/100g sln.

AUXILIARY INFORMATION

METHOD/APPARATUS/PROCEDURE:

In a round bottomed flask, 50 mL of water and a sufficient quantity of alcohol were introduced until two separate layers were formed. The flask assembly was equilibrated by agitation for at least 3 h in a constant temperature bath. Equilibrium solubility was attained by first supersaturating at a slightly lower temperature (solubility of alcohols in water decreases with increasing temperature) and then equilibrating at the desired temperature. The aqueous layer was separated after an overnight storage in a bath. The alcohol content was determined by reacting the aqueous solution with potassium dichromate and titrating the excess dichromate with ferrous sulfate solution in the presence of phosphoric acid and diphenylamine barium sulfonate as an indicator.

SOURCE AND PURITY OF MATERIALS:

(1) Fluka A.G. Buchs S.G. redistilled with 10:1 reflux ratio;

b.p. 121.2 °C.

n_D^{25} = 1.41330

(2) twice distilled from silica apparatus or ion-exchanged with Sagei A20.

ESTIMATED ERROR:

Solubility: relative error of 2 determinations less than 1%.

Temperature: ± 0.05°C

REFERENCES:

COMPONENTS:

(1) 2-Ethyl-1-butanol; $C_6H_{14}O$; [97-95-0]

(2) Water; H_2O; [7732-18-5]

EVALUATOR:

A. Maczynski, Institute of Physical Chemistry of the Polish Academy of Sciences, Poland.

November 1982

CRITICAL EVALUATION:

Solubilities in the system comprising 2-ethyl-1-butanol (1) and water (2) have been reported in only two publications. Crittenden and Hixon (ref 1) determined the mutual solubilities of the two components at one temperature (298 K), presumably by the titration method. Ratouis and Dodé (ref 2) measured the solubility of (1) in the water-rich phase at 298 and 303 K by an analytical method.

The values for the solubility of (1) in (2) at 298 K are in disagreement (0.4 and 1.0 g(1)/100g sln, respectively, in ref 1 and 2). As the results in ref 2 refer to two temperatures, and are accompanied by an error estimate, these form the basis of the tentative data below.

Tentative values for the mutual solubilities of 2-ethyl-1-butanol (1) and water (2).

T/K	Water-rich phase		Alcohol-rich phase	
	g(1)/100g sln	10^3x_1	g(2)/100g sln	x_2
298	1.0	1.8	5.6	0.25
303	0.9	1.6		

References

1. Crittenden, E.D.,Jr.; Hixon, A.N. *Ind. Eng. Chem.* 1954, *46*, 265.

2. Ratouis, M.; Dodé, M. *Bull. Soc. Chim. Fr.* 1965, 3318.

COMPONENTS:

(1) 2-Ethyl-1-butanol; $C_6H_{14}O$; [97-95-0]

(2) Water; H_2O; [7732-18-5]

ORIGINAL MEASUREMENTS:

Crittenden, E.D, Jr.; Hixon, A.N.;

Ind. Eng. Chem. 1954, *46*, 265-8.

VARIABLES:

One temperature: 25°C

PREPARED BY:

A. Maczynski

EXPERIMENTAL VALUES:

The solubility of 2-ethyl-1-butanol in water at 25°C was reported to be 0.4 g(1)/100 sln.

The corresponding mole fraction, x_1, calculated by the compiler is 0.0007.

The solubility of water in 2-ethyl-1-butanol at 25°C was reported to be 5.6 g(2)/100g sln.

The corresponding mole fraction, x_2, calculated by the compiler is 0.25.

AUXILIARY INFORMATION

METHOD/APPARATUS/PROCEDURE:

Presumably the titration method described for ternary systems containing HCl was used. In this method the solubility was determined by bringing 100 mL samples of (1) or (2) to a temperature 25.0 ± 0.1°C and the second component was then added from a calibrated buret, with vigorous stirring, until the solution became permanently cloudy.

SOURCE AND PURITY OF MATERIALS:

(1) source not specified;
purified;
purity not specified.

(2) not specified.

ESTIMATED ERROR:

Solubility: 2% (alcohol-rich) - 10% (water-rich).

Temperature: ± 0.10°C

REFERENCES:

COMPONENTS:

(1) 2-Ethyl-1-butanol; $C_6H_{14}O$; [97-95-0]

(2) Water; H_2O; [7732-18-5]

ORIGINAL MEASUREMENTS:

Ratouis, M.; Dodé, M.;

Bull. Soc. Chim. Fr. 1965, 3318-22.

VARIABLES:

Temperature: 25°C and 30°C

Ringer solution also studied

PREPARED BY:

S.C. Valvani; S.H. Yalkowsky; A.F.M. Barton

EXPERIMENTAL VALUES:

Proportion of 2-ethyl-1-butanol (1) in water-rich phase

$t/°C$	g(1)/100g sln	$10^3\ x_1$ (compilers)
25	1.00	1.78
30	0.92	1.63

The proportion of (1) in the water-rich phase of a mixture with Ringer solution at equilibrium at 30°C was reported to be 0.90 g(1)/100g sln.

AUXILIARY INFORMATION

METHOD/APPARATUS/PROCEDURE:

In a round bottomed flask, 50 mL of water and a sufficient quantity of alcohol were introduced until two separate layers were formed. The flask assembly was equilibrated by agitation for at least 3 h in a constant temperature bath. Equilibrium solubility was attained by first supersaturating at a slightly lower temperature (solubility of alcohols in water decreases with increasing temperature) and then equilibrating at the desired temperature. The aqueous layer was separated after an overnight storage in a bath. The alcohol content was determined by reacting the aqueous solution with potassium dichromate and titrating the excess dichromate with ferrous sulfate solution in the presence of phosphoric acid and diphenylamine barium sulfonate as an indicator.

SOURCE AND PURITY OF MATERIALS:

(1) Fluka A.G., Buchs S.G. redistilled with 10:1 reflux ratio

b.p. 147°C/763.5 mm Hg

n_D^{25} = 1.42055

(2) twice distilled from silica apparatus or ion-exchanged with Sagei A20.

ESTIMATED ERROR:

Solubility: relative error of 2 determinations less than 1%.

Temperature: ±0.05°C

REFERENCES:

COMPONENTS:	EVALUATOR:
(1) 2-Methyl-1-pentanol; $C_6H_{14}O$; [105-30-6] (2) Water; H_2O; [7732-18-5]	A. Maczynski, Institute of Physical Chemistry of the Polish Academy of Sciences, Warsaw, Poland; and A.F.M. Barton, Murdoch University, Perth, Western Australia. November 1982

CRITICAL EVALUATION:

Solubilities in the system comprising 2-methyl-1-pentanol (1) and water (2) have been reported in only two publications. Crittenden and Hixon (ref 1) determined the mutual solubility of the two components at 298 K, presumably by the titration method. Ratouis and Dodé (ref 2) used an anlytical method for the solubility of (1) in the water-rich phase at 298 and 303 K. Their value of 0.81 g(1)/100g sln at 298 K is not in agreement with the value of 0.6 g(1)/100g sln reported in ref 1, so it is possible to report only tentative values.

Tentative values for the mutual solubilities of 2-methyl-1-pentanol (1) and water (2)

T/K	Water-rich phase		Alcohol-rich phase	
	g(1)/100g sln	$10^3 x_1$	g(2)/100g sln	x_2
298	0.81	1.4	5.4	0.24
303	0.76	1.4		

References

1. Crittenden, E.D.,Jr.; Hixon, A.N. *Ind. Eng. Chem.* 1954, *46*, 265.

2. Ratouis, M.; Dodé, M. *Bull. Soc. Chim. Fr.* 1965, 3318.

COMPONENTS:

(1) 2-Methyl-1-pentanol; $C_6H_{14}O$; [105-30-6]

(2) Water; H_2O; [7732-18-5]

ORIGINAL MEASUREMENTS:

Crittenden, E.D., Jr., Hixon, A.N.

Ind. Eng. Chem. 1954, *46*, 265-8.

VARIABLES:

One temperature: 25°C

PREPARED BY:

A. Maczynski

EXPERIMENTAL VALUES:

The solubility of 2-methyl-1-pentanol in water at 25°C was reported to be 0.6 g(1)/100g sln.

The corresponding mole fraction, x_1, calculated by the compiler is 0.0011.

The solubility of water in 2-methyl-1-pentanol at 25°C was reported to be 5.4 g(2)/100g sln.

The corresponding mole fraction, x_2, calculated by the compiler is 0.24.

AUXILIARY INFORMATION

METHOD/APPARATUS/PROCEDURE:

Presumably the titration method described for ternary systems containing HCl was used. In this method the solubility was determined by bringing 100-mL samples of (1) or (2) to a temperature 25.0 ± 0.1°C and the second component was then added from a calibrated buret, with vigorous stirring, until the solution became permanently cloudy.

SOURCE AND PURITY OF MATERIALS:

(1) source not specified; purified; purity not specified.

(2) not specified.

ESTIMATED ERROR:

Solubility: 2% (alcohol-rich) - 10% (water-rich).

Temperature: ± 0.10°C

REFERENCES:

COMPONENTS:

(1) 2-Methyl-1-pentanol; $C_6H_{14}O$; [105-30-6]

(2) Water; H_2O; [7732-18-5]

ORIGINAL MEASUREMENTS:

Ratouis, M.; Dodé, M.;
Bull. Soc. Chim. Fr. 1965, 3318-22.

VARIABLES:

Temperature: 25-30°C
Ringer solution also studied

PREPARED BY:

S.C. Valvani; S.H. Yalkowsky; A.F.M. Barton

EXPERIMENTAL VALUES:

Proportion of 2-methyl-1-pentanol (1) in water-rich phase

t/°C	g(1)/100g sln	x_1 (compilers)
25	0.81	0.00144
30	0.76	0.00135

The proportion of (1) in the water-rich phase of a mixture with Ringer solution at equilibrium at 30°C was reported to be 0.72 g(1)/100g sln.

AUXILIARY INFORMATION

METHOD/APPARATUS/PROCEDURE:

In a round bottomed flask, 50 mL of water and a sufficient quantity of alcohol were introduced until two separate layers were formed. The flask assembly was equilibrated by agitation for at least 3 h in a constant temperature bath. Equilibrium solubility was attained by first supersaturating at a slightly lower temperature (solubility of alcohols in water decreases with increasing temperature) and then equilibrating at the desired temperature. The aqueous layer was separated after an overnight storage in a bath. The alcohol content was determined by reacting the aqueous solution with potassium dichromate and titrating the excess dichromate with ferrous sulfate solution in the presence of phosphoric acid and diphenylamine barium sulfonate as an indicator.

SOURCE AND PURITY OF MATERIALS:

(1) Fluka A.G. Buchs S.G. redistilled with 10:1 reflux

b.p. 148-148.1°C/760 mm Hg

n_D^{25} = 1.41631

(2) twice distilled from silica apparatus or ion-exchanged with Sagei A20

ESTIMATED ERROR:

Solubility: relative error of 2 determinations less than 1%.

Temperature: ± 0.05°C

REFERENCES:

COMPONENTS:

(1) 2-Methyl-2-pentanol
(dimethyl-n-propylcarbinol, tert-hexyl alcohol); $C_6H_{14}O$; [590-36-3]

(2) Water; H_2O; [7732-18-5]

EVALUATOR:

Z. Maczynska, Institute of Physical Chemistry of the Polish Academy of Sciences, Warsaw, Poland; and A.F.M. Barton, Murdoch University, Perth, Western Australia.
November 1982

CRITICAL EVALUATION:

Solubilities in the system comprising 2-methyl-2-pentanol (1) and water (2) have been reported in three publications. Ginnings and Webb (ref 1) carried out measurements of the mutual solubilities of the two components at 293, 298 and 303 K by the volumetric method (Figures 1 and 2). Addison (ref 2) determined the solubility of (1) in the water-rich phase at one temperature (293 K) by a surface tension method. Ratouis and Dodé (ref 3) used an analytical method for (1) in (2), also at one temperature (303 K). The last two reported values are 10% lower than those of ref 1 for the water-rich phase, so the following figures, based on ref 1, must be regarded as tentative only.

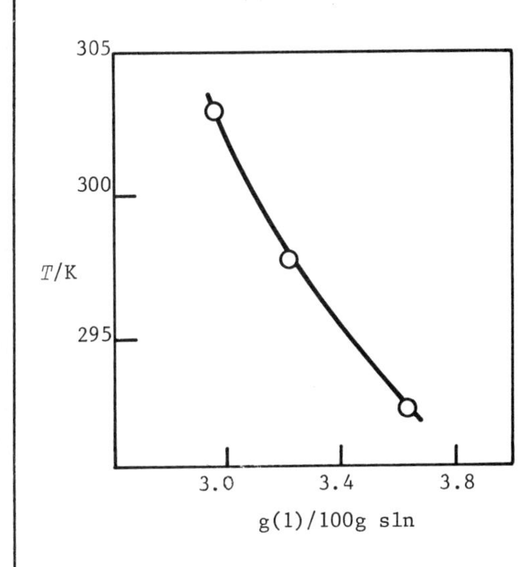

Fig. 1. Water-rich phase

305
300
295
T/K
89.8
89.9
90.0
g(1)/100g sln

Fig. 2. Alcohol-rich phase

Tentative values for the mutual solubilities of 1-methyl-2-pentanol (1) and water (2)

T/K	Water-rich phase		Alcohol-rich phase	
	g(1)/100g sln	$10^3 x_1$	g(2)/100g sln	x_2
293	3.6	6.6	10.1	0.390
298	3.2	5.9	10.0	0.388
303	2.9	5.3	10.0	0.386

References

1. Ginnings, P.M.; Webb, R. *J. Am. Chem. Soc.* 1938, *60*, 1388.

2. Addison, C.C. *J. Chem. Soc.* 1945, 98.

3. Ratouis, M.; Dodé, M. *Bull. Soc. Chim. Fr.* 1965, 3318.

COMPONENTS:

(1) 2-Methyl-2-pentanol
(dimethyl-n-propylcarbinol, tert-hexyl alcohol), $C_6H_{14}O$; [590-36-3]

(2) Water; H_2O; [7732-18-5]

ORIGINAL MEASUREMENTS:

Ginnings, P.M.; Webb, R.
J. Am. Chem. Soc. 1938, *60*, 1388-9.

VARIABLES:

Temperature: 20-30°C

PREPARED BY:

A. Maczynski and Z. Maczynska

EXPERIMENTAL VALUES:

Mutual solubility of 2-methyl-2-pentanol (1) and water (2)

$t/°C$	g(1)/100g sln		x_1 (compiler)	
	(2)-rich phase	(1)-rich phase	(2)-rich phase	(1)-rich phase
20	3.63	89.87	0.00660	0.6099
25	3.24	89.95	0.00586	0.6120
30	2.96	89.99	0.00534	0.6131

Relative density, d_4

$t/°C$	Water-rich phase	Alcohol-rich phase
20	0.9935	0.8321
25	0.9929	0.8280
30	0.9921	0.8237

AUXILIARY INFORMATION

METHOD/APPARATUS/PROCEDURE:

The volumetric method was used as described in ref 1.

Both components were introduced in known amounts into a two-bulb graduated and calibrated flask and shaken mechanically in a water-bath at constant temperature. After sufficient time the liquids were allowed to separate and the total volume was measured. Upon centrifucation, the phase separation line was read, and phase volumes were calcualted. From the total weights of the components, the total volume, individual phase volumes, and component concentrations in either phase were evaluated.

SOURCE AND PURITY OF MATERIALS:

(1) Eastman best grade;
distilled, and redistilled from calcium oxide
b.p. range 122.4-122.6°C
d_4^{25} 0.8053

(2) not specified

ESTIMATED ERROR:

Temperature: ± 0.1°C

Solubility: better than 0.1 wt % (type of error not specified)

REFERENCES:

1. Ginnings, P.M.; Baum, R.J.
J. Am. Chem. Soc. 1937, *59*, 1111.

COMPONENTS:

(1) 2-Methyl-2-pentanol, *(dimethyl-n-propylcarbinol, tert-hexyl alcohol)*; $C_6H_{14}O$; [590-36-3]

(2) Water; H_2O; [7732-18-5]

ORIGINAL MEASUREMENTS:

Addison, C.C.

J. Chem. Soc. 1945, 98-106.

VARIABLES:

One tempertature: 20°C

PREPARED BY:

S.H. Yalkowsky; S.C. Valvani; A.F.M. Barton

EXPERIMENTAL VALUES:

The proportion of 2-methyl-2-pentanol (1) in the water-rich phase at equilibrium at 20°C was reported to be 3.23 g(1)/100g sln.

The corresponding mole fraction solubility calculated by the compilers is x_1 = 0.00585.

AUXILIARY INFORMATION

METHOD/APPARATUS/PROCEDURE:

A surface tension method was used. Sufficient excess of (1) was added to 100mL of (2) in a stoppered flask to form a separate lens on the surface. The mixture was swirled gently, too vigorous an agitation giving a semi-permanent emulsion and incorrect readings. After settling, a small sample of the clear aqueous solution was withdrawn into a drop weight pipet and the surface tension determined. The swirling was continued until a constant value was obtained. The surface tension-concentration curve was known, and only a slight extrapolation (logarithmic scale) was necessary to find the concentration corresponding to the equilibrium value.

SOURCE AND PURITY OF MATERIALS:

(1) impure alcohols were purified by fractional distillation, the middle fraction from a distillation being redistilled;

b.p. 122.0°C d_4^{20} 0.8048

n_D^{20} 1.4120

(2) not stated

ESTIMATED ERROR:

Solubility: ± 0.5%

REFERENCES:

COMPONENTS:

(1) 2-Methyl-2-pentanol *(dimethyl-n-propylcarbinol, tert-hexyl alcohol)*; $C_6H_{14}O$; [590-36-3].

(2) Water, H_2O [7732-18-5].

ORIGINAL MEASUREMENTS:

Ratouis, M.; Dodé, M.; *Bull. Soc. Chim. Fr.* 1965, 3318-22.

VARIABLES:

One temperature: 30°C

Ringer solution also studied

PREPARED BY:

S.C. Valvani; S.H. Yalkowsky; A.F.M. Barton

EXPERIMENTAL VALUES:

The proportion of 2-methyl-2-pentanol (1) in the water-rich phase at equilibrium at 30°C was reported to be 2.82 g(1)/100g sln.

The corresponding mole fraction solubility, calculated by the compilers, is x_1 = 0.00509.

The proportion of (1) in the water-rich phase of a mixture with Ringer solution at 30°C was reported to be 2.51 g(1)/100g sln.

AUXILIARY INFORMATION

METHOD/APPARATUS/PROCEDURE:

In a round bottomed flask, 50 mL of water and a sufficient quantity of alochol were introduced until two separate layers were formed. The flask assembly was equilibrated by agitation for at least 3 h in a constant temperature bath. Equilibrium solubility was attained by first supersaturation at a slightly lower temperature (solubility of alcohols in water decreases with increasing temperature) and then equilibrating at the desired temperature. Aqueous layer was separated after an overnight storage in a bath. The alcohol content was determined by reacting the aqueous solution with potassium dichromate and titrating the excess dichromate with ferrous sulfate solution in the presence of phosphoric acid and diphenylamine barium sulfonate as an indicator.

SOURCE AND PURITY OF MATERIALS:

(1) laboratory preparation; redistilled with 10:1 reflux ratio;
b.p. 121.6-121.8°C/757.8 mm Hg
n_D^{25} = 1.40888

(2) twice distilled from silica apparatus or ion-exchanged with Sagei A20.

ESTIMATED ERROR:

Solubility: relative error of 2 determinations less than 1%.

Temperature: ± 0.05°C

REFERENCES:

COMPONENTS:	EVALUATOR:
(1) 2-Methyl-3-pentanol *(isopropylethylcarbinol)*; $C_6H_{14}O$; [565-67-3] (2) Water; H_2O; [7732-18-5]	Z. Maczynska, Institute of Physical Chemistry of the Polish Academy of Sciences, Warsaw, Poland; and A.F.M. Barton, Murdoch University, Perth, Western Australia. November 1982.

CRITICAL EVALUATION:

Solubilities in the system comprising 2-methyl-3-pentanol (1) and water (2) have been reported in two publications (Figure 1 and 2). Ginnings and Webb (ref 1) carried out measurements of the mutual solubilities of the two components at 293, 298 and 303 K by the volumetric method. Ratouis and Dodé (ref 2) determined the solubility of (1) in the water-rich phase at one temperature (303 K) by an analytical method. Their value of 1.85 g(1)/100g soln at this temperature is in excellent agreement with the value 1.82 ± 0.1 g(1)/100g sln of ref 1. The 303 K value is recommended, but the other five points are derived from a single source and are regarded as tentative.

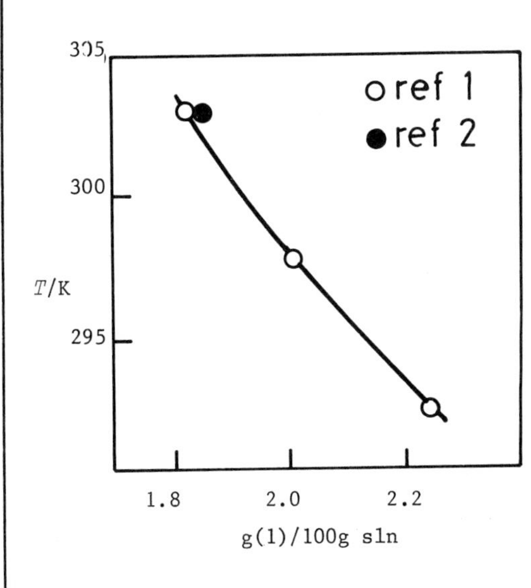

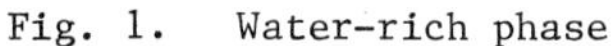
Fig. 1. Water-rich phase

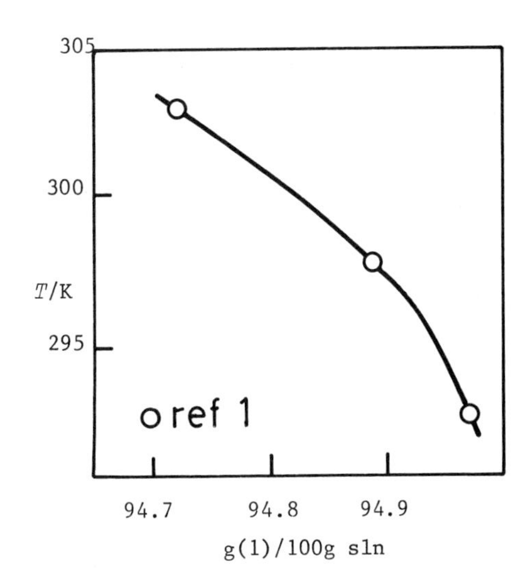

Fig. 2. Alcohol-rich phase

<u>Tentative values for the mutual solubilities of 2-methyl-3-pentanol (1) and water (2)</u>

T/K	Water-rich phase g(1)/100g sln		10^3x_1	Alcohol-rich phase g(2)/100g sln		x_2
293	2.2	(tentative)	4.0	5.0	(tentative)	0.231
298	2.0	(tentative)	3.6	5.1	(tentative)	0.234
303	1.8	(recommended)	3.3	5.3	(tentative)	0.240

<u>References</u>

1. Ginnings, P.M.; Webb, R. *J. Am. Chem. Soc.* <u>1938</u>, *60*, 1388.

2. Ratouis, M.; Dodé, M. *Bull. Soc. Chim. Fr.* <u>1965</u>, 3318.

COMPONENTS:

(1) 2-Methyl-3-pentanol *(isopropylethylcarbinol)*; $C_6H_{14}O$; [565-67-3]

(2) Water; H_2O; [7732-18-5]

ORIGINAL MEASUREMENTS:

Ginnings, P.M.; Webb, R.

J. Am. Chem. Soc. 1938, *60*, 1388-9.

VARIABLES:

Temperature: 20-30°C

PREPARED BY:

A. Maczynski and Z. Maczynska

EXPERIMENTAL VALUES:

Mutual solubility of 2-methyl-3-pentanol(1) and water(2)

t/°C	g(1)/100g sln		x_1(compiler)	
	(2)-rich phase	(1)-rich phase	(2)-rich phase	(1)-rich phase
20	2.24	94.97	0.00402	0.7689
25	2.01	94.89	0.00360	0.7660
30	1.82	94.72	0.00326	0.7597

Relative density, d_4

t/°C	Water-rich phase	Alcohol-rich phase
20	0.9950	0.8323
25	0.9941	0.8288
30	0.9940	0.8249

AUXILIARY INFORMATION

METHOD/APPARATUS/PROCEDURE:

The volumetric method was used as described in ref 1.

Both components were introduced in known amounts into a two-bulb graduated and calibrated flask and shaken mechanically in a water-bath at constant temperature. After sufficient time the liquids were allowed to separate and the total volume was measured. Upon centrifugation, the phase separation line was read, and phase volumes were calculated. From the total weights of the components, the total volume, individual phase volumes, and component concentrations in either phase were evaluated.

SOURCE AND PURITY OF MATERIALS:

(1) prepared by Grignard synthesis; distilled from calcium oxide; b.p. range 126.3-127.3°C, d_4^{25} 0.8186; purity not specified.

(2) not specified.

ESTIMATED ERROR:

Temperature: ± 0.1°C
Solubility: better than 0.1 wt % (type of error not specified)

REFERENCES:

1. Ginnings, P.M.; Baum, R.J. *J. Am. Chem. Soc.* 1937, *59*, 1111.

COMPONENTS:	ORIGINAL MEASUREMENTS:
(1) 2-Methyl-3-pentanol *(isopropylethylcarbinol)*; $C_6H_{14}O$; [565-67-3] (2) Water; H_2O; [7732-18-5]	Ratouis, M.; Dodé, M.; *Bull. Soc. Chim. Fr.* 1965, 3318-22.
VARIABLES:	**PREPARED BY:**
One temperature: 30°C Ringer solution also studied	S.C. Valvani; S.H. Yalkowsky; A.F.M. Barton

EXPERIMENTAL VALUES:

The proportion of 2-methyl-3-pentanol (1) in the water-rich phase at equilibrium at 30°C was reported to be 1.85 g(1)/100g sln.

The corresponding mole fraction solubility, calculated by the compilers, is x_1 = 0.00331.

The proportion of (1) in the water-rich phase of a mixture with Ringer solution at 30°C was reported to be 1.77 g(1)/100g sln.

AUXILIARY INFORMATION

METHOD/APPARATUS/PROCEDURE:

In a round bottomed flask, 50 mL of water and a sufficient quantity of alcohol were introduced until two separate layers were formed. The flask assembly was equilibrated by agitation for at least 3 h in a constant temperature bath. Equilibrium solubility was attained by first supersaturation at a slightly lower temperature (solubility of alcohols in water decreases with increasing temperature) and then equilibrating at the desired temperature. The aqueous layer was separated after an overnight storage in a bath. The alcohol content was determined by reacting the aqueous solution with potassium dichromate and titrating the excess dichromate with ferrous sulfate solution in the presence of phosphoric acid and diphenylamine barium sulfonate as an indicator.

SOURCE AND PURITY OF MATERIALS:

(1) laboratory preparation; redistilled with 10:1 reflux ratio,

b.p. 126.4°C/753 mm Hg

n_D^{25} = 1.41504

(2) twice distilled from silica apparatus or ion-exchanged with Sagei A20.

ESTIMATED ERROR:

Solubility: relative error of 2 determinations less than 1%.

Temperature: ±0.05°C

REFERENCES:

COMPONENTS:	EVALUATOR:
(1) 3-Methyl-2-pentanol *(sec-butylmethylcarbinol)*; $C_6H_{14}O$; [565-60-6] (2) Water; H_2O; [7732-18-5]	Z. Maczynska, Institute of Physical Chemistry of the Polish Academy of Sciences, Warsaw, Poland. November 1982

CRITICAL EVALUATION:

Solubilities in the system comprising 3-methyl-2-pentanol (1) and water (2) have been reported in two publications. Ginnings and Webb (ref 1) carried out measurements of the mutual solubilities of the two components at 293, 298 and 303 K by the volumetric method (Figure 1). Ratouis and Dodé (ref 2) determined the solubility of (1) in the water-rich phase at one temperature (303 K) by an analytical method. Their value of 1.65 g(1)/100g sln at this temperature is in reasonable agreement with the value 1.79 ± 0.1 g(1)/100g sln of ref 1. The data are regarded as tentative, since comparison can be made only at a single temperature and the other five points are derived from one source.

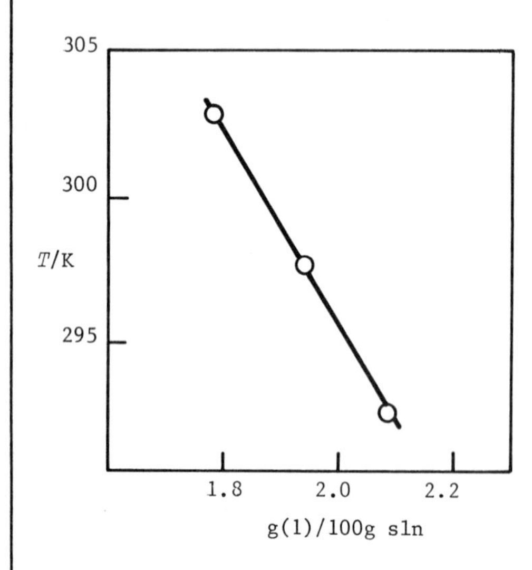

Fig. 1. Water-rich phase (ref 1)

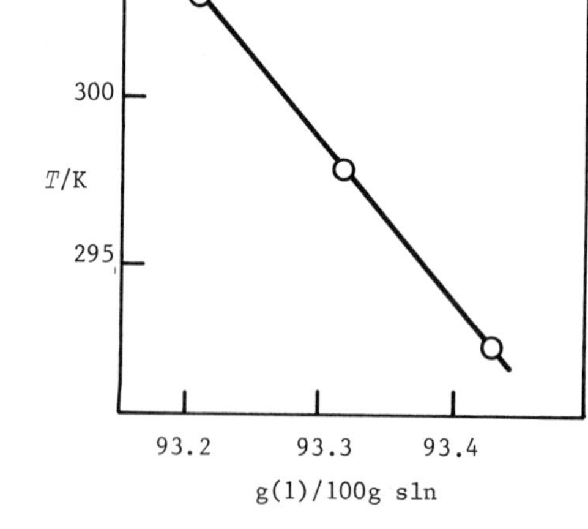

Fig. 2. Alcohol-rich phase (ref 1)

Tentative values for the mutual solubilities of 3-methyl-2-pentanol (1) and water (2)

T/K	Water-rich phase		Alcohol-rich phase	
	g(1)/100g sln	10^3x_1	g(2)/100g sln	x_2
293	2.1	3.8	6.6	0.285
298	1.9	3.5	6.7	0.289
303	1.8	3.2	6.8	0.292

References

1. Ginnings, P.M.; Webb, R. *J. Am. Chem. Soc.* 1938, *60*, 1388.

2. Ratouis, M.; Dodé, M. *Bull. Soc. Chim. Fr.* 1965, 3318.

COMPONENTS:

(1) 3-Methyl-2-pentanol
(sec-butylmethylcarbinol); $C_6H_{14}O$; [565-60-6]

(2) Water; H_2O; [7732-18-5]

ORIGINAL MEASUREMENTS:

Ginnings, P.M.; Webb, R.

J. Am. Chem. Soc. 1938, *60*, 1388-9.

VARIABLES:

Temperature: 20-30 °C

PREPARED BY:

A. Maczynski and Z. Maczynska

EXPERIMENTAL VALUES:

Mutual solubility of 3-methyl-2-propanol(1) and water(2)

t/°C	g(1)/100g sln		x_1 (compiler)	
	(2)-rich phase	(1)-rich phase	(2)-rich phase	(1)-rich phase
20	2.09	93.43	0.00375	0.7148
25	1.94	93.32	0.00347	0.7112
30	1.79	93.21	0.00320	0.7076

Relative density, d_4

t/°C	Water-rich phase	Alcohol-rich phase
20	0.9960	0.8390
25	0.9950	0.8356
30	0.9939	0.8316

AUXILIARY INFORMATION

METHOD/APPARATUS/PROCEDURE:

The volumetric method was used as described in ref 1.

Both components were introduced in known amounts into a two-bulb graduated and calibrated flask and shaken mechanically in a water-bath at constant temperature. After sufficient time the liquids were allowed to separate and the total volume was measured. Upon centrifugation, the phase separation line was read, and phase volumes were calculated. From the total weights of the components, the total volume, individual phase volumes, and component concentrations in either phase were evaluated.

SOURCE AND PURITY OF MATERIALS:

(1) prepared by Grignard synthesis; distilled from calcium oxide; b.p. range 133.5-134.5 °C/760 mm Hg d_4^{25} 0.8231; purity unspecified

(2) not specified.

ESTIMATED ERROR:

Temperature: ± 0.1 °C
Solubility: better than 0.1 wt %
(type of error not specified)

REFERENCES:

1. Ginnings, P.M.; Baum, R.J. *J. Am. Chem. Soc.* 1937, *59*, 1111.

COMPONENTS:	ORIGINAL MEASUREMENTS:
(1) 3-Methyl-2-pentanol, *(sec-butylmethylcarbinol)*; $C_6H_{14}O$; [565-60-6] (2) Water; H_2O; [7732-18-5]	Ratouis, M.; Dodé, M.; *Bull. Soc. Chim. Fr.* 1965, 3318-22.
VARIABLES:	**PREPARED BY:**
One temperature: 30°C Ringer solution also studied	S.C. Valvani; S.H. Yalkowsky; A.F.M. Barton.

EXPERIMENTAL VALUES:

The proportion of 3-methyl-2-pentanol (1) in the water-rich phase at equilibrium at 30°C was reported to be 1.65 g(1)/100g sln.

The corresponding mole fraction solubility, calculated by the compilers, is x_1 = 0.00295.

The proportion of (1) in the water-rich phase of a mixture with Ringer solution at 30°C was reported to be 1.52 g(1)/100g sln.

AUXILIARY INFORMATION

METHOD/APPARATUS/PROCEDURE:

In a round bottomed flask, 50 mL of water and a sufficient quantity of alcohol were introduced until two separate layers were formed. The flask assembly was equilibrated by agitation for at least 3 h in a constant temperature bath. Equilibrium solubility was attained by first supersaturation at a slightly lower temperature (solubility of alcohols in water decrease with increasing temperature) and then equilibrating at the desired temperature. The aqueous layer was separated after an overnight storage in a bath. The alcohol content was determined by reacting the aqueous solution with potassium dichromate and titrating the excess dichromate with ferrous sulfate solution in the presence of phosphoric acid and diphenylamine barium sulfonate as an indicator.

SOURCE AND PURITY OF MATERIALS:

(1) Laboratory preparation; redistilled with 10:1 reflux ratio;
b.p. 132.4-132.5°C/750 mm Hg
n_D^{25} = 1.41827

(2) twice distilled from silica apparatus or ion-exchanged with Sagei A20.

ESTIMATED ERROR:

Solubility: relative error of 2 determinations less than 1%.

Temperature: ±0.05°C.

REFERENCES:

COMPONENTS:	EVALUATOR:
(1) 3-Methyl-3-pentanol *(diethylmethylcarbinol)*; $C_6H_{14}O$; [77-74-7] (2) Water; H_2O; [7732-18-5]	Z. Maczynska, Institute of Physical Chemistry of the Polish Academy of Sciences, Warsaw, Poland. November 1982

CRITICAL EVALUATION:

Solubilities in the system comprising 3-methyl-3-pentanol (1) and water (2) have been reported in two publications (Figures 1 and 2). Ginnings and Webb (ref 1) carried out measurements of the mutual solubilities of the two components at 293, 298 and 303 K by the volumetric method. Ratouis and Dodé (ref 2) determined the solubility of (1) in the water-rich phase at one temperature (303 K) by an analytical method. Their value of 3.57 g(1)/100g sln at this temperature is in reasonable agreement with the value 3.81 ± 0.1 g(1)/100g sln of ref 1. The data are regarded as tentative, since comparison can be made only at a single temperature and the other five points are derived from one source.

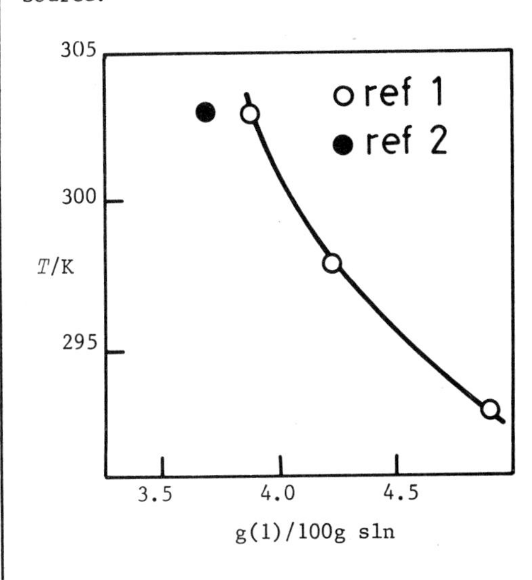

Fig. 1. Water-rich phase

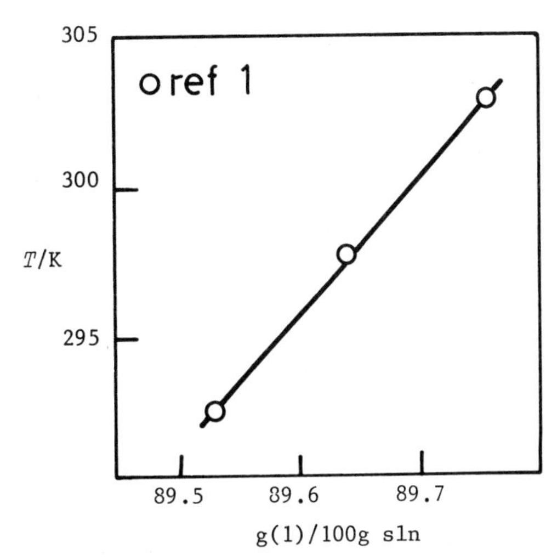

Fig. 2. Alcohol-rich phase

Tentative values for the mutual solubilities of 3-methyl-3-pentanol (1) and water (2)

T/K	Water-rich phase		Alcohol-rich phase	
	g(1)/100g sln	10^3x_1	g(2)/100g sln	x_2
293	4.8	8.9	10.5	0.399
298	4.3	7.8	10.4	0.396
303	3.8	6.9	10.2	0.393

References

1. Ginnings, P.M.; Webb, R. *J. Am. Chem. Soc.* 1938, *60*, 1388.
2. Ratouis, M.; Dodé, M. *Bull. Soc. Chim. Fr.* 1965, 3318.

COMPONENTS:

(1) 3-Methyl-3-pentanol *(diethylmethylcarbinol)*; $C_6H_{14}O$; [77-74-7]

(2) Water; H_2O; [7731-18-5]

ORIGINAL MEASUREMENTS:

Ginnings, P.M.; Webb, R.

J. Am. Chem. Soc. 1938, *60*, 1388-9.

VARIABLES:

Temperature: 20-30°C

PREPARED BY:

A. Maczynski and Z. Maczynska

EXPERIMENTAL VALUES:

Mutual solubility of 3-methyl-3-pentanol(1) and water(2)

$t/^\circ C$	g(1)/100g sln		x_1 (compiler)	
	(2)-rich phase	(1)-rich phase	(2)-rich phase	(1)-rich phase
20	4.82	89.53	0.00885	0.6011
25	4.26	89.64	0.00778	0.6040
30	3.81	89.76	0.00693	0.6071

Relative density, d_4

$t/^\circ C$	Water-rich phase	Alcohol-rich phase
20	0.9933	0.8498
25	0.9926	0.8454
30	0.9919	0.8410

AUXILIARY INFORMATION

METHOD/APPARATUS/PROCEDURE:

The volumetric method was used as described in ref 1.

Both components were introduced in known amounts into a two-bulb graduated and calibrated flask and shaken mechanically in a water-bath at constant temperature. After sufficient time the liquids were allowed to separate and the total volume was measured. Upon centrifugation, the phase separation line was read, and phase volumes were calculated. From the total weights of the components, the total volume, individual phase volumes, and component concentrations in either phase were evaluated.

SOURCE AND PURITY OF MATERIALS:

(1) prepared by Grignard synthesis; distilled from calcium oxide; b.p. range 122.6-122.8°C/760 mm Hg d_4^{25} 0.8242; purity not specified.

(2) not specified.

ESTIMATED ERROR:

Temperature: ± 0.1°C
Solubility: better than 0.1 wt % (type of error not specified)

REFERENCES:

1. Ginnings, P.M.; Baum, R.J. *J. Am. Chem Soc.* 1937, *59*, 1111.

COMPONENTS:	ORIGINAL MEASUREMENTS:
(1) 3-Methyl-3-pentanol *(diethylmethylcarbinol)*; $C_6H_{14}O$; [77-74-7] (2) Water; H_2O; [7732-18-5]	Ratouis, M.; Dodé, M.; *Bull. Soc. Chim. Fr.* 1965, 3318-22.
VARIABLES:	PREPARED BY:
One temperature: 30°C Ringer solution also studied	S.C. Valvani; S.H Yalkowsky; A.F.M. Barton

EXPERIMENTAL VALUES:

The proportion of 3-methyl-3-pentanol (1) in the water-rich phase at equilibrium at 30°C was reported to be 3.57 g(1)/100g sln.

The corresponding mole fraction solubility, calculated by the compilers, is x_1 = 0.00649

The proportion of (1) in the water-rich phase of a mixture with Ringer solution at 30°C was reported to be 3.35 g(1)/100g sln.

AUXILIARY INFORMATION

METHOD/APPARATUS/PROCEDURE:

In a round bottomed flask, 50 mL of water and a sufficient quantity of alcohol were introduced until two separate layers were formed. The flask assembly was equilibrated by agitation for at least 3 h in a constant temperature bath. Equilibrium solubility was attained by first supersaturation at a slightly lower temperature (solubility of alcohols in water decreases with increasing temperature) and then equilibrating at the desired temperature. The aqueous layer was separated after an overnight storage in a bath. The alcohol content was determined by reacting the aqueous solution with potassium dichromate and titrating the excess dichromate with ferrous sulfate solution in the presence of phosphoric acid and diphenylamine barium sulfonate as an indicator.

SOURCE AND PURITY OF MATERIALS:

(1) laboratory preparation; redistilled with 10:1 reflux ratio

b.p. 123-123.1°C/770.4 mm Hg

n_D^{25} = 1.41665

(2) twice distilled from silica apparatus or ion-exchanged with Sagei A20.

ESTIMATED ERROR:

Solubility: relative error of 2 determinations less than 1%.

Temperature: ±0.05°C

REFERENCES:

COMPONENTS:

(1) 4-Methyl-1-pentanol; $C_6H_{14}O$; [626-89-1]

(2) Water; H_2O; [7732-18-5]

ORIGINAL MEASUREMENTS:

Ratouis, M.; Dodé, M.;

Bull. Soc. Chim. Fr. 1965, 3318-22.

VARIABLES:

Temperature: 25°C and 30°C

Ringer solution also studied

PREPARED BY:

S.C. Valvani; S.H. Yalkowsky; A.F.M. Barton

EXPERIMENTAL VALUES:

Proportion of 4-methyl-1-pentanol (1) in water-rich phase

t/°C	g(1)/100g sln	x_1 (compilers)
25	0.76	0.00135
30	0.715	0.00127

The proportion of (1) in the water-rich phase of a mixture with Ringer solution at equilibrium at 30°C was reported to be 0.70 g(1)/100g sln.

AUXILIARY INFORMATION

METHOD/APPARATUS/PROCEDURE:

In a round bottomed flask, 50 mL of water and a sufficient quantity of alcohol were introduced until two separate layers were formed. The flask assembly was equilibrated by agitation for at least 3 h in a constant temperature bath. Equilibrium solubility was attained by first supersaturation at a slightly lower temperature (solubility of alcohols in water decreases with increasing temperature) and then equilibrating at the desired temperature. The aqueous layer was separated after an overnight storage in a bath. The alcohol content was determined by reacting the aqueous solution with potassium dichromate and titrating the excess dichromate with ferrous sulfate solution in the presence of phosphoric acid and diphenylamine barium sulfonate as an indicator.

SOURCE AND PURITY OF MATERIALS:

(1) laboratory preparation; redistilled with 10:1 reflux ratio

b.p. 151.6-152°C/760.2 mm Hg

n_D^{25} = 1.41392

(2) twice distilled from silica apparatus or ion-exchanged with Sagei A20.

ESTIMATED ERROR:

Solubility: relative error of 2 determinations less than 1%.

Temperature: ±0.05°C

REFERENCES:

COMPONENTS:	EVALUATOR:
(1) 4-Methyl-2-pentanol *(isobutylmethylcarbinol)*; $C_6H_{14}O$; [108-11-2] (2) Water; H_2O; [7732-18-5]	A. Maczynski, Institute of Physical Chemistry of the Polish Academy of Sciences, Warsaw, Poland; and A.F.M. Barton, Murdoch University, Perth, Western Australia. November 1982

CRITICAL EVALUATION:

Solubilities in the system comprising 4-methyl-2-pentanol (1) and water (2) have been reported in four publications. Ginnings and Webb (ref 1) carried out measurements of the mutual solubilities of the two components at 293, 298 and 303 K by the volumetric method. Crittenden and Hixon (ref 2) used (presumably) the titration method at one temperature (298 K) for both phases and Ratouis and Dodé (ref 3) also determined only one temperature point (303 K) of the solubility in the water-rich phase by an analytical method. Dakshinamurty *et al.* (ref 4) carried out determination of the mutual solubility of (1) and (2) at 283, 301, and 323 K by the synthetic method.

The data on the water-rich phase are collected in Figure 1. The results of ref 1, 2, 3, and of ref 4 at 301 K are in reasonable agreement. The value at 300 K is recommended, with tentative values being proposed at 280 and 320 K.

Results on the alcohol-rich phase are in poor agreement, and the proposed value at 300 K is tentative only.

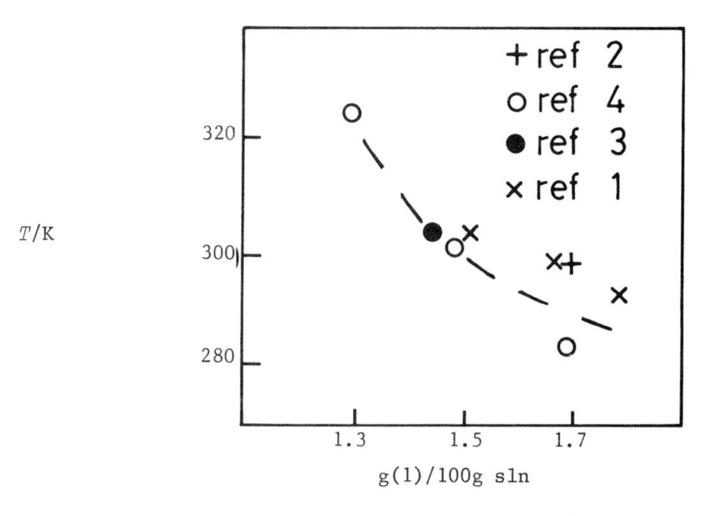

Fig. 1. Solubility of (1) in (2)

Recommended and tentative values for the mutual solubilities of 4-methyl-2-pentanol (1) and water (2)

T/K	Water-rich phase g(1)/100g sln	10^3x_1	Alcohol-rich phase g(2)/100g sln	x_2
280	2 (tentative)	4		
300	1.5 (recommended)	2.7	6 (tentative)	0.7
320	1 (tentative)	2		

(continued next page)

COMPONENTS:	EVALUATOR:
(1) 4-Methyl-2-pentanol (*isobutylmethylcarbinol*); $C_6H_{14}O$; [108-11-2] (2) Water; H_2O; [7732-18-5]	A. Maczynski, Institute of Physical Chemistry of the Polish Academy of Sciences, Warsaw, Poland; and A.F.M. Barton, Murdoch University, Perth, Western Australia. November 1982

CRITICAL EVALUATION: (continued)

References

1. Ginnings, P.M.; Webb, R. *J. Am. Chem. Soc.* 1938, *60*, 1388.

2. Grittenden, E.D.,Jr.; Hixon, A.N. *Ind. Eng. Chem.* 1954, *46*, 265.

3. Ratouis, M.; Dodé, M. *Bull. Soc. Chim. Fr,* 1965, 3318.

4. Dakshinamurty, P.; Chiranjivi, C.; Rao, P.V.; Subrahmanyam, V. *J. Chem. Eng. Data* 1972, *17*, 379.

COMPONENTS:

(1) 4-Methyl-2-pentanol *(isobutylmethylcarbinol)*; $C_6H_{14}O$; [108-11-2];

(2) Water; H_2O; [7732-18-5]

ORIGINAL MEASUREMENTS:

Ginnings, P.M.; Webb, R.

J. Am. Chem. Soc. 1938, *60*, 1388-9.

VARIABLES:

Temperature: 20-30°C

PREPARED BY:

A. Maczynski and Z. Maczynska

EXPERIMENTAL VALUES:

Mutual solubility of 4-methyl-2-pentanol(1) and water(2)

$t/^{\circ}C$	g(1)/100g sln		x_1 (compiler)	
	(2)-rich phase	(1)-rich phase	(2)-rich phase	(1)-rich phase
20	1.79	93.79	0.00320	0.7269
25	1.64	93.65	0.00293	0.7222
30	1.52	93.45	0.00271	0.7155

Relative density, d_4

$t/^{\circ}C$	Water-rich phase	Alcohol-rich phase
20	0.9956	0.8186
25	0.9948	0.8149
30	0.9939	0.8114

AUXILIARY INFORMATION

METHOD/APPARATUS/PROCEDURE:

The volumetric method was used as described in ref 1.

Both components were introduced in known amounts into a two-bulb graduated and calibrated flask and shaken mechanically in a water-bath at constant temperature. After sufficient time the liquids were allowed to separate and the total volume was measured. Upon centrifugation, the phase separation line was read, and phase volumes were calculated. From the total weights of the components, the total volume, individual phase volumes, and component concentrations in either phase were evaluated.

SOURCE AND PURITY OF MATERIALS:

(1) Eastman practical grade; fractionally distilled, and redistilled from calcium oxide; b.p. range 131.6-131.8°C/760 mm Hg d_4^{25} 0.8034.

(2) not specified.

ESTIMATED ERROR:

Temperature: ± 0.1°C
Solubility: better than 0.1 wt % (type of error not specified)

REFERENCES:

1. Ginnings, P.M.; Baum, R.J. *J. Am. Chem. Soc.* 1937, *59*, 1111.

COMPONENTS:

(1) 4-Methyl-2-pentanol *(isobutylmethylcarbinol)*; $C_6H_{14}O$; [108-11-2]

(2) Water; H_2O; [7732-18-5]

ORIGINAL MEASUREMENTS:

Crittenden, E.D., Jr.; Hixon, A.N.

Ind. Eng. Chem. 1954, *46*, 265-8.

VARIABLES:

One temperature: 25°C

PREPARED BY:

A. Maczynski

EXPERIMENTAL VALUES:

The solubility of 4-methyl-2-pentanol in water at 25°C was reported to be 1.7 g(1)/100g sln.

The corresponding mole fraction, x_1, calculated by the compiler is 0.0030.

The solubility of water in 4-methyl-2-pentanol at 25°C was reported to be 5.8 g(2)/100 sln.

The corresponding mole fraction, x_2, calculated by the compiler is 0.26

AUXILIARY INFORMATION

METHOD/APPARATUS/PROCEDURE:

Presumably the titration method described for ternary systems containing HCl was used. In this method the solubility was determined by bringing 100-mL samples of (1) or (2) to a temperature 25.0 ± 0.1°C and the second component was then added from a calibrated buret, with vigorous stirring, until the solution became permanently cloudy.

SOURCE AND PURITY OF MATERIALS:

(1) source not specified;
purified;
purity not specified.

(2) not specified.

ESTIMATED ERROR:

Solubility: 2% (alcohol-rich) - 10% (Water-rich)

Temperature: ± 0.10°C

REFERENCES:

COMPONENTS:	ORIGINAL MEASUREMENTS:
(1) 4-Methyl-2-pentanol *(isobutylmethylcarbinol)*; $C_6H_{14}O$; [108-11-2] (2) Water; H_2O; [7732-18-5]	Ratouis, M.; Dodé, M.; *Bull. Soc. Chim. Fr.* 1965, 3318-22.
VARIABLES:	**PREPARED BY:**
One temperature: 30°C Ringer solution also studied	S.C. Valvani; S.H. Yalkowsky; A.F.M. Barton

EXPERIMENTAL VALUES:

The proportion of 4-methyl-2-pentanol (1) in the water-rich phase at equilibrium at 30°C was reported to be 1.43 g(1)/100g sln.

The corresponding mole fraction solubility, calculated by the compilers, is x_1 = 0.00255.

The proportion of (1) in the water-rich phase of a mixture with Ringer solution at 30°C was reported to be 1.39 g(1)/100g sln.

AUXILIARY INFORMATION

METHOD/APPARATUS/PROCEDURE:

In a round bottomed flask, 50 mL of water and a sufficient quantity of alcohol were introduced until two separate layers were formed. The flask assembly was equilibrated by agitation for at least 3 h in a constant temperature bath. Equilibrium solubility was attained by first supersaturation at a slightly lower temperature (solubility of alcohols in water decreases with increasing temperature) and then equilibrating at the desired temperature. The aqueous layer was separated after an overnight storage in a bath. The alcohol content was determined by reacting the aqueous solution with potassium dichromate and titrating the excess dichromate with ferrous sulfate solution in the presence of phosphoric acid and diphenylamine barium sulfonate as an indicator.

SOURCE AND PURITY OF MATERIALS:

(1) Fluka A.G. Buchs S.G.
b.p. 129.6-129.7°C.
n_D^{25} = 1.40919

(2) twice distilled from silica apparatus or ion-exchanged with Sagei A20.

ESTIMATED ERROR:

Solubility: relative error of 2 determinations less than 1%.

Temperature: ±0.05°C

REFERENCES:

COMPONENTS:

(1) 4-Methyl-2-pentanol; $C_6H_{14}O$; [108-11-2]

(2) Water; H_2O; [7732-18-5]

ORIGINAL MEASUREMENTS:

Dakshinamurty, P.; Chiranjivi, C.; Rao, P.V.; Subrahmanyam, V.

J. Chem. Eng. Data, 1972, *17*, 379-83.

VARIABLES:

Temperature: 10-50°C

PREPARED BY:

A. Maczynski

EXPERIMENTAL VALUES:

Mutual solubility of 4-methyl-2-pentanol and water

$t/^{\circ}C$	g(1)/100g sln		x_1 (compiler)	
	(2)-rich phase	(1)-rich phase	(2)-rich phase	(1)-rich phase
10	1.7	96.7	0.0030	0.838
28	1.5	95.7	0.0027	0.797
50	1.3	94.7	0.0023	0.759

AUXILIARY INFORMATION

METHOD/APPARATUS/PROCEDURE:

The synthetic method was used.

Weighed mixtures of (1) and (2) were placed in a borosilicate glass test tube (20 ml) with a ground glass joint, and fitted with a thermometer. The test tube was slowly heated or cooled in a water bath and the temperature, corresponding to the appearance or disappearance of turbidity was noted. The end points were reproducible to ± 0.3°C either on heating or cooling. The solubilities at required temperatures were interpolated graphically.

SOURCE AND PURITY OF MATERIALS:

(1) British Drug House; distilled; b.p. range 131.5-132°C, n_D^{30} 1.4090, d^{30} 0.8003 g cm^{-3}

(2) not specified.

ESTIMATED ERROR:

Temperature: ± 0.5°C.
Solubility: not specified.

REFERENCES:

COMPONENTS:	EVALUATOR:
(1) 1-Hexanol (*n-hexyl alcohol*); $C_6H_{14}O$; [111-27-3] (2) Water; H_2O; [7732-18-5]	G.T. Hefter and A.F.M. Barton, Murdoch University, Perth, Western Australia. June, 1983

CRITICAL EVALUATION:

Solubilities in the 1-hexanol (1)-water (2) system have been reported in the following publications.

Reference	*T*/K	Phase	Method
Führer (ref 1)	273-378	(1) in (2)	synthetic
Butler *et al.* (ref 2)	298	(1) in (2)	interferometric
Addison (ref 3)	293	(1) in (2)	surface tension
Laddha and Smith (ref 4)	293	mutual	titration
Donahue (ref 5)	298	mutual	analytical
Erichsen (ref 6)	273-495	mutual	synthetic
Erichsen (ref 7)	273-323	(1) in (2)	synthetic
Crittenden and Hixon (ref 8)	298	mutual	titration
Kinoshita *et al.* (ref 9)	298	(1) in (2)	surface tension
Venkataratnam and Rao (ref 10)	303	mutual	turbidimetric
Ababi and Popa (ref 11)	298	mutual	turbidimetric
Chandy and Rao (ref 12)	303	mutual	turbidimetric
Hanssens (ref 13)	298	(1) in (2)	interferometric
Ratouis and Dodé (ref 14)	303	(1) in (2)	analytical
Krasnov and Gartseva (ref 15)	285-313	(2) in (1)	analytical
Vochten and Petre (ref 16)	288	(1) in (2)	surface tension
Hill and White (ref 17)	278-306	(1) in (2)	interferometric
Korenman *et al.* (ref 18)	298	(1) in (2)	analytical
Lavrova and Lesteva (ref 19)	313-333	mutual	titration
Filippov and Markuzin (ref 20)	294-313	(2) in (1)	not specified
Nishino and Nakamura (ref 21)	280-350	mutual	synthetic
Tokunaga *et al.* (ref 22)	288-308	(2) in (1)	analytical

The original data are compiled in the data sheets immediately following this Critical Evaluation.

In this Critical Evaluation the data of Hanssens (ref 13), Vochten and Petre (ref 16) and Korenman *et al.* (ref 18) given in weight/volume fractions are excluded from consideration as density information was not included in the original references. The graphical data of Nishino and Nakamura (ref 21) are also excluded.

In the water-rich phase the data of Addison (ref 3), Laddha and Smith (ref 4), Donahue (ref 5), Venkataratnam and Rao (ref 10), Ababi and Popa (ref 11) and Chandy and Rao (ref 12), all at either 298 or 303 K, disagree markedly from all other studies and are rejected.

In the alcohol-rich phase the low temperature (≤ 283 K) data of Erichsen (ref 6 & 7) and Krasnov and Gartseva (ref 15) disagree markedly and both are rejected. The data of Erichsen (ref 6 & 7) between 293 and 303 K are in poor agreement with all other studies (ref 5, 8, 1, 15, 20, 22) and have also been rejected as has the data of Ababi and Popa

(continued next page)

COMPONENTS:	EVALUATOR:
(1) Hexanol *(n-hexyl alcohol)*; $C_6H_{14}O$; [111-27-3] (2) Water; H_2O; [7732-18-5]	G.T. Hefter and A.F.M. Barton, Murdoch University, Perth, Western Australia. June, 1983

CRITICAL EVALUATION: (continued)

(ref 11). At higher temperatures Erichsen's values (ref 6 & 7) are in somewhat better agreement with other studies and have been included.

All other data are included in the Tables below. Values obtained by the Evaluators by graphical interpolation or extrapolation from the data sheets are indicated by an asterisk (*). "Best" values have been obtained by simple averaging. The uncertainty limits (σ_n) attached to the "best" values do not have statistical significance and should be regarded only as a convenient representation of the spread of reported values and not as error limits. The letter (*R*) indicates "Recommended" data. Data are "Recommended" if two or more apparently reliable studies are in reasonable agreement (≤ ± 5% relative).

For convenience the two phases will be further discussed separately.

The solubility of 1-hexanol (1) and water (2)

There is excellent agreement between the data of Führer (ref 1) and Erichsen (ref 6 & 7) over a very wide temperature range and also with other studies (ref 17 & 19) over more limited ranges (Figure 1). This has enabled values to be recommended over an unusually wide range of temperatures (see Table below).

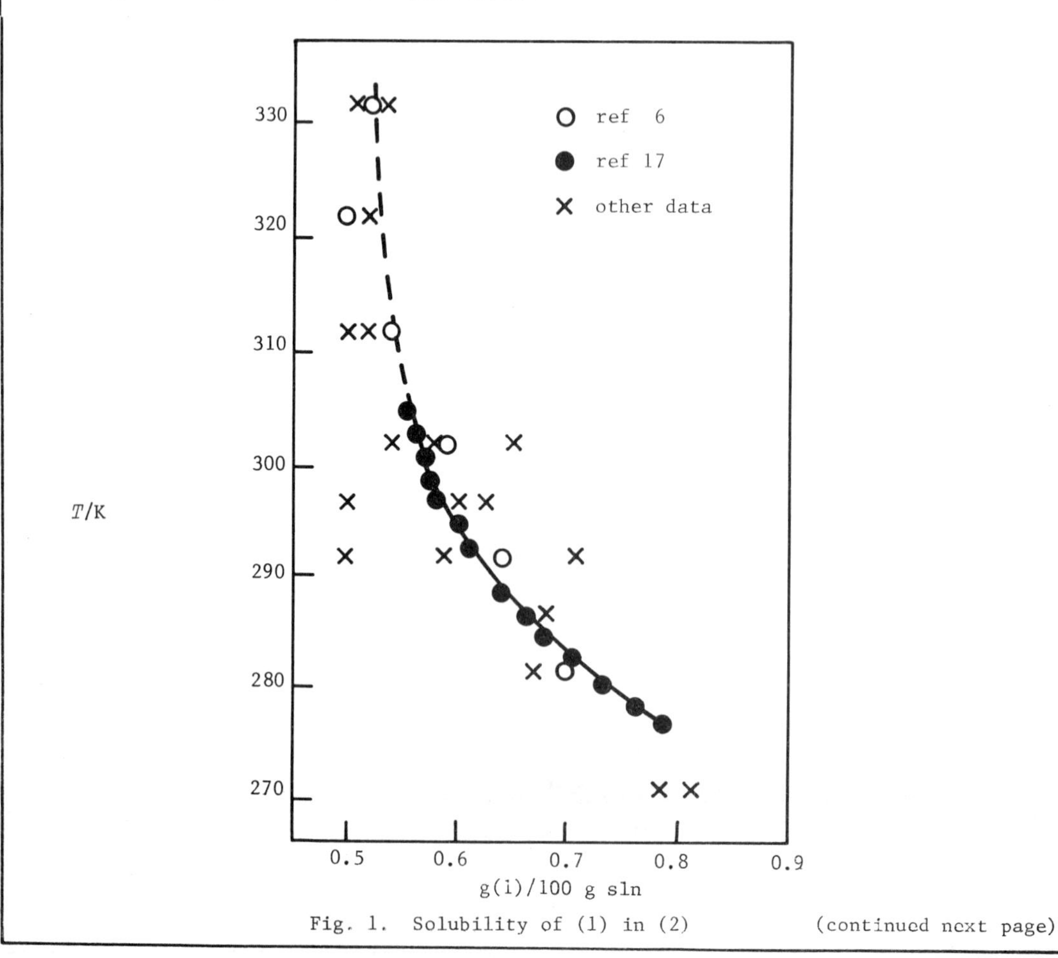

Fig. 1. Solubility of (1) in (2)

(continued next page)

COMPONENTS:	EVALUATOR:
(1) Hexanol (*n-hexyl alcohol*); $C_6H_{14}O$; [111-27-3] (2) Water; H_2O; [7732-18-5]	G.T. Hefter and A.F.M. Barton, Murdoch University, Perth, Western Australia June, 1983

CRITICAL EVALUATION (continued)

Recommended (R) and tentative solubilities of 1-hexanol (1) in water (2)

T/K	Solubility, g(1)/100g sln: Reported values	"Best" value ($\pm\sigma_n$)
273	0.78 (ref 1), 0.81 (ref 6), 0.79 (ref 7)	0.79 ± 0.01 (R)
283	0.67 (ref 1), 0.70 (ref 6), 0.68 (ref 7), 0.72* (ref 17)	0.69 ± 0.02 (R)
293	0.59 (ref 1), 0.64 (ref 6), 0.62 (ref 7), 0.62* (ref 17)	0.62 ± 0.02 (R)
298	0.57* (ref 1), 0.624 (ref 2), 0.62 (ref 6), 0.6 (ref 8), 0.60 (ref 9), 0.585* (ref 17)	0.60 ± 0.02 (R)
303	0.545 (ref 1), 0.59 (ref 6), 0.56 (ref 7), 0.58 (ref 14), 0.566 (ref 17)	0.57 ± 0.02 (R)
313	0.52 (ref 1), 0.54 (ref 6), 0.56 (ref 7), 0.55 (ref 17), 0.50 (ref 19)	0.53 ± 0.02 (R)
323	0.51 (ref 1), 0.50 (ref 6), 0.51 (ref 7)	0.51 ± 0.01 (R)
333	0.53 (ref 1), 0.52 (ref 6), 0.51 (ref 19)	0.52 ± 0.01 (R)
343	0.565 (ref 1), 0.56 (ref 6)	0.56 ± 0.01 (R)
353	0.62 (ref 1), 0.61 (ref 6)	0.62 ± 0.01 (R)
363	0.68 (ref 1), 0.69 (ref 6)	0.69 ± 0.01 (R)
373	0.785 (ref 1), 0.80 (ref 6)	0.79 ± 0.01 (R)
383	0.89 (ref 1), 0.91 (ref 6)	0.90 ± 0.01 (R)
393	1.04 (ref 6)	1.0
403	1.19 (ref 6)	1.2
413	1.37 (ref 6)	1.4
423	1.63 (ref 6)	1.6
433	2.05 (ref 6)	2.1
443	2.70 (ref 6)	2.7
453	3.61 (ref 6)	3.6
463	4.87 (ref 6)	4.9
473	6.75 (ref 6)	6.8
483	9.70 (ref 6)	9.7
493	16.30 (ref 6)	16.3

Solubility of water (2) in 1-hexanol (1)

Numerous studies (ref 5, 6, 8, 10, 12, 15, 19, 20, 22) are in good agreement over the range 298 - 313 K enabling values to be recommended. Agreement at all other temperatures is only fair and values are considered tentative only. In particular above 343 K only Erichsen's data (ref 6) are available and for reasons noted above they cannot be considered reliable without confirmatory studies.

COMPONENTS:	EVALUATOR:
(1) Hexanol (*n-hexyl alcohol*); $C_6H_{14}O$; [111-27-3]	G.T. Hefter and A.F.M. Barton, Murdoch University, Perth, Western Australia
(2) Water; H_2O; [7732-18-5]	June, 1983

CRITICAL EVALUATION: (continued)

Recommended (*R*) and tentative values of the solubility of water (2) in 1-hexanol (1)

T/K	Solubility, g(2)/100g sln	
	Reported values	"Best" values ($\pm\sigma_n$)
293	7.1* (ref 15), 6.2* (ref 20), 7.30 (ref 22)	6.9 ± 0.5
298	6.7 (ref 5), 7.2 (ref 8), 7.16 (ref 15), 6.7* (ref 20) 7.38 (ref 22)	7.0 ± 0.3 (*R*)
303	6.8 (ref 10), 7.0 (ref 12), 7.2* (ref 15), 6.9* (ref 20), 7.43 (ref 22)	7.1 ± 0.2 (*R*)
313	6.70 (ref 6), 7.3 (ref 15), 7.0 (ref 19), 7.2 (ref 20)	7.1 ± 0.2 (*R*)
323	7.35 (ref 6)	7.4
333	8.05 (ref 6, 9.1 (ref 19)	8.6 ± 0.5
343	8.80 (ref 6)	8.8
353	9.65 (ref 6)	9.7
363	10.45 (ref 6)	10.5
373	11.30 (ref 6)	11.3
383	12.25 (ref 6)	12.3
393	13.30 (ref 6)	13.3
403	14.50 (ref 6)	14.5
413	15.85 (ref 6)	15.9
423	17.50 (ref 6)	17.5
433	19.35 (ref 6)	19.4
443	21.50 (ref 6)	21.5
453	24.40 (ref 6)	24.4
463	28.20 (ref 6)	28.2
473	33.40 (ref 6)	33.4
483	41.25 (ref 6)	41.3
493	55.35 (ref 6)	55.4

The upper critical solution temperature

The UCST has been reported as 495.4 K (222.2 °C) by Erichsen (ref 6).

The "best" values for the mutual solubility of 1-hexanol and water, taken from the Tables above, are plotted in Figure 2.

(continued next page)

COMPONENTS:

(1) Hexanol (*n-hexyl alcohol*); $C_6H_{14}O$; [111-27-3]

(2) Water; H_2O; [7732-18-5]

EVALUATOR:

G.T. Hefter and A.F.M. Barton, Murdoch University, Perth, Western Australia.
June, 1983

CRITICAL EVALUATION: (continued)

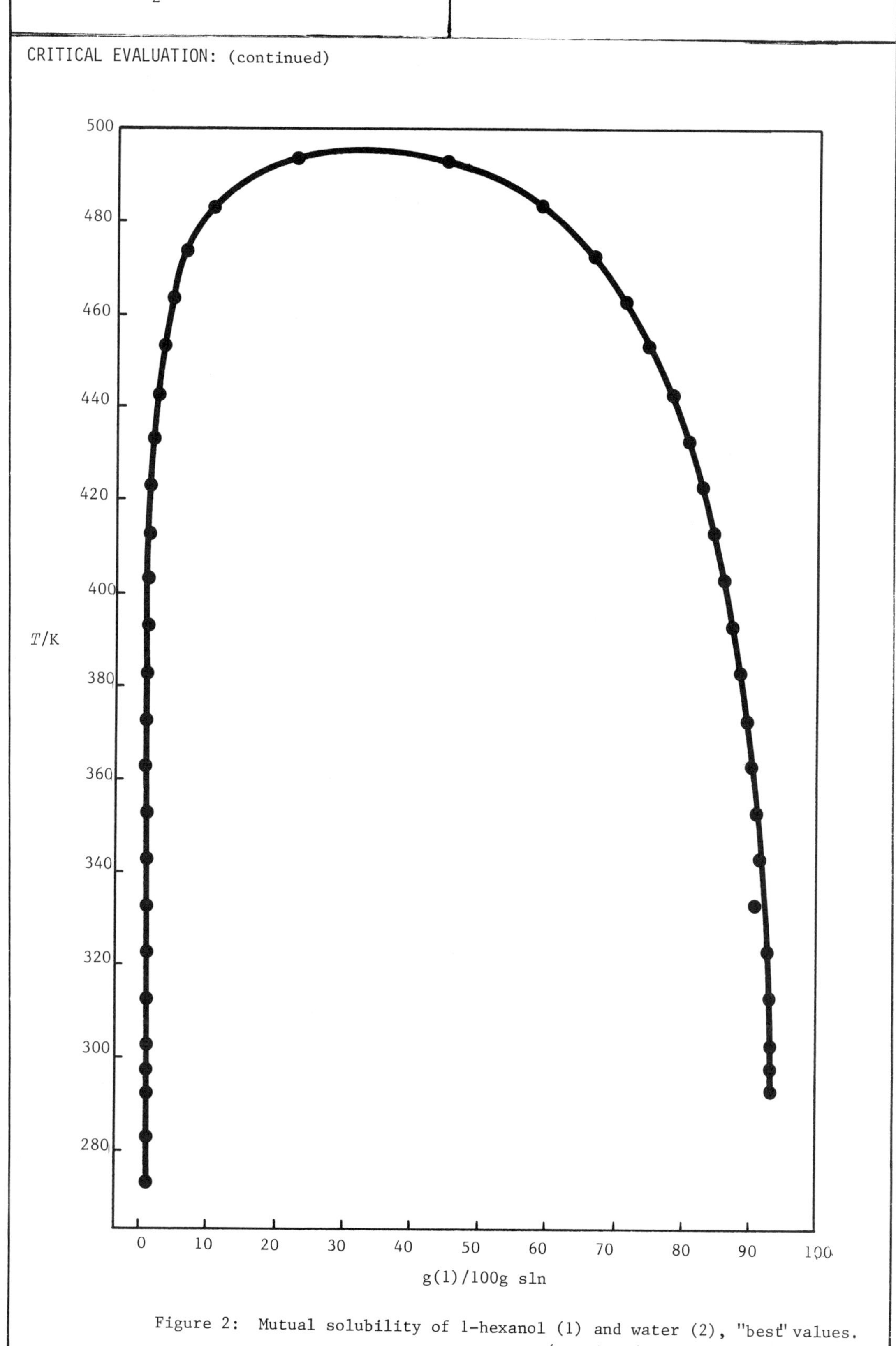

Figure 2: Mutual solubility of 1-hexanol (1) and water (2), "best" values.

(continued next page)

COMPONENTS:	EVALUATOR:
(1) Hexanol (*n-hexyl alcohol*); $C_6H_{14}O$; [111-27-3] (2) Water; H_2O; [7732-18-5]	G.T. Hefter and A.F.M. Barton, Murdoch University, Perth, Western Australia. June, 1983

CRITICAL EVALUATION (continued)

References

1. Fühner, H. *Ber. Dtsch. Chem. Ges.* 1924, *57*, 510
2. Butler, J.A.V.; Thomson, D.W.; Maclennan, W.H. *J. Chem. Soc.* 1933, 674.
3. Addison, C.C. *J. Chem. Soc.* 1945, 98.
4. Laddha, G.S.; Smith, J.M. *Ind. Eng. Chem.* 1948, *40*, 494.
5. Donahue, D.J.; Bartell, F.E. *J. Phys. Chem.* 1952, *56*, 480.
6. Erichsen, L. von. *Brennst. Chem.* 1952, *33*, 166
7. Erichsen, L. von. *Naturwissenschaften* 1952, *39*, 41.
8. Crittenden, E.D.,Jr.; Hixon, A.N. *Ind. Eng. Chem.* 1954, *46*, 265.
9. Kinoshita, K.; Ishikawa, H.; Shinoda, K. *Bull. Chem. Soc. Jpn.* 1958, *31*, 1081.
10. Venkataratnam, A.; Rao, R.I. *J. Sci. Ind. Res.* 1958, *17B*, 108.
11. Ababi, V.; Popa, A. *An. Stiint. Univ. "Al. I. Cuza" Iasi.* 1960, *6*, 929.
12. Chandy, C.A.; Rao, M.R. *J. Chem. Eng. Data* 1962, *7*, 473.
13. Hanssens, I. *Associatie van normale alcoholen en hun affinifeit voor water en organische solventen*, Doctoraatsproefschrift, Leuven, Belgium, 1969; Huyskens, P.; Mullens, J.; Gomez, A.; Tack, J. *Bull. Soc. Chim. Belg.* 1975, *84*, 253.
14. Ratouis, M.; Dodé, M. *Bull. Soc. Chim. Fr.* 1965, 3318.
15. Krasnov, K.S.; Gartseva, L.A. *Izv. Vysshykh Uchebn. Zavednii, Khim. Khim. Tekhnol.* 1970, *13*, 952.
16. Vochten, R.; Petre, G. *J. Colloid Interface Sci.* 1973, *42*, 320.
17. Hill, D.J.T.; White, L.R. *Aust. J. Chem.* 1974, *27*, 1905.
18. Korenman, I.M.; Gorokhov, A.A.; Polozenko, G.N. *Zh. Fiz. Khim.* 1974, *48*, 1810; *Russ. J. Phys. Chem.* 1974, *48*, 1065.
19. Lavrova, O.A.; Lesteva, T.M. *Zh. Fiz. Khim.* 1976, *50*, 1617; *Dep. Doc. VINITI* 3813-75.
20. Filippov, V.V.; Markuzin, N.P.; Sazonov, V.P. *Zh. Prikl. Khim.* 1977, *6*, 1321.
21. Nishino, N.; Nakamura, M. *Bull. Chem. Soc. Jpn.* 1978, *51*, 1617; *54*, 545.
22. Tokunaga, S.; Manabe, M.; Koda, M. *Niihama Kogyo Koto Semmon Gakko Kiyo, Rikogaku Hen (Memoirs Niihama Technical College, Sci. and Eng.)* 1980, *16*, 96.

COMPONENTS:

(1) 1-Hexanol; $C_6H_{14}O$; [111-27-3]

(2) Water; H_2O; [7732-18-5]

ORIGINAL MEASUREMENTS:

Fühner, H.

Ber. Dtsch. Chem. Ges. 1924, *57*, 510-5.

VARIABLES:

Temperature: 0-110°C

PREPARED BY:

A. Maczynski; Z. Maczynska; A. Szafranski.

EXPERIMENTAL VALUES:

Solubility of 1-hexanol (1) in water (2)

$t/°C$	g(1)/100g sln	x_1 (compiler)
0	0.78	0.00138
10	0.67	0.00119
20	0.59	0.00105
30	0.545	0.00096
40	0.52	0.00092
50	0.515	0.00091
60	0.53	0.00094
70	0.565	0.00100
80	0.62	0.00110
90	0.68	0.00121
100	0.785	0.00139
110	0.89	0.00158

AUXILIARY INFORMATION

METHOD/APPARATUS/PROCEDURE:

Rothmund's synthetic method (ref 1) was used.

Small amounts of (1) and (2) were sealed in a glass tube and heated with shaking in an oil bath to complete dissolution. The solution was cooled until a milky turbidity appeared and this temperature was adopted as the equilibrium temperature.

SOURCE AND PURITY OF MATERIALS:

(1) source not specified; specially purified, but no details provided.

(2) not specified.

ESTIMATED ERROR:

Not specified.

REFERENCES:

1. Rothmund, V. *Z. Physik. Chem.* 1898, *26*, 433.

COMPONENTS:	ORIGINAL MEASUREMENTS:
(1) 1-Hexanol; $C_6H_{14}O$; [111-27-3] (2) Water; H_2O; [7732-18-5]	Butler, J.A.V.; Thomson, D.W.; Maclennan, W.H. *J. Chem. Soc.* 1933, 674-86.
VARIABLES:	PREPARED BY:
One temperature: 25°C	S.H. Yalkowsky; S.C. Valvani; A.F.M. Barton

EXPERIMENTAL VALUES:

The proportion of 1-hexanol (1) in the water-rich phase at equilibrium at 25°C was reported to be 0.624 g(1)/100g sln, the mean of seven determinations (0.637, 0.625, 0.622, 0.615, 0.611, 0.627, 0.634 g(1)/100g sln).

The corresponding mole fraction solubility was reported as x_1 = 0.00111.

AUXILIARY INFORMATION

METHOD/APPARATUS/PROCEDURE:

An analytical method was used, with a U-tube apparatus having two internal stoppers. Suitable quantities of (1) and (2) were placed in one of the connected vessels and shaken in a thermostat for some hours. The liquid was allowed to separate into two layers, the heavier aqueous layer being separated by raising the stoppers and allowing part of the liquid to run into the connected vessel. A weighed portion of the separated sln was diluted with about an equal quantity of (2) and the resulting sln compared with calibration slns in an interferometer. To avoid the possibility of reading the position of the wrong fringe, 2 cells (1 cm and 5 cm) were used. The method was unsuitable for analysis of alcohol-rich slns, as no stoppered interferometer cell was available.

SOURCE AND PURITY OF MATERIALS:

(1) BDH;
repeatedly fractionated in vacuum with a Hempel column, the middle fractions being refluxed with Ca and refractionated;
b.p. 80.8-80.9°C/12 mm Hg, 155.7°C/760 mm Hg,
d_4^{25} 0.81648, n_D^{20} 1.41778

(2) not stated.

ESTIMATED ERROR:

Solubility: the result is the mean of seven determinations agreeing within 0.013 g(1)/100g sln.
Temperature: not stated (but in related experiments it was ± 0.03°C).

REFERENCES:

COMPONENTS:

(1) 1-Hexanol; $C_6H_{14}O$; [111-27-3]

(2) Water; H_2O; [7732-18-5]

ORIGINAL MEASUREMENTS:

Addison, C.C.

J. Chem. Soc. 1945, 98-106.

VARIABLES:

One temperature: 20°C

PREPARED BY:

S.H. Yalkowsky; S.C. Valvani; A.F.M. Barton

EXPERIMENTAL VALUES:

The proportion of 1-hexanol (1) in the water-rich phase at equilibrium at 20°C was reported to be 0.706 g(1)/100g sln.

The corresponding mole fraction solubility, calculated by the compilers, is x_1 = 0.00125.

AUXILIARY INFORMATION

METHOD/APPARATUS/PROCEDURE:

A surface tension method was used. Sufficient excess of (1) was added to 100 mL of (2) in a stoppered flask to form a separate lens on the surface. The mixture was swirled gently, too vigorous an agitation being avoided as this gave a semi-permanent emulsion and incorrect readings. After settling, a small sample of the clear aqueous solution was withdrawn into a drop weight pipet and the surface tension determined. The swirling was continued until a constant value was obtained. The surface tension - concentration curve was known, and only a slight extrapolation (logarithmic scale) was necessary to find the concentration corresponding to the equilibrium value.

SOURCE AND PURITY OF MATERIALS:

(1) impure alcohols were purified by fractional distillation, the middle fraction from a distillation being redistilled;

b.p. 157.5°C d_4^{20} 0.8194

n_D^{20} 1.4182

(2) not stated

ESTIMATED ERROR:

Solubility: ± 0.5%

REFERENCES:

COMPONENTS:	ORIGINAL MEASUREMENTS:
(1) 1-Hexanol; $C_6H_{14}O$; [111-27-3] (2) Water; H_2O; [7732-18-5]	Laddha, G.S.; Smith, J.M. *Ind. Eng. Chem.* 1948, *40*, 494-6.
VARIABLES:	PREPARED BY:
One temperature: 20°C	A. Maczynski

EXPERIMENTAL VALUES:

The solubility of 1-hexanol in water at 20°C was reported to be 0.5 g(1)/100g sln.

The corresponding mole fraction, x_1, calculated by the compiler is 9 x 10^{-4}.

The solubility of water in 1-hexanol at 20°C was reported to be 5.7 g(2)/100 sln.

The corresponding mole fraction, x_2 calculated by the compiler is 0.26.

AUXILIARY INFORMATION

METHOD/APPARATUS/PROCEDURE:

The titration method was used.

One component was placed in a 20°C constant-temperature bath for 1 h . Then titration was carried out in several steps, in order that the mixture could be frequently returned to the constant-temperature bath to ensure maintenance of the 20°C temperature. The end point was taken when turbidity appeared over the entire solution.

SOURCE AND PURITY OF MATERIALS:

(1) Carbide and Carbon Chemicals Co., technical grade; b.p. range 156.5-157°C, d^{20} 0.820.

(2) distilled.

ESTIMATED ERROR:

Not specified.

REFERENCES:

COMPONENTS:

(1) 1-Hexanol; $C_6H_{14}O$; [111-27-3]

(2) Water; H_2O; [7732-18-5]

ORIGINAL MEASUREMENTS:

Donahue, D.J.; Bartell, F.E.

J. Phys. Chem. 1952, *56*, 480-4.

VARIABLES:

One temperature: 25°C

PREPARED BY:

A.F.M. Barton

EXPERIMENTAL VALUES:

	Density	Mutual solubilities	
	g mL^{-1}	x_1	g(1)/100g sln (compiler)
Alcohol-rich phase	0.8284	0.712	93.3
Water-rich phase	0.9962	0.00110[a]	0.62

[a] From ref 1.

AUXILIARY INFORMATION

METHOD/APPARATUS/PROCEDURE:

Mixtures were placed in glass stoppered flasks and were shaken intermittently for at least 3 days in a water bath. The organic phase was analyzed for water content by the Karl Fischer method and the aqueous phase was analyzed interferometrically. The solubility measurements formed part of a study of water-organic liquid interfacial tensions.

SOURCE AND PURITY OF MATERIALS:

(1) "best reagent grade"; fractional distillation

(2) purified

ESTIMATED ERROR:

Temperature: ±0.1°C

REFERENCES:

(1) Butler, J.A.V.; Thomson, D.W.; Maclennan, W.H. *J. Chem. Soc.* 1933, 674

COMPONENTS:

(1) 1-Hexanol; $C_6H_{14}O$; [111-27-3]

(2) Water; H_2O; [7732-18-5]

ORIGINAL MEASUREMENTS:

Erichsen, L. von

Brennst. Chem. 1952, *33*, 166-72.

VARIABLES:

Temperature: 0-220 °C

PREPARED BY:

S.H. Yalkowsky and Z. Maczynska

EXPERIMENTAL VALUES:

Mutual solubility of 1-hexanol (1) and water (2)

$t/°C$	(2)-rich phase		(1)-rich phase	
	g(1)/100g sln	x_1	g(1)/100g sln	x_1
0	0.81	0.0014	95.55	0.7911
10	0.70	0.0012	95.05	0.7720
20	0.64	0.0011	94.50	0.7524
30	0.59	0.0010	93.95	0.7326
40	0.54	0.0010	93.30	0.7106
50	0.50	0.0009	92.65	0.6898
60	0.52	0.0009	91.95	0.6688
70	0.56	0.0010	91.20	0.6464
80	0.61	0.0011	90.35	0.6228
90	0.69	0.0012	89.55	0.6018
100	0.80	0.0014	88.70	0.5806
110	0.91	0.0016	87.75	0.5582
120	1.04	0.0019	86.70	0.5348
130	1.19	0.0021	85.50	0.5098
140	1.37	0.0024	84.15	0.4836
150	1.63	0.0027	82.50	0.4540
160	2.05	0.0037	80.65	0.4237
170	2.70	0.0050	78.50	0.3917
180	3.61	0.0066	75.60	0.3534
190	4.87	0.0090	71.80	0.3099
200	6.75	0.0128	66.60	0.2602
210	9.70	0.0189	58.75	0.2008
220	16.30	0.0342	44.65	0.1245

The UCST is 222.2 °C

AUXILIARY INFORMATION

METHOD/APPARATUS/PROCEDURE:

The synthetic method was used.

The measurements were carried out in 2 mL glass ampules. These were placed in an aluminium block equipped with two glass windows. Cloud points were measured with a thermocouple wound around the ampule. Each measurement was repeated twice.

SOURCE AND PURITY OF MATERIALS:

(1) Merck, or Ciba, or industrial product; distilled and chemically free from isomers; b.p. 156.4-156.5 °C (751 mm Hg) n_D^{20} 1.4184.

(2) not specified.

ESTIMATED ERROR:

Not specified.

REFERENCES:

COMPONENTS:

(1) 1-Hexanol; $C_6H_{14}O$; [111-27-3]

(2) Water; H_2O; [7732-18-5]

ORIGINAL MEASUREMENTS:

Erichsen, L. von

Naturwissenschaften 1952, *39*, 41-2.

VARIABLES:

Temperature: 0-50°C

PREPARED BY:

A. Maczynski and Z. Maczynska

EXPERIMENTAL VALUES:

Solubility of 1-hexanol (1) in water (2)

$t/^{\circ}C$	x_1	g(1)/100g sln (compiler)
0	0.0014	0.79
10	0.0012	0.68
20	0.0011	0.62
30	0.0010	0.56
40	0.0010	0.56
50	0.0009	0.51

AUXILIARY INFORMATION

METHOD/APPARATUS/PROCEDURE:

The synthetic method was used.

No details were reported in the paper.

SOURCE AND PURITY OF MATERIALS:

(1) not specified.

(2) not specified.

ESTIMATED ERROR:

Not specified.

REFERENCES:

COMPONENTS:

(1) 1-Hexanol; $C_6H_{14}O$; [111-27-3]

(2) Water; H_2O; [7732-18-5]

ORIGINAL MEASUREMENTS:

Crittenden, E.D., Jr.; Hixon, A.N.;

Ind. Eng. Chem. 1954, 46, 265-8.

VARIABLES:

One temperature: 25°C

PREPARED BY:

A. Maczynski

EXPERIMENTAL VALUES:

The solubility of 1-hexanol in water at 25°C was reported to be 0.6 g(1)/100g sln.

The corresponding mole fraction, x_1, calculated by the compiler is 0.0011.

The solubility of water in 1-hexanol at 25°C was reported to be 7.2 g(2)/100g sln.

The corresponding mole fraction, x_2, calculated by the compiler is 0.31.

AUXILIARY INFORMATION

METHOD/APPARATUS/PROCEDURE:

Presumably the titration method described for ternary systems containing HCl was used. In this method the solubility was determined by bringing 100-mL samples of (1) or (2) to a temperature 25.0 ± 0.1°C and the second component was then added from a calibrated buret, with vigorous stirring, until the solution became permanently cloudy.

SOURCE AND PURITY OF MATERIALS:

(1) source not specified;
purified;
purity not specified.

(2) not specified.

ESTIMATED ERROR:

Temperature: ± 0.10°C.
Solubility: 2% (alcohol-rich)-10% (water-rich)

REFERENCES:

COMPONENTS:

(1) 1-Hexanol; $C_6H_{14}O$; [111-27-3]

(2) Water; H_2O; [7732-18-5]

ORIGINAL MEASUREMENTS:

Kinoshita, K.; Ishikawa, H.; Shinoda, K.

Bull. Chem. Soc. Jpn. 1958, *31*, 1081-4.

VARIABLES:

One temperature: 25°C

PREPARED BY:

S.H. Yalkowsky; S.C. Valvani; A.F.M. Barton

EXPERIMENTAL VALUES:

The equilibrium concentration of 1-hexanol (1) in the water-rich phase at 25°C was reported to be 0.059 mol(1)/L sln. The mass percentage solubility was reported as 0.60 g(1)/100g sln, and the corresponding mole fraction solubility, calculated by the compilers, is x_1 = 0.00106.

AUXILIARY INFORMATION

METHOD/APPARATUS/PROCEDURE:

The surface tension in aqueous solutions of alcohols monotonously decreases up to their saturation concentration and remains constant in the heterogeneous region (ref 1-4). Surface tension was measured by the drop weight method, using a tip 6 mm in diameter, the measurements being carried out in a water thermostat. From the (surface tension) - (logarithm of concentration) curves the saturation points were determined as the intersections of the curves with the horizontal straight lines passing through the lowest experimental points.

SOURCE AND PURITY OF MATERIALS:

(1) purified by vacuum distillation through 50-100 cm column;

b.p. 155.7 - 156°C

(2) not stated

ESTIMATED ERROR:

Temperature: ± 0.05%

Solubility: within 4%

REFERENCES:

(1) Motylewski, S. *Z. Anorg. Chem.* 1904, *38*, 410

(2) Taubamann, A. *Z. physik. Chem.* 1932, *A161*, 141

(3) Zimmerman, H.K, Jr. *Chem. Rev.* 1952, *51*, 25

(4) Shinoda, K.; Yamanaka, T.; Kinoshita, K. *J. Phys. Chem.* 1959, *63*, 648

COMPONENTS:

(1) 1-Hexanol; $C_6H_{14}O$; [111-27-3]

(2) Water; H_2O; [7732-18-5]

ORIGINAL MEASUREMENTS:

Venkataratnam, A.; Rao, R.I.

J. Sci. Ind. Res. 1958, *17B*, 108-10.

VARIABLES:

One temperature: 30°C

PREPARED BY:

A. Maczynski

EXPERIMENTAL VALUES:

The solubility of 1-hexanol in water at 30°C was reported to be 0.4 g(1)/100 sln.

The corresponding mole fraction, x_1, calculated by the compiler is 7×10^{-4}.

The solubility of water in 1-hexanol at 30°C was reported to be 6.8 g(2)/100g sln.

The corresponding mole fraction, x_2, calculated by the compiler is 0.29.

AUXILIARY INFORMATION

METHOD/APPARATUS/PROCEDURE:

The method of appearance and disappearance of turbidity described in ref 1 was used.

No details were reported in the paper.

SOURCE AND PURITY OF MATERIALS:

(1) source not specified;
distilled;
b.p. 157°C, n^{30} 1.4188, d^{30} 0.8179 g/ml

(2) distilled;
free from carbon dioxide.

ESTIMATED ERROR:

Not specified.

REFERENCES:

1. Othmer, D.F.; White, R.E.; Trueger, E. *Ind. Eng. Chem.* 1941, *33*, 1240.

COMPONENTS:

(1) 1-Hexanol; $C_6H_{14}O$; [111-27-3]

(2) Water; H_2O; [7732-18-5]

ORIGINAL MEASUREMENTS:

Ababi, V.; Popa, A.

An. Stiint. Univ. "Al. I. Cuza" Iasi. 1960, *6*, 929-42.

VARIABLES:

One temperature: 25°C

PREPARED BY:

A. Maczynski

EXPERIMENTAL VALUES:

The solubility of 1-hexanol in water at 25°C was reported to be 0.5 g(1)/100g sln.

The corresponding mole fraction, x_1, calculated by the compiler is 9×10^{-4}.

The solubility of water in 1-hexanol at 25°C was reported to be 7.9 g(2)/100g sln.

The corresponding mole fraction, x_2, calculated by the compiler is 0.33.

AUXILIARY INFORMATION

METHOD/APPARATUS/PROCEDURE:

The turbidimetric method was used.

Ternary solubilities were described in the paper but nothing was reported on the method for binary solubilities.

SOURCE AND PURITY OF MATERIALS:

(1) Merck analytical reagent; used as received.

(2) not specified.

ESTIMATED ERROR:

Not specified.

REFERENCES:

COMPONENTS:

(1) 1-Hexanol; $C_6H_{14}O$; [111-27-3]

(2) Water; H_2O; [7732-18-5]

ORIGINAL MEASUREMENTS:

Chandy, C.A.; Rao, M.R.

J. Chem. Eng. Data 1962, *7*, 473-5.

VARIABLES:

One temperature: 30°C

PREPARED BY:

A. Maczynski

EXPERIMENTAL VALUES:

The solubility of 1-hexanol in water at 30°C was reported to be 0.65 g(1)/100g sln.

The corresponding mole fraction, x_1, calculated by the compiler is 1.1×10^{-3}.

The solubility of water in 1-hexanol at 30°C was reported to be 7.0 g(2)/100g sln.

The corresponding mole fraction, x_2, calculated by the compiler is 0.30.

AUXILIARY INFORMATION

METHOD/APPARATUS/PROCEDURE:

The method of appearance and disappearance of turbidity described in ref 1 was used.

No details were reported in the paper.

SOURCE AND PURITY OF MATERIALS:

(1) Jean A. du Crocq, Jr. Ltd., (Holland); distilled; b.p. range 156-156.5°C.

(2) distilled; free from carbon dioxide.

ESTIMATED ERROR:

Not specified.

REFERENCES:

1. Othmer, D.F.; White, R.E.; Trueger, E. *Ind. Eng. Chem.* 1941, 33, 1240.

COMPONENTS:

(1) 1-hexanol; $C_6H_{14}O$; [111-27-3]

(2) Water; H_2O; [7732-18-5]

ORIGINAL MEASUREMENTS:

*Hanssens, I.
Associatie van normal alcoholen en hun affiniteit voor water en organische solventen
Doctoraatsproefschrift, Leuven, 1969.

Huyskens, P.; Mullens, J.; Gomez, A.; Tack, J.
Bull. Soc. Chim. Belg. 1975, *84*, 253-62.

VARIABLES:

One temperature; 298 K

PREPARED BY:

M.C. Haulait-Pirson; A.F.M. Barton

EXPERIMENTAL VALUES:

The concentration of 1-hexanol (1) in the water-rich phase was reported as 0.0581 mol(1)/L sln, (0.0585 in the 1975 published paper), and the concentration of water (2) in the alcohol-rich phase was reported as 7.493 mol(2)/L sln.

The corresponding solubilities on a mass/volume basis, calculated by the compilers, are 5.9 g(1)/L sln, and 135.0 g(2)/L sln, respectively.

(The temperature was unspecified in the Thesis, but reported as 298 K in the 1975 published paper).

AUXILIARY INFORMATION

METHOD/APPARATUS/PROCEDURE:

(1) and (2) were equilibrated using a cell described in ref 1. The Rayleigh M75 interference refractometer with the cell M160 for liquids was used for the determination of the concentrations. Cell thicknesses were 1, 3 and 10 cm depending on the concentration range. Standard solutions covering the whole range of concentrations investigated were used for the calibration.

SOURCE AND PURITY OF MATERIALS:

(1) Merck p.a.

(2) distilled

ESTIMATED ERROR:

Solubility: ± 0.00036 - 0.05 mol(1)/L sln, depending on the concentration

REFERENCES:

1. Meeussen, E.; Huyskens, P.
J. Chim. Phys. 1966, *63*, 845.

COMPONENTS:

(1) 1-Hexanol; $C_6H_{14}O$; [111-27-3]

(2) Water; H_2O; [7732-18-5]

ORIGINAL MEASUREMENTS:

Ratouis, M.; Dodé, M.;

Bull. Soc. Chim. Fr. 1965, 3318-22.

VARIABLES:

One temperature: 30°C

Ringer solution also studied

PREPARED BY:

S.C. Valvani; S.H. Yalkowsky; A.F.M. Barton

EXPERIMENTAL VALUES:

The proportion of 1-hexanol (1) in the water-rich phase at equilibrium at 30°C was reported to be 0.58 g(1)/100g sln.

The corresponding mole fraction solubility, calculated by the compilers, is x_1 = 0.00103.

The proportion of (1) in the water-rich phase of a mixture with Ringer solution at 30°C was reported to be 0.53 g(1)/100g sln.

AUXILIARY INFORMATION

METHOD/APPARATUS/PROCEDURE:

In a round bottomed flask, 50 mL of water and a sufficient quantity of alcohol were introduced until two separate layers were formed. The flask assembly was equilibrated by agitation for at least 3 h in a constant temperature bath. Equilibrium solubility was attained by first supersaturation at a slightly lower temperature (solubility of alcohols in water decreases with increasing temperature) and then equilibrating at the desired temperature. The aqueous layer was separated after an overnight storage in a bath. The alcohol content was determined by reacting the aqueous solution with potassium dichromate and titrating the excess dichromate with ferrous sulfate solution in the presence of phosphoric acid and diphenylamine barium sulfonate as an indicator.

SOURCE AND PURITY OF MATERIALS:

(1) Fluka A.G. Buchs S.G.; redistilled with 10:1 reflux ratio;

b.p. 157°C/760 mm Hg

n_D^{25} = 1.41607

(2) twice distilled from silica apparatus or ion-exchanged with Sagei A20.

ESTIMATED ERROR:

Solubility: relative error of 2 determinations less than 1%.

Temperature: ±0.05°C.

REFERENCES:

COMPONENTS:

(1) 1-Hexanol; $C_6H_{14}O$; [111-27-3]

(2) Water; H_2O; [7732-18-5]

ORIGINAL MEASUREMENTS:

Krasnov, K.S., Gartseva, L.A.

Izv. Vysshykh Uchebn. Zavedenii, Khim. Khim. Tekhnol. 1970, *13*, 952-6.

VARIABLES:

Temperature: 12-40 °C

PREPARED BY:

A. Maczynski and Z. Maczynska

EXPERIMENTAL VALUES:

Solubility of water (2) in 1-hexanol (1)

t/°C	g(2)/100g sln	x_2(compiler)
12	7.12	0.303
25	7.16	0.304
40	7.31	0.309

AUXILIARY INFORMATION

METHOD/APPARATUS/PROCEDURE:

The analytical method was used.

A saturated mixture of (1) and (2) was placed in a thermostat and the phases were allowed to separate. Then (2) was determined in the organic layer by Karl Fischer analysis.

SOURCE AND PURITY OF MATERIALS:

(1) CP reagent;
source not specified;
distilled;
no isomers by GLC;
d_4^{25} 0.8147.

(2) not specified

ESTIMATED ERROR:

Temperature: ± 0.05 °C
Solubility: ± 0.05 wt %
(type of error not specified)

REFERENCES:

COMPONENTS:

(1) 1-Hexanol; $C_6H_{14}O$; [111-27-3]

(2) Water; H_2O; [7732-18-5]

ORIGINAL MEASUREMENTS:

Vochten, R.; Petre, G.

J. Colloid Interface Sci. 1973, *42*, 320-7.

VARIABLES:

One temperature: 15°C

PREPARED BY:

S.H. Yalkowsky; S.C. Valvani; A.F.M. Barton

EXPERIMENTAL VALUES:

The equilibrium concentration of 1-hexanol (1) in the water-rich phase at 15°C was reported to be 0.066 mol(1)/L sln.

The corresponding mass/volume solubility, calculated by the compilers, is 6.75 g(1)/L sln.

AUXILIARY INFORMATION

METHOD/APPARATUS/PROCEDURE:

The solubility was obtained from the surface tension of saturated solutions, measured by the static method of Wilhelmy (platinum plate). The apparatus consisted of an electrobalance (R.G. Cahn) connected with a high impedance null detector (Fluke type 845 AR). An all-Pyrex vessel was used.

SOURCE AND PURITY OF MATERIALS:

(1) purified by distillation and preparative gas chromatography; b.p. 157.0°C/760 mm Hg

(2) triply distilled from permanganate solution

ESTIMATED ERROR:

Temperature: ± 0.1°C

Solubility: (probably standard deviation) ± 0.001 mol(1) /L sln.

REFERENCES:

COMPONENTS:

(1) 1-Hexanol; $C_6H_{14}O$; [111-27-3]

(2) Water; H_2O; [7732-18-5]

ORIGINAL MEASUREMENTS:

Hill, D.J.T.; White, L.R.

Aust. J. Chem., 1974, *27*, 1905-16.

VARIABLES:

Temperature: 279-306 K

PREPARED BY:

A. Maczynski

EXPERIMENTAL VALUES:

Solubility of 1-hexanol (1) in water (2)

T/K	x_1	g(1)/100g sln (compiler)
278.66	0.001398	0.7879
280.00	0.001358	0.7655
281.83	0.001308	0.7375
284.15	0.001252	0.7061
286.09	0.001209	0.6820
287.83	0.001179	0.6651
290.19	0.001139	0.6427
293.86	0.001085	0.6124
296.14	0.001060	0.5984
298.16	0.001034	0.5838
300.14	0.001021	0.5764
302.09	0.001010	0.5702
304.07	0.0009986	0.5639
306.24	0.0009846	0.5560

AUXILIARY INFORMATION

METHOD/APPARATUS/PROCEDURE:

The interferometric method was used.

The saturated solutions of (1) were prepared in an equilibrium flask and solution concentrations were determined from their refractive index using a Zeiss interferometer and appropriate calibration curves. Duplicate determinations were always within the error limits calculated from the curve fit to the calibration measurements (0.5% at the 95% confidence level).

Numerous technical details were reported in the paper.

SOURCE AND PURITY OF MATERIALS:

(1) Fluka puriss grade; dried over molecular sieves and purified by vacuum distillation; b.p. 335.7 K (20 mm Hg), n_D^{25} 1.4150, 99.5% purity.

(2) freshly double-distilled.

ESTIMATED ERROR:

Temperature: ± 0.02°C
Solubility: < 0.5% (accuracy at 95% confidence level)

REFERENCES:

COMPONENTS:

(1) 1-Hexanol; $C_6H_{14}O$; [111-27-3]

(2) Water; H_2O; [7732-18-5]

ORIGINAL MEASUREMENTS:

Korenman, I.M.; Gorokhov, A.A.; Polozenko, G.N.

Zhur. Fiz. Khim. 1974, *48*, 1810-2;

**Russ. J. Phys. Chem.* 1974, *48*, 1065-7.

VARIABLES:

One temperature: 25°C

PREPARED BY:

A.F.M. Barton

EXPERIMENTAL VALUES:

The concentration of 1-hexanol (1) in the water-rich phase at equilibrium at 25.0°C was reported to be 0.058 mol(1)/L sln and the concentration of water (2) in the alcohol-rich phase was reported to be 3.59 mol(2)/L sln.

The corresponding solubilities on a mass/volume basis, calculated by the compiler, are 5.9 g(1)/L sln and 64.7 g(2)/L sln respectively.

AUXILIARY INFORMATION

METHOD/APPARATUS/PROCEDURE:

The two liquids were shaken in a closed vessel at 25.0±0.1°C until equilibrium was established. The soly of the alcohol in the aqueous phase was determined on a Tsvet-1 chromatograph with a flame-ionisation detector. The sorbent was a polyethylene glycol adipate deposited on Polychrom-1 (10% of the mass of the carrier). The 1 m column had an internal diameter 4 mm, its temp 140°C, and the flow of the carrier gas (nitrogen) 50 mL min^{-1}. The soly of water in the alcohol was determined on a UKh-2 universal chromatograph under isothermal conditions (150°C) with a heat-conductivity detector. The 1 m by 6 mm column was filled with Polysorb. The carrier gas was helium (50 mL min^{-1}). The study formed part of an investigation of salting-out by alkali halides of higher alcohol-water systems.

SOURCE AND PURITY OF MATERIALS:

Not stated

ESTIMATED ERROR:

Temperature: ±0.1°C

Solubility: not stated; the results reported are the arithmetic means from four sets of experiments.

REFERENCES:

COMPONENTS:	ORIGINAL MEASUREMENTS:
(1) 1-Hexanol; $C_6H_{14}O$; [111-27-3] (2) Water; H_2O; [7732-18-5]	Lavrova, O.A.; Lesteva, T.M. *Zh. Fiz. Khim.*, 1976, *50*, 1617. *Dep. Doc. VINITI*, 3813-75
VARIABLES:	PREPARED BY:
Temperature: 40 and 60°C	A. Maczynski

EXPERIMENTAL VALUES:

Mutual solubility of 1-hexanol (1) and water (2)

$t/^{\circ}C$	g(1)/100g sln		x_1 (compiler)	
	(2)-rich phase	(1)-rich phase	(2)-rich phase	(1)-rich phase
40	0.50	93.0	0.0008	0.701
60	0.51	90.9	0.00090	0.638

AUXILIARY INFORMATION

METHOD/APPARATUS/PROCEDURE:

The titration method was used.

No details were reported in the paper.

SOURCE AND PURITY OF MATERIALS:

(1) source not specified; distilled with heptane; purity 99.95 wt % with 0.05 wt % water, n_D^{20} 1.4133, b.p. 157.0°C, d_4^{20} 0.8186

(2) not specified.

ESTIMATED ERROR:

Not specified.

REFERENCES:

COMPONENTS:

(1) 1-Hexanol; $C_6H_{14}O$; [111-27-3]

(2) Water; H_2O; [7732-18-5]

ORIGINAL MEASUREMENTS:

Filippov, V.V.; Markuzin, N.P.; Sazonov, V.P.

Zh. Prikl. Khim. 1977, *6*, 1321-4.

VARIABLES:

Temperature: 21-40°C

PREPARED BY:

A. Maczynski and Z. Maczynska

EXPERIMENTAL VALUES:

Solubility of water in 1-hexanol

$t/^\circ C$	x_2	g(2)/100g sln
21.0	0.276	6.30
23.0	0.283	6.50
40.0	0.305	7.18

AUXILIARY INFORMATION

METHOD/APPARATUS/PROCEDURE:

Not specified

SOURCE AND PURITY OF MATERIALS:

(1) pure grade reagent;
purified;
purity not specified.

(2) twice distilled

ESTIMATED ERROR:

Not specified.

REFERENCES:

COMPONENTS:

(1) 1-Hexanol; $C_6H_{14}O$; [111-27-3]

(2) Water; H_2O; [7732-18-5]

ORIGINAL MEASUREMENTS:

Nishino, N.; Nakamura, M.

Bull. Chem. Soc. Jpn. 1978, *51*, 1617-20; 1981, *54*, 545-8.

VARIABLES:

Temperature: 275-360 K

PREPARED BY:

G.T. Hefter

EXPERIMENTAL VALUES:

The mutual solubility of (1) and (2) in mole fractions are reported over the temperature range in graphical form. Graphical data are also presented for the heat of evaporation of (1).

AUXILIARY INFORMATION

METHOD/APPARATUS/PROCEDURE:

The turbidimetric method was used. Twenty to thirty glass ampoules containing aqueous solutions of *ca.* 5 cm^3 of various concentrations near the solubility at room temperature were immersed in a water thermostat. The distinction between clear and turbid ampoules was made after equilibrium was established (*ca.* 2h). The smooth curve drawn to separate the clear and turbid regions was regarded as the solubility curve.

SOURCE AND PURITY OF MATERIALS:

(1) G.R. grade (various commercial sources given); dried over calcium oxide; kept in ampoules over magnesium powder.

(2) Deionized, refluxed for 15h with potassium permanganate then distilled.

ESTIMATED ERROR:

Not stated

REFERENCES:

COMPONENTS:

(1) 1-Hexanol; $C_6H_{14}O$; [111-27-3]

(2) Water; H_2O; [7732-18-5]

ORIGINAL MEASUREMENTS:

Tokunaga, S.; Manabe, M.; Koda, M.
Niihama Kogyo Koto Semmon Gakko Kiyo, Rikogaku Hen (Memoirs Niihama Technical College, Sci. and Eng.) 1980, *16*, 96-101.

VARIABLES:

Temperature: 15-35°C

PREPARED BY:

A.F.M. Barton

EXPERIMENTAL VALUES:

Solubility of water (2) in the alcohol-rich phase

t/°C	g(2)/100g sln	x_2	mol(1)/mol(2)
15	7.13	0.303	2.29
20	7.30	0.309	2.25
25	7.38	0.311	2.22
30	7.43	0.313	2.19
35	7.57	0.317	2.16

AUXILIARY INFORMATION

METHOD/APPARATUS/PROCEDURE:

The mixtures of 1-hexanol (-5mL) and water (-10mL) were stirred magnetically in a stoppered vessel and allowed to stand for 10-12h in a water thermostat. The alcohol-rich phase was analyzed for water by Karl Fischer titration.

SOURCE AND PURITY OF MATERIALS:

(1) distilled;
no impurities detectable by gas chromatography

(2) deionized;
distilled prior to use

ESTIMATED ERROR:

Temperature: ± 0.1°C

Solubility: each result is the mean of three determinations.

REFERENCES:

COMPONENTS:	EVALUATOR:
(1) 2-Hexanol (*n-butylmethylcarbinol*); $C_6H_{14}O$; [626-93-7] (2) Water; H_2O; [7732-18-5]	Z. Maczynska, Institute of Physical Chemistry of the Polish Academy of Sciences, Warsaw, Poland; and A.F.M. Barton, Murdoch University, Perth, Western Australia. November 1982.

CRITICAL EVALUATION:

Solubilities in the system comprising 2-hexanol (1) and water (2) have been reported in four publications. Ginnings and Webb (ref 1) carried out measurements of the mutual solubilities in the two phases at 293, 298 and 303 K by the volumetric method (Figures 1 and 2). Ratouis and Dodé (ref 2) determined the solubility of (1) in the water-rich phase at one temperature (303 K) by an analytical method. Their value of 1.19 g(1)/100g sln is in good agreement with the value 1.28 ± 0.1 g(1)/100g sln of ref 1. Nishino and Nakamura provided graphical information only of the solubility of (1) in (2) (ref 3) and (2) in (1) (ref 4). Since direct comparison is possible for only a single data point, the data of ref 1 are regarded as tentative.

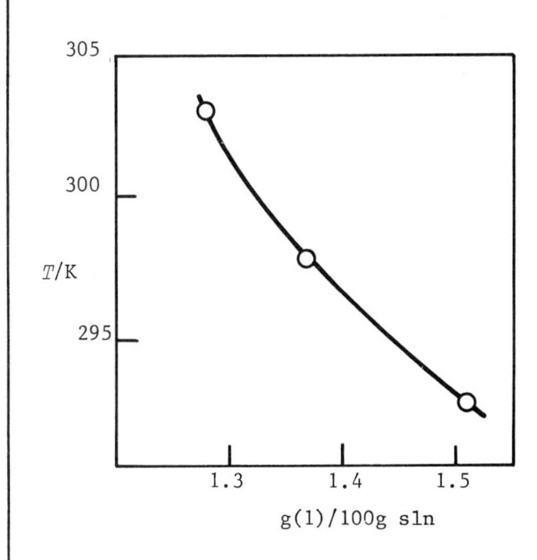

Fig 1. Water-rich phase (ref 1)

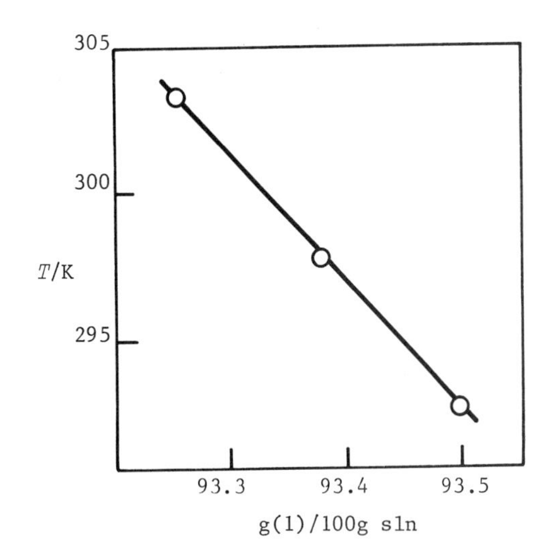

Fig. 2. Alcohol-rich phase (ref 1)

Tentative values for the mutual solubilities of 2-hexanol (1) and water (2)

T/K	Water-rich phase		Alcohol-rich phase	
	g(1)/100g sln	10^3x_1	g(2)/100g sln	x_2
293	1.5	2.7	6.5	0.283
298	1.4	2.4	6.6	0.287
303	1.3	2.3	6.7	0.292

References

1. Ginnings, P.M.; Webb, R. *J. Am. Chem. Soc.* 1938, *60*, 1388.
2. Ratouis, M.; Dodé, M. *Bull. Soc. Chim. Fr.* 1965, 3318.
3. Nishino, N.; Nakamura, M. *Bull. Chem. Soc. Jpn.* 1978, *51*, 1617.
4. Nishino, N.; Nakamura, M. *Bull. Chem. Soc. Jpn.* 1981, *54*, 545.

COMPONENTS:

(1) 2-Hexanol *(n-butylmethylcarbinol)*; $C_6H_{14}O$ [623-93-7]

(2) Water; H_2O; [7732-18-5]

ORIGINAL MEASUREMENTS:

Ginnings, P.M.; Webb, R.

J. Am. Chem. Soc. 1938, *60*, 1388-9.

VARIABLES:

Temperature: 20-30°C

PREPARED BY:

A. Maczynski and Z. Maczynska

EXPERIMENTAL VALUES:

Mutual solubility of 2-hexanol(1) and water(2)

$t/^{\circ}C$	g(1)/100g sln		x_1 (compiler)	
	(2)-rich phase	(1)-rich phase	(2)-rich phase	(1)-rich phase
20	1.51	93.50	0.00270	0.7171
25	1.37	93.38	0.00244	0.7132
30	1.28	93.25	0.00228	0.7089

Relative density, d_4

$t/^{\circ}C$	Water-rich phase	Alcohol-rich phase
20	0.9965	0.8264
25	0.9954	0.8231
30	0.9942	0.8194

AUXILIARY INFORMATION

METHOD/APPARATUS/PROCEDURE:

The volumetric method was used as described in ref 1.

Both components were introduced in known amounts into a two-bulb graduated and calibrated flask and shaken mechanically in a water-bath at constant temperature. After sufficient time the liquids were allowed to separate and the total volume was measured. Upon centrifugation, the phase separation line was read, and phase volumes were calculated. From the total weights of the components, the total volume, individual phase volumes, and component concentrations in either phase were evaluated.

SOURCE AND PURITY OF MATERIALS:

(1) Eastman best grade; distilled, and redistilled from calcium oxide; b.p. range 139.0-140.0°C, d_4^{25} 0.8108,

(2) not specified.

ESTIMATED ERROR:

Temperature: ± 0.1°C
Solubility: better than 0.1 wt % (type of error not specified)

REFERENCES:

1. Ginnings, P.M.; Baum, R.J. *J. Am. Chem. Soc.* 1937, *59*, 1111.

COMPONENTS:

(1) 2-Hexanol *(n-butylmethylcarbinol)*; $C_6H_{14}O$; [626-93-7]

(2) Water; H_2O; [7732-18-5]

ORIGINAL MEASUREMENTS:

Ratouis, M.; Dodé, M.;

Bull. Soc. Chim. Fr. 1965, 3318-22.

VARIABLES:

One temperature: 30°C

Ringer solution also studied

PREPARED BY:

S.C. Valvani; S.H. Yalkowsky; A.F.M. Barton

EXPERIMENTAL VALUES:

The proportion of 2-hexanol (1) in the water-rich phase at equilibrium at 30°C was reported to be 1.19 g(1)/100g sln.

The corresponding mole fraction solubility, calculated by the compilers is x_1 = 0.00212.

The proportion of (1) in the water-rich phase of a mixture with Ringer solution at 30°C was reported to be 1.12 g(1)/100g sln.

AUXILIARY INFORMATION

METHOD/APPARATUS/PROCEDURE:

In a round bottomed flask, 50 mL of water and a sufficient quantity of alcohol were introduced until two separate layers were formed. The flask assembly was equilibrated by agitation for at least 3 h in a constant temperature bath. Equilibrium solubility was attained by first supersaturation at a slightly lower temperature (solubility of alcohols in water decreases with increasing temperature) and then equilibrating at the desired temperature. The aqueous layer was separated after an overnight storage in a bath. The alcohol content was determined by reacting the aqueous solution with potassium dichromate and titrating the excess dichromate with ferrous sulfate solution in the presence of phosphoric acid and diphenylamine barium sulfonate as an indicator.

SOURCE AND PURITY OF MATERIALS:

(1) Fluka A.G. Buchs S.G. redistilled with 10:1 reflux ratio

b.p. 138.6-138.8°C/770 mm Hg

n_D^{25} = 1.41269

(2) twice distilled from silica apparatus or ion-exchanged with Sagei A20.

ESTIMATED ERROR:

Solubility: relative error of 2 determinations less than 1%.

Temperature: ±0.05°C.

REFERENCES:

COMPONENTS:	ORIGINAL MEASUREMENTS:
(1) 2-Hexanol (*n-butylmethylcarbinol*) $C_6H_{14}O$; [626-93-7] (2) Water; H_2O; [7732-18-5]	Nishino, N.; Nakamura, M. *Bull. Chem. Soc. Jpn.* 1978, *51*, 1617-20; 1981, *54*, 545-8.
VARIABLES:	PREPARED BY:
Temperature: 275 - 360 K	G.T. Hefter

EXPERIMENTAL VALUES:

The mutual solubility of (1) and (2) in mole fractions are reported over the temperature range in graphical form. Graphical data are also presented for the heat of evaporation of (1)

AUXILIARY INFORMATION

METHOD/APPARATUS/PROCEDURE:

The turbidimetric method was used. Twenty to thirty glass ampoules containing aqueous solutions of *ca.* 5 cm^3 of various concentrations near the solubility at room temperature were immersed in a water thermostat. The distinction between clear and turbid ampoules was made after equilibrium was established (*ca.* 2h). The smooth curve drawn to separate the clear and turbid regions was regarded as the solubility curve.

SOURCE AND PURITY OF MATERIALS:

(1) G.R. grade (various commercial sources given); dried over calcium oxide; kept in ampoules over magnesium powder.

(2) Deionized, refluxed for 15h with potassium permanganate then distilled.

ESTIMATED ERROR:

Not stated

REFERENCES:

COMPONENTS:

(1) 3-Hexanol (*n-propylethylcarbinol*)
$C_6H_{14}O$; [623-37-0]

(2) Water; H_2O; [7732-18-5]

EVALUATOR:

Z. Maczynska, Institute of Physical Chemistry of the Polish Academy of Sciences, Warsaw, Poland; and A.F.M. Barton, Murdoch University, Perth, Western Australia.
November 1982

CRITICAL EVALUATION:

Solubilities in the system comprising 3-hexanol (1) and water (2) have been reported in four publications. Ginnings and Webb (ref 1) carried out measurements of the mutual solubilities in the two phases at 293, 298 and 303 K by the volumetric method (Figures 1 and 2). Ratouis and Dodé (ref 2) determined the solubility of (1) in the water-rich phase at one temperature (303 K) by an analytical method. Their value of 1.36 g(1)/100g sln is in good agreement with the value 1.49 ± 0.1 g(1)/100g sln of ref 1. Nishino and Nakamura provided graphical information only of the solubility of (1) in (2) (ref 3) and of (2) in (1) (ref 4). Since direct comparison is possible for only a single data point, the data of ref 1 are regarded as tentative.

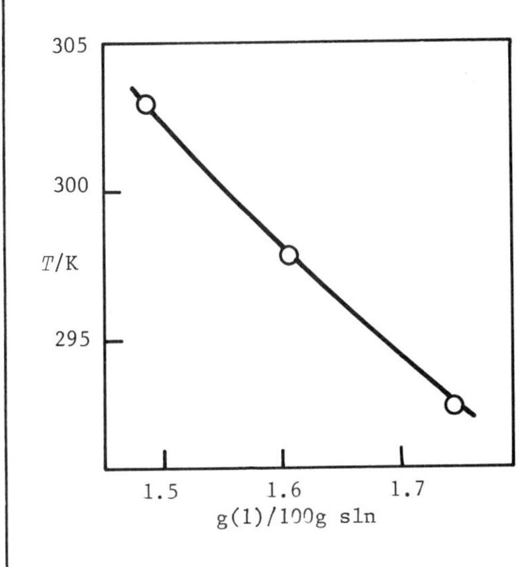

Fig. 1. Water-rich phase (ref 1)

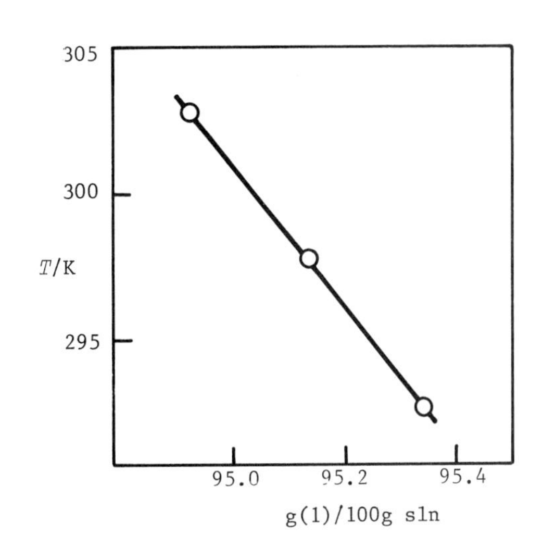

Fig. 2. Alcohol-rich phase (ref 2)

Tentative values for the mutual solubilities of 3-hexanol (1) and water (2)

T/K	Water-rich phase		Alcohol-rich phase	
	g(1)/100g sln	$10^3 x_1$	g(2)/100g sln	x_2
293	1.8	3.1	4.7	0.217
298	1.6	2.9	4.9	0.225
303	1.5	2.7	5.1	0.232

References

1. Ginnings, P.M.; Webb, R. *J. Am. Chem. Soc.* 1938, *60*, 1388.
2. Ratouis, M.; Dodé, M. *Bull. Soc. Chim. Fr.* 1965, 3318.
3. Nishino, N.; Nakamura, M. *Bull. Chem. Soc. Jpn.* 1978, *51*, 1617.
4. Nishino, N.; Nakamura, M. *Bull. Chem. Soc. Jpn.* 1981, *54*, 545.

COMPONENTS:

(1) 3-Hexanol *(n-propylethylcarbinol)*; $C_6H_{14}O$; [623-37-0]

(2) Water; H_2O; [7732-18-5]

ORIGINAL MEASUREMENTS:

Ginnings, P.M.; Webb, R.

J. Am. Chem. Soc. 1938, *60*, 1388-9.

VARIABLES:

Temperature: 20-30°C

PREPARED BY:

A. Maczynski and Z. Maczynska

EXPERIMENTAL VALUES:

Mutual solubility of 3-hexanol(1) and water(2)

$t/^{\circ}C$	g(1)/100g sln		x_1 (compiler)	
	(2)-rich phase	(1)-rich phase	(2)-rich phase	(1)-rich phase
20	1.75	95.34	0.00313	0.7829
25	1.61	95.14	0.00286	0.7753
30	1.49	94.93	0.00266	0.7675

Relative density, d_4

$t/^{\circ}C$	Water-rich phase	Alcohol-rich phase
20	0.9952	0.8264
25	0.9951	0.8225
30	0.9939	0.8190

AUXILIARY INFORMATION

METHOD/APPARATUS/PROCEDURE:

The volumetric method was used as described in ref. 1.

Both components were introduced in known amounts into a two-bulb graduated and calibrated flask and shaken mechanically in a water-bath at constant temperature. After sufficient time the liquids were allowed to separate and the total volume was measured. Upon centrifugation, the phase separation line was read, and phase volumes were calculated. From the total weights of the components, the total volume, individual phase volumes, and component concentrations in either phase were evaluated.

SOURCE AND PURITY OF MATERIALS:

(1) prepared by Grignard synthesis; distilled from calcium oxide; b.p. range 134.5-135.0°C, d_4^{25} 0.8143; purity not specified.

(2) not specified.

ESTIMATED ERROR:

Temperature: ± 0.1°C
Solubility: better than 0.1 wt % (type of error not specified)

REFERENCES:

1. Ginnings. P.M.; Baum, R.J. *J. Am. Chem. Soc.* 1937, *59*, 1111.

COMPONENTS:

(1) 3-Hexanol *(n-propylethylcarbinol)*
$C_6H_{14}O$; [623-37-0]

(2) Water; H_2O [7732-18-5]

ORIGINAL MEASUREMENTS:

Ratouis, M.; Dodé, M.;
Bull. Soc. Chim. Fr. 1965, 3318-22.

VARIABLES:

One temperature: 30°C
Ringer solution also studied

PREPARED BY:

S.C. Valvani; S.H. Yalkowsky; A.F.M. Barton

EXPERIMENTAL VALUES:

The proportion of 3-hexanol (1) in the water-rich phase at equilibrium at 30°C was reported to be 1.36 g(1)/100g sln.

The corresponding mole fraction solubility, calculated by the compilers, is x_1 = 0.00243.

The proportion of (1) in the water-rich phase of a mixture with Ringer solution at 30°C was reported to be 1.25 g(1)/100g sln.

AUXILIARY INFORMATION

METHOD/APPARATUS/PROCEDURE:

In a round bottomed flask, 50 mL of water and a sufficient quantity of alcohol were introduced until two separate layers were formed. The flask assembly was equilibrated by agitation for at least 3 h in a constant temperature bath. Equilibrium solubility was attained by first supersaturation at a slightly lower temperature (solubility of alcohols in water decreases with increasing temperature) and then equilibrating at the desired temperature. The aqueous layer was separated after an overnight storage in a bath. The alcohol content was determined by reacting the aqueous solution with potassium dichromate and titrating the excess dichromate with ferrous sulfate solution in the presence of phosphoric acid and diphenylamine barium sulfonate as an indicator.

SOURCE AND PURITY OF MATERIALS:

(1) Fluka A.G., Buchs S.G.
redistilled with 10:1 reflux ratio
b.p. 136.1-136.2°C
n_D^{25} = 1.41392

(2) twice distilled from silica apparatus or ion-exchanged with sagei A20.

ESTIMATED ERROR:

Solubility: relative error of 2 determinations less than 1%.

Temperature: ±0.05°C

REFERENCES:

COMPONENTS:

(1) 3-Hexanol (*n-propylethylcarbinol*)
$C_6H_{14}O$; [623-37-0]

(2) Water; H_2O; [7732-18-5]

ORIGINAL MEASUREMENTS:

Nishino, N.; Nakamura, M.
Bull. Chem. Soc. Jpn. 1978, *51*, 1617-20; 1981, *54*, 545-8

VARIABLES:

Temperature: 275 - 360 K

PREPARED BY:

G.T. Hefter

EXPERIMENTAL VALUES:

The mutual solubility of (1) and (2) in mole fractions are reported over the temperature range in graphical form. Graphical data are also presented for the heat of evaporation of (1).

AUXILIARY INFORMATION

METHOD/APPARATUS/PROCEDURE:

The tubidimetric method was used. Twenty to thirty glass ampoules containing aqueous solutions of *ca.* 5 cm^3 of various concentrations near the solubility at room temperature were immersed in a water thermostat. The distinction between clear and turbid ampoules was made after equilibrium was established (*ca.* 2h). The smooth curve drawn to separate the clear and turbid regions was regarded as the solubility curve.

SOURCE AND PURITY OF MATERIALS:

(1) G.R. grade (various commercial sources given); dried over calcium oxide; kept in ampoules over magnesium powder.

(2) Deionized, refluxed for 15h with potassium permanganate then distilled.

ESTIMATED ERROR:

Not stated

REFERENCES:

COMPONENTS:

(1) 2,3,3-Trimethyl-2-butanol; $C_7H_{16}O$; [594-83-2]

(2) Water; H_2O; [7732-18-5]

ORIGINAL MEASUREMENTS:

Ginnings, P.M.; Hauser, M.

J. Am. Chem. Soc. 1938, *60*, 2581-2.

VARIABLES:

One temperature: 40°C

PREPARED BY:

A. Maczynski and Z. Maczynska

EXPERIMENTAL VALUES:

The mutual solubility of 2,3,3-trimethyl-2-butanol and water at 40°C was reported to be 2.20 and 94.72 g(1)/100g sln.

The corresponding mole fractions x_1 calculated by the compilers are 3.47×10^{-3} and 0.7355.

The relative densities of the water-rich and alcohol-rich phases were 0.9902 and 0.8352 respectively.

AUXILIARY INFORMATION

METHOD/APPARATUS/PROCEDURE:

The volumetric method was used as described in ref 1.

Both components were introduced in known amounts into a two-bulb graduated and calibrated flask and shaken mechanically in a water-bath at constant temperature. After sufficient time the liquids were allowed to separate and the total volume was measured. Upon centrifugation, the phase separation line was read, and phase volumes were calculated. From the total weights of the components, the total volume, individual phase volumes, and component concentrations in either phase were evaluated.

SOURCE AND PURITY OF MATERIALS:

(1) prepared by Grignard synthesis; distilled from calcium oxide; b.p. range 130.3-130.8°C, d_4^{25} 0.8380; purity not specified.

(2) not specified.

ESTIMATED ERROR:

Solubility: better than 0.1 wt % (type of error not specified)

REFERENCES:

1. Ginnings, P.M.; Baum, R.J. *J. Am. Chem. Soc.* 1937, *59*, 1111.

COMPONENTS:

(1) 2,2-Dimethyl-1-pentanol; $C_7H_{16}O$; [2370-12-9]

(2) Water: H_2O; [7732-18-5]

ORIGINAL MEASUREMENTS:

Ratouis, M.; Dodé, M.

Bull. Soc. Chim. Fr. 1965, 3318-22

VARIABLES:

Temperature: 25°C and 30°C

Ringer solution also studied

PREPARED BY:

S.C. Valvani; S.H. Yalkowsky; A.F.M. Barton

EXPERIMENTAL VALUES:

Proportion of 2,2-dimethyl-1-pentanol (1) in water-rich phase

$t/^{\circ}C$	g(1)/100g sln	$10^4\ x_1$ (compilers)
25	0.36	5.6
30	0.34	5.3

The proportion of (1) in the water-rich phase of a mixture with Ringer solution at equilibrium at 30°C was reported to be 0.30 g(1)/100g sln.

AUXILIARY INFORMATION

METHOD/APPARATUS/PROCEDURE:

In a round bottomed flask, 50 mL of water and a sufficient quantity of alcohol were introduced until two separate layers were formed. The flask assembly was equilibrated by agitation for at least 3h in a constant temperature bath. Equilibrium solubility was attained by first supersaturation at a slightly lower temperature (solubility of alcohols decreases with increasing temperature) and then equilibrating at the desired temperature. The aqueous layer was separated after an overnight storage in a bath. The alcohol content was determined by reacting the aqueous solution with potassium dichromate and titrating the excess dichromate with ferrous sulfate solution in the presence of phosphoric acid and diphenylamine barium sulfonate as an indicator.

SOURCE AND PURITY OF MATERIALS:

(1) Fluka A.G. Buchs S.G.; redistilled with 10:1 reflux ratio

b.p. 153.8-154°C/758.4 mm Hg

$n_D^{25} = 1.42556$

(2) twice distilled from silica apparatus or ion-exchanged with Sagei A20

ESTIMATED ERROR:

Solubility: relative error of 2 determinations less than 1%

Temperature: ± 0.05°C

REFERENCES:

COMPONENTS:	EVALUATOR:
(1) 2, 2-Dimethyl-3-pentanol; $C_7H_{16}O$; [3970-62-5] (2) Water; H_2O; [7732-18-5]	Z. Maczynska, Institute of Physical Chemistry of the Polish Academy of Sciences, Warsaw, Poland. November 1982

CRITICAL EVALUATION:

Solubilities in the system comprising 2,2-dimethyl-3-pentanol (1) and water (2) have been reported in two publications. Ginnings and Hauser (ref 1) carried out measurements of the mutual solubilities in the two phases at 293, 298 and 303 K by the volumetric method (Figures 1 and 2). Ratouis and Dodé (ref 2) determined the solubility of (1) in the water-rich phase at one temperature (303 K) by an analytical method. Their value of 0.72 g(1)/100g sln is in good agreement with the value 0.79 ± 0.1 g(1)/100g sln of ref 1. However, as the comparison involves a single point and as the other five points are derived from only one source, the data are regarded as tentative.

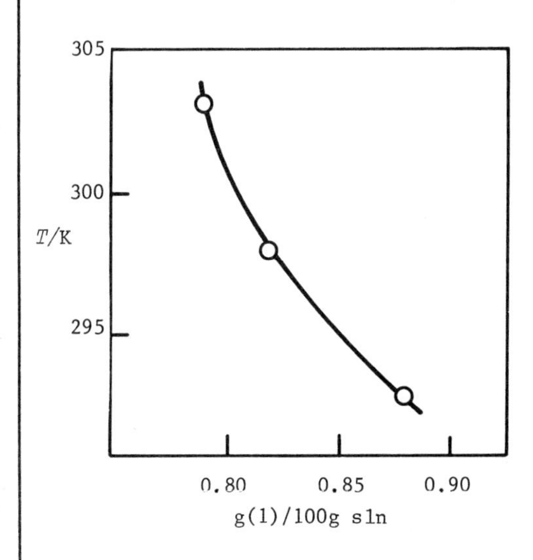

Fig. 1. Water-rich phase (ref 1)

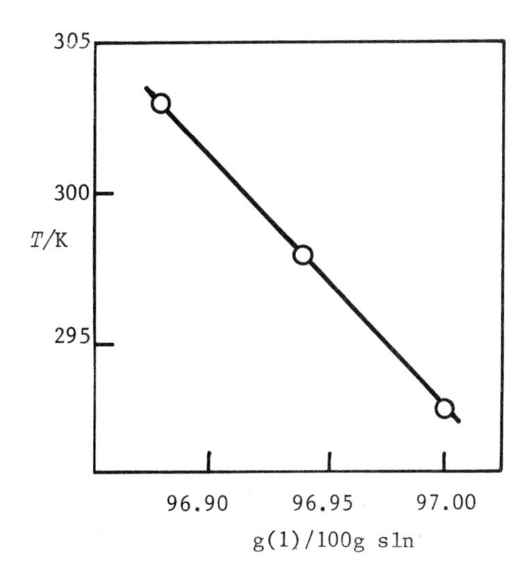

Fig. 2. Alcohol-rich phase (ref 1)

Tentative values for the mutual solubilities of 2,2-dimethyl-3-pentanol (1) and water (2)

T/K	Water-rich phase		Alcohol-rich phase	
	g(1)/100g sln	10^3x_1	g(2)/100g sln	x_2
293	0.88	1.4	3.0	0.166
298	0.82	1.3	3.1	0.169
303	0.79	1.2	3.1	0.172

References:

1. Ginnings, P.M.; Hauser, M. *J. Am. Chem. Soc.* 1938, *60*, 2581.

2. Ratouis, M.; Dodé, M. *Bull. Soc. Chim. Fr.* 1965, 3318.

AWW-K

COMPONENTS:

(1) 2,2-Dimethyl-3-pentanol; $C_7H_{16}O$; [3970-62-5]

(2) Water; H_2O; [7732-18-5]

ORIGINAL MEASUREMENTS:

Ginnings, P.M.; Hauser, M.

J. Am. Chem. Soc. 1938, *60*, 2581-2.

VARIABLES:

Temperature: 20-30°C

PREPARED BY:

A. Maczynski and Z. Maczynska

EXPERIMENTAL VALUES:

Mutual solubility of 2,2-dimethyl-3-pentanol(1) and water(2)

t/°C	g(1)/100g sln		x_1 (compiler)	
	(2)-rich phase	(1)-rich phase	(2)-rich phase	(1)-rich phase
20	0.88	97.00	0.00137	0.8336
25	0.82	96.94	0.00128	0.8308
30	0.79	96.88	0.00123	0.8280

Relative density, d_4

t/°C	Water-rich phase	Alcohol-rich phase
20	0.9971	0.8329
25	0.9962	0.8289
30	0.9950	0.8253

AUXILIARY INFORMATION

METHOD/APPARATUS/PROCEDURE:

The volumetric method was used as described in ref 1.

Both components were introduced in known amounts into a two-bulb graduated and calibrated flask and shaken mechanically in a water-bath at constant temperature. After sufficient time the liquids were allowed to separate and the total volume was measured. Upon centrifugation, the phase separation line was read, and phase volumes were calculated. From the total weights of the components, the total volume, individual phase volumes, and component concentrations in either phase was evaluated.

SOURCE AND PURITY OF MATERIALS:

(1) prepared by Grignard synthesis; distilled from calcium oxide; b.p. range 134.7-135.1°C, d_4^{25} 0.8224; purity not specified.

(2) not specified.

ESTIMATED ERROR:

Solubility: better than 0.1 wt %
(type of error not specified)

REFERENCES:

1. Ginnings, P.M.; Baum, R.J. *J. Am. Chem. Soc.* 1937, *59*, 1111.

COMPONENTS:

(1) 2,2-Dimethyl-3-pentanol; $C_7H_{16}O$; [3970-62-5]

(2) Water; H_2O; [7732-18-5]

ORIGINAL MEASUREMENTS:

Ratouis, M.; Dodé, M.

Bull. Soc. Chim. Fr. 1965, 3318-22.

VARIABLES:

One temperature: 30°C

Ringer solution also studied

PREPARED BY:

S.C. Valvani; S.H. Yalkowsky; A.F.M. Barton

EXPERIMENTAL VALUES:

The proportion of 2,2-dimethyl-3-pentanol (1) in the water-rich phase at equilibrium at 30°C was reported to be 0.72 g(1)/100g sln.

The corresponding mole fraction solubility, calculated by the compilers, is x_1 = 0.00112.

The proportion of (1) in the water-rich phase of a mixture with Ringer solution at 30°C was reported to be 0.64 g(1)/100g sln.

AUXILIARY INFORMATION

METHOD/APPARATUS/PROCEDURE:

In a round bottomed flask, 50 mL of water and a sufficient quantity of alcohol were introduced until two separate layers were formed. The flask assembly was equilibrated by agitation for at least 3h in a constant temperature bath. Equilibrium solubility was attained by first supersaturation at a slightly lower temperature (solubility of alcohols in water decreases with increasing temperature) and then equilibrating at the desired temperature. The aqueous layer was separated after an overnight storage in a bath. The alcohol content was determined by reacting the aqueous solution with potassium dichromate and titrating the excess dichromate with ferrous sulfate solution in the presence of phosphoric acid and diphenylamine barium sulfonate as an indicator.

SOURCE AND PURITY OF MATERIALS:

(1) laboratory preparation; redistilled with 10:1 reflux ratio

b.p. 135.4°C/768.2 mm Hg

n_D^{25} = 1.42358

(2) twice distilled from silica apparatus or ion-exchanged with Sagei A20.

ESTIMATED ERROR:

Solubility: relative error of 2 determinations less than 1%

Temperature: ± 0.05°C

REFERENCES:

COMPONENTS:	EVALUATOR:
(1) 2,3-Dimethyl-2-pentanol; $C_7H_{16}O$; [4911-70-0] (2) Water; H_2O; [7732-18-5]	Z. Maczynska, Institute of Physical Chemistry of the Polish Academy of Sciences, Warsaw, Poland November 1982

CRITICAL EVALUATION:

Solubilities in the system comprising 2,3-dimethyl-2-pentanol (1) and water (2) have been reported in two publications. Ginnings and Hauser (ref 1) carried out measurements of the mutual solubilities in the two phases at 293, 298 and 303 K by the volumetric method (Figure 1). Ratouis and Dodé (ref 2) determined the solubility of (1) in the water-rich phase at one temperature (303 K) by an anlytical method. Their value of 1.15 g(1)/100g sln agrees poorly with the value 1.40 ± 0.1 g(1)/100g sln of ref 1. The results are regarded as tentative.

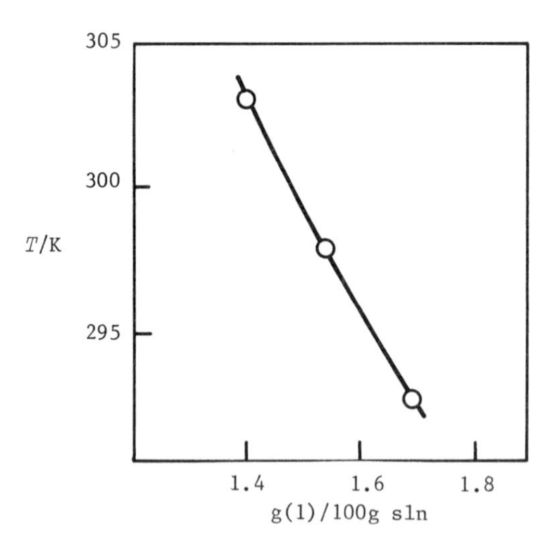

Fig. 1 Solubility of (1) in (2) (ref 1)

Tentative values for the mutual solubilities of 2,3-dimethyl-2-pentanol (1) and water (2)

T/K	Water-rich phase		Alcohol-rich phase	
	g(1)/100g sln	10^3x_1	g(2)/100g sln	x_2
293	1.7	2.7	6.3	0.30
298	1.5	2.4	6.3	0.30
303	1.4	2.2	6.3	0.30

References

1. Ginnings, P.M.; Hauser, M. *J. Am. Chem. Soc.* 1938, *60*, 2581

2. Ratouis, M.; Dodé, M. *Bull. Soc. Chim. Fr.* 1965, 3318

COMPONENTS:

(1) 2,3-Dimethyl-2-pentanol; $C_7H_{16}O$; [4911-70-0]

(2) Water; H_2O; [7732-18-5]

ORIGINAL MEASUREMENTS:

Ginnings, P.M.; Hauser, M.

J. Am. Chem. Soc. 1938, *60*, 2581-2.

VARIABLES:

Temperature: 20-30°C

PREPARED BY:

A. Maczynski and Z. Maczynska

EXPERIMENTAL VALUES:

Mutual solubility of 2,3-dimethyl-2-pentanol(1) and water(2)

$t/^{\circ}C$	g(1)/100g sln		x_1 (compiler)	
	(2)-rich phase	(1)-rich phase	(2)-rich phase	(1)-rich phase
20	1.69	93.69	0.00265	0.6971
25	1.54	93.69	0.00242	0.6971
30	1.40	93.71	0.00219	0.6978

Relative density, d_4

$t/^{\circ}C$	Water-rich phase	Alcohol-rich phase
20	0.9964	0.8477
25	0.9955	0.8441
30	0.9943	0.8404

AUXILIARY INFORMATION

METHOD/APPARATUS/PROCEDURE:

The volumetric method was used as described in ref 1.

Both components were introduced in known amounts into a two-bulb graduated and calibrated flask and shaken mechanically in a water-bath at constant temperature. After sufficient time the liquids were allowed to separate and the total volume was measured. Upon centrifugation, the phase separation line was read, and phase volumes were calculated. From the total weights of the components, the total volume, individual phase volumes, and component concentrations in either phase were evaluated.

SOURCE AND PURITY OF MATERIALS:

(1) prepared by Grignard synthesis; distilled from calcium oxide; b.p. range 138.5-139.5°C, d_4^{25} 0.8307; purity not specified.

(2) not specified.

ESTIMATED ERROR:

Solubility: better than 0.1 wt %
(type of error not specified)

REFERENCES:

1. Ginnings, P.M.; Baum, R.J. *J. Am. Chem. Soc.* 1937, *59*, 1111.

COMPONENTS:

(1) 2,3-Dimethyl-2-pentanol; $C_7H_{16}O$; [4911-70-0]

(2) Water; H_2O; [7732-18-5]

ORIGINAL MEASUREMENTS:

Ratouis, M.; Dodé, M.

Bull. Soc. Chim. Fr. 1965, 3318-22.

VARIABLES:

One temperature: 30°C

Ringer solution also studied

PREPARED BY:

S.C. Valvani; S.H. Yalkowsky; A.F.M. Barton

EXPERIMENTAL VALUES:

The proportion of 2,3-dimethyl-2-pentanol (1) in the water-rich phase at equilibrium at 30°C was reported to be 1.15 g(1)/100g sln.

The corresponding mole fraction solubility, calculated by the compilers, is x_1 = 0.00180.

The proportion of (1) in the water-rich phase of a mixture with Ringer solution at 30°C was reported to be 1.02 g(1)/100g sln.

AUXILIARY INFORMATION

METHOD/APPARATUS/PROCEDURE:

In a round bottomed flask, 50 mL of water and a sufficient quantity of alcohol were introduced until two separate layers were formed. The flask assembly was equilibrated by agitation for at least 3h in a constant temperature bath. Equilibrium solubility was attained by first supersaturation at a slightly lower temperature (solubility of alcohols in water decreases with increasing temperature) and then equilibrating at the desired temperature. The aqueous layer was separated after an overnight storage in a bath. The alcohol content was determined by reacting the aqueous solution with potassium dichromate and titrating the excess dichromate with ferrous sulfate solution in the presence of phosphoric acid and diphenylamine barium sulfonate as an indicator.

SOURCE AND PURITY OF MATERIALS:

(1) laboratory preparation; redistilled with 10:1 reflux ratio

b.p. 84.4°C/97 mm Hg

n_D^{25} = 1.42348

(2) twice distilled from silica apparatus or ion-exchanged with Sagei A20

ESTIMATED ERROR:

Solubility: relative error of 2 determinations less than 1%

Temperature: ± 0.05°C

REFERENCES:

COMPONENTS:	EVALUATOR:
(1) 2,3-Dimethyl-3-pentanol; $C_7H_{16}O$; [595-41-5]	Z. Maczynska, Institute of Physical Chemistry of the Polish Academy of Sciences, Warsaw, Poland
(2) Water; H_2O; [7732-18-5]	November 1982

CRITICAL EVALUATION:

Solubilities in the system comprising 2,3-dimethyl-3-pentanol (1) and water (2) have been reported in two publications (Figure 1). Ginnings and Hauser (ref 1) carried out measurements of the mutual solubilities in the two phases at 293, 298 and 303 K by the volumetric method. Ratouis and Dodé (ref 2) determined the solubility of (1) in the water-rich phase at one temperature (303K) by an analytical method. Their value of 1.41 g(1)/100g sln is in excellent agreement with the value 1.43 ± 0.1 g(1)/100g sln of ref 1. However, the comparison involves a single point and as the remaining five points are derived from only one source, the data are regarded as tentative.

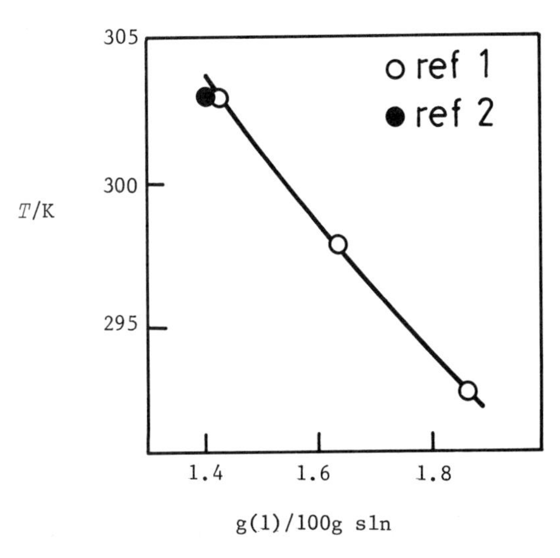

Fig. 1 Solubility of (1) in (2)

Tentative values for the mutual solubilties of 2,3-dimethyl-3-pentanol (1) and water (2)

T/K	Water-rich phase		Alcohol-rich phase	
	g(1)/100g sln	10^3x_1	g(2)/100g sln	x_2
293	1.9	2.9	5.9	0.29
298	1.6	2.6	5.9	0.29
303	1.4	1.4	5.9	0.29

References

1. Ginnings, P.M.; Hauser, M. *J. Am. Chem. Soc.* 1938, *60*, 2581.

2. Ratouis, M.; Dodé, M. *Bull. Soc. Chim. Fr.* 1965, 3318.

COMPONENTS:

(1) 2,3-Dimethyl-3-pentanol; $C_7H_{16}O$; [595-41-5]

(2) Water; H_2O; [7732-18-5]

ORIGINAL MEASUREMENTS:

Ginnings, P.M.; Hauser, M.

J. Am. Chem. Soc. 1938, *60*, 2581-2.

VARIABLES:

Temperature: 20-30°C

PREPARED BY:

A. Maczynski and Z. Maczynska

EXPERIMENTAL VALUES:

Mutual solubility of 2,3-dimethyl-3-pentanol(1) and water(2)

$t/^{\circ}C$	g(1)/100g sln		x_1 (compiler)	
	(2)-rich phase	(1)-rich phase	(2)-rich phase	(1)-rich phase
20	1.87	94.11	0.00294	0.7123
25	1.64	94.12	0.00258	0.7127
30	1.43	94.12	0.00224	0.7127

Relative density, d_4

$t/^{\circ}C$	Water-rich phase	Alcohol-rich phase
20	0.9965	0.8513
25	0.9961	0.8470
30	0.9945	0.8430

AUXILIARY INFORMATION

METHOD/APPARATUS/PROCEDURE:

The volumetric method was used as described in ref 1.

Both components were introduced in known amounts into a two-bulb graduated and calibrated flask and shaken mechanically in a water-bath at constant temperature. After sufficient time the liquids were allowed to separate and the total volume was measured. Upon centrifugation, the phase separation line was read, and phase volumes were calculated. From the total weights of the components, the total volume, individual phase volumes, and component concentrations in either phase were evaluated.

SOURCE AND PURITY OF MATERIALS:

(1) prepared by Grignard synthesis; distilled from calcium oxide; b.p. range 139.6-139.8°C, d_4^{25} 0.8365; purity not specified.

(2) not specified.

ESTIMATED ERROR:

Solubility: better than 0.1 wt % (type of error not specified)

REFERENCES:

1. Ginnings. P.M.; Baum, R.J. *J. Am. Chem. Soc.* 1937, *59*, 1111.

COMPONENTS:

(1) 2,3-Dimethyl-3-pentanol; $C_7H_{16}O$; [595-41-5]

(2) Water; H_2O; [7732-18-5]

ORIGINAL MEASUREMENTS:

Ratouis, M.; Dodé, M.

Bull. Soc. Chim. Fr. 1965, 3318-22.

VARIABLES:

One temperature: 30°C

Ringer solution also studied

PREPARED BY:

S.C. Valvani; S.H. Yalkowsky; A.F.M. Barton

EXPERIMENTAL VALUES:

The proportion of 2,3-dimethyl-3-pentanol (1) in the water-rich phase at equilibrium at 30°C was reported to be 1.41 g(1)/100g sln.

The corresponding mole fraction solubility, calculated by the compilers, is x_1 = 0.00221.

The proportion of (1) in the water-rich phase of a mixture with Ringer solution at 30°C was reported to be 1.31 g(1)/100g sln.

AUXILIARY INFORMATION

METHOD/APPARATUS/PROCEDURE:

In a round bottomed flask, 50 mL of water and a sufficient quantity of alcohol were introduced until two separate layers were formed. The flask assembly was equilibrated by agitation for at least 3h in a constant temperature bath. Equilibrium solubility was attained by first supersaturation at a slightly lower temperature (solubility of alcohols in water decreases with increasing temperature) and then equilibrating at the desired temperature. The aqueous layer was separated after an overnight storage in a bath. The alcohol content was determined by reacting the aqueous solution with potassium dichromate and titrating the excess dichromate with ferrous sulfate solution in the presence of phosphoric acid and diphenylamine barium sulfonate as an indicator.

SOURCE AND PURITY OF MATERIALS:

(1) laboratory preparation; redistilled with 10:1 reflux ratio

b.p. 60.2-60.3°C/40 mm Hg

n_D^{25} = 1.42671

(2) twice distilled from silica apparatus or ion-exchanged with Sagei A20

ESTIMATED ERROR:

Solubility: relative error of 2 determinations less than 1%

Temperature: ± 0.05°C

REFERENCES:

COMPONENTS:

(1) 2,4-Dimethyl-1-pentanol; $C_7H_{16}O$; [6305-71-1]

(2) Water; H_2O; [7732-18-5]

ORIGINAL MEASUREMENTS:

Ratouis, M.; Dodé, M.

Bull. Soc. Chim. Fr. 1965, 3318-22.

VARIABLES:

Temperature: 25°C and 30°C

Ringer solution also studied

PREPARED BY:

S.C. Valvani; S.H. Yalkowsky; A.F.M. Barton

EXPERIMENTAL VALUES:

Proportion of 2,4-dimethyl-1-pentanol (1) in water-rich phase

$t/^{\circ}C$	g(1)/100g sln	$10^4\ x_1$ (compilers)
25	0.30	4.7
30	0.285	4.42

The proportion of (1) in the water-rich phase of a mixture with Ringer solution at equilibrium at 30°C was reported to be 0.265 g(1)/100g sln.

AUXILIARY INFORMATION

METHOD/APPARATUS/PROCEDURE:

In a round bottomed flask, 50 mL of water and a sufficient quantity of alcohol were introduced until two separate layers were formed. The flask assembly was equilibrated by agitation for at least 3h in a constant temperature bath. Equilibrium solubility was attained by first supersaturation at a slightly lower temperature (solubility of alcohols in water decreases with increasing temperature) and then equilibrating at the desired temperature. The aqueous layer was separated after an overnight storage in a bath. The alcohol content was determined by reacting the aqueous solution with potassium dichromate and titrating the excess dichromate with ferrous sulfate solution in the presence of phosphoric acid and diphenylamine barium sulfonate as an indicator.

SOURCE AND PURITY OF MATERIALS:

(1) Fluka A.G. Buchs S.G.;
redistilled with 10:1 reflux ratio

b.p. 158.2-159°C/753.6 mm Hg

n_D^{25} = 1.42009

(2) twice distilled from silica apparatus or ion-exchanged with Sagei A20

ESTIMATED ERROR:

Solubility: relative error of 2 determinations less than 1%

Temperature: ± 0.05°C

REFERENCES:

COMPONENTS:	EVALUATOR:
(1) 2,4-Dimethyl-2-pentanol; $C_7H_{16}O$; [625-06-9] (2) Water; H_2O; [7732-18-5]	Z. Maczynska, Institute of Physical Chemistry of the Polish Academy of Sciences, Warsaw, Poland November 1982

CRITICAL EVALUATION:

Solubilities in the system comprising 2,4-dimethyl-2-pentanol (1) and water (2) have been reported in two publications. Ginnings and Hauser (ref 1) carried out measurements of the mutual solubilities at 293, 298 and 303 K by the volumetric method (Figure 1). Ratouis and Dodé (ref 2) determined the solubility of (1) in the water-rich phase at one temperature (303 K) by an analytical method. Their value of 0.98 g(1)/100g sln is in reasonable agreement with the value 1.22 ± 0.1 g(1)/100g sln of ref 1. The data are regarded as tentative, since comparison can be made only at a single temperature and the other five points are derived from one reference.

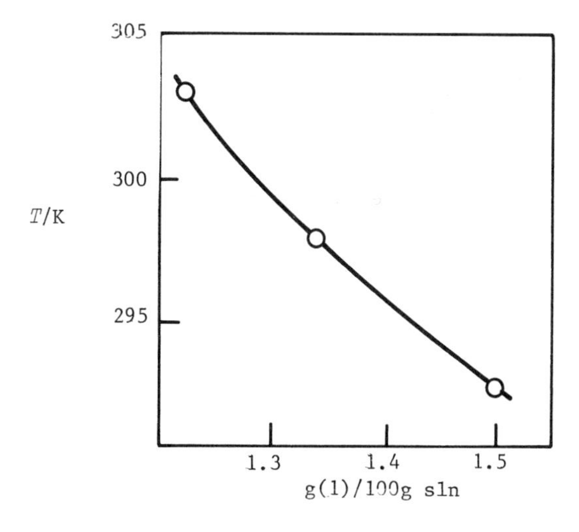

Fig. 1 Solubility of (1) in (2) (ref 1)

Tentative values for the mutual solubilities of 2,4-dimethyl-2-pentanol (1) and water (2)

T/K	Water-rich phase		Alcohol-rich phase	
	g(1)/100g sln	$10^3 x_1$	g(2)/100g sln	x_2
293	1.5	2.4	6.5	0.31
298	1.3	2.1	6.5	0.31
303	1.2	1.9	6.5	0.31

References

1. Ginnings, P.M.; Hauser, M. *J. Am. Chem. Soc.* 1938, *60*, 2581.
2. Ratouis, M.; Dodé, M. *Bull. Soc. Chim. Fr.* 1965, *5*, 3318

COMPONENTS:

(1) 2,4-Dimethyl-2-pentanol; $C_7H_{16}O$; [625-06-9]

(2) Water; H_2O; [7732-18-5]

ORIGINAL MEASUREMENTS:

Ginnings, P.M.; Hauser, M.

J. Am. Chem. Soc. 1938, *60*, 2581-2.

VARIABLES:

Temperature: 20-30°C

PREPARED BY:

A. Maczynski and Z. Maczynska

EXPERIMENTAL VALUES:

Mutual solubility of 2,4-dimethyl-2-pentanol(1) and water(2)

t/°C	g(1)/100g sln		x_1 (compiler)	
	(2)-rich phase	(1)-rich phase	(2)-rich phase	(1)-rich phase
20	1.50	93.49	0.00235	0.6900
25	1.34	93.48	0.00210	0.6897
30	1.22	93.49	0.00191	0.6900

Relative density, d_4

t/°C	Water-rich phase	Alcohol-rich phase
20	0.9962	0.8279
25	0.9954	0.8240
30	0.9943	0.8199

AUXILIARY INFORMATION

METHOD/APPARATUS/PROCEDURE:

The volumetric method was used as described in ref 1.

Both components were introduced in known amounts into a two-bulb graduated and calibrated flask and shaken mechanically in a water-bath at constant temperature. After sufficient time the liquids were allowed to separate and the total volume was measured. Upon centrifugation, the phase separation line was read, and phase volumes were calculated. From the total weights of the components, the total volume individual phase volumes, and component concentrations in either phase were evaluated.

SOURCE AND PURITY OF MATERIALS:

(1) prepared by Grignard synthesis; distilled from calcium oxide; b.p. range 132.5-133.5°C, d_4^{25} 0.8100; purity not specified.

(2) not specified.

ESTIMATED ERROR:

Solubility: better than 0.1 wt % (type of error not specified)

REFERENCES:

1. Ginnings, P.M.; Baum, R.J. *J. Am. Chem. Soc.* 1937, *59*, 1111.

COMPONENTS:

(1) 2,4-Dimethyl-2-pentanol; $C_7H_{16}O$; [625-06-9]

(2) Water; H_2O; [7732-18-5]

ORIGINAL MEASUREMENTS:

Ratouis, M.; Dodé, M.

Bull. Soc. Chim. Fr. 1965, 3318-22.

VARIABLES:

One temperature: 30°C

Ringer solution also studied

PREPARED BY:

S.C. Valvani; S.H. Yalkowsky; A.F.M. Barton

EXPERIMENTAL VALUES:

The proportion of 2,4-dimethyl-2-pentanol (1) in the water-rich phase at equilibrium at 30°C was reported to be 0.98 g(1)/100g sln.

The corresponding mole fraction solubility, calculated by the compilers, is x_1 = 0.00153.

The proportion of (1) in the water-rich phase of a mixture with Ringer solution at 30°C was reported to be 0.89 g(1)/100g sln.

AUXILIARY INFORMATION

METHOD/APPARATUS/PROCEDURE:

In a round bottomed flask, 50 mL of water and a sufficient quantity of alcohol were introduced until two separate layers were formed. The flask assembly was equilibrated by agitation for at least 3h in a constant temperature bath. Equilibrium solubility was attained by first supersaturation at a slightly lower temperature (solubility of alcohols in water decreases with increasing temperature) and then equilibrating at the desired temperature. The aqueous layer was separated after an overnight storage in a bath. The alcohol content was determined by reacting the aqueous solution with potassium dichromate and titrating the excess dichromate with ferrous sulfate solution in the presence of phosphoric acid and diphenylamine barium sulfonate as an indicator.

SOURCE AND PURITY OF MATERIALS:

(1) laboratory preparation; redistilled with 10:1 reflux ratio

b.p. 132.3-133.6°C/752.4 mm Hg

n_D^{25} = 1.41216

(2) twice distilled from silica apparatus or ion-exchanged with Sagei A20

ESTIMATED ERROR:

Solubility: relative error of 2 determinations less than 1%

Temperature: ± 0.05°C

REFERENCES:

COMPONENTS:	EVALUATOR:
(1) 2,4-Dimethyl-3-pentanol (*diisopropylcarbinol*); $C_7H_{16}O$; [600-36-2] (2) Water; H_2O; [7732-18-5]	Z. Maczynska, Institute of Physical Chemistry of the Polish Academy of Sciences, Warsaw, Poland November 1982.

CRITICAL EVALUATION:

Solubilities in the system comprising 2,4-dimethyl-3-pentanol (1) and water (2) have been reported in two publications. Ginnings and Hauser (ref 1) carried out measurements of the mutual solubilities in the two phases at 293, 298 and 303 K by the volumetric method (Figures 1 and 2). Ratouis and Dodé (ref 2) determined the solubility of (1) in the water-rich phase at one temperature (303 K) by an analytical method, their value of 0.61 g(1)/100g sln being in good agreement with the value of 0.67 ± 0.1 g(1)/100g sln of ref 1. However, as the comparison involves a single point and as the remaining five points are derived from only one source, the data are regarded as tentative.

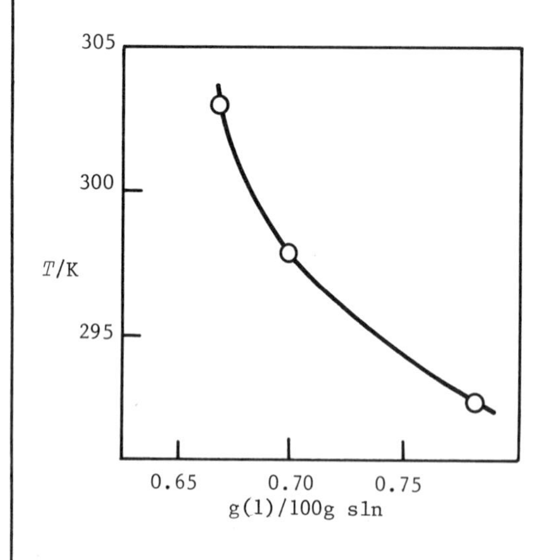

Fig. 1. Water-rich phase (ref 1)

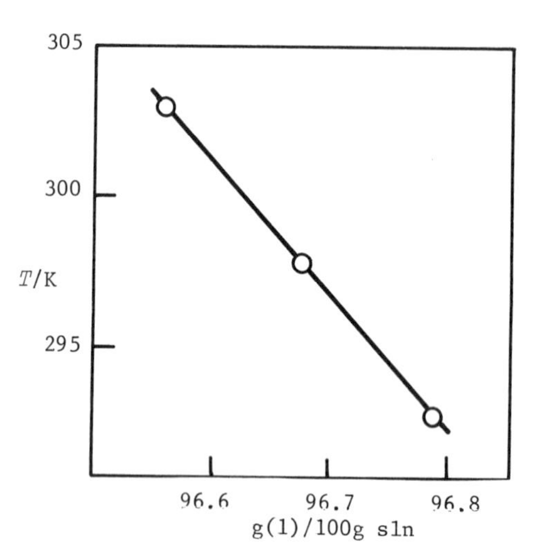

Fig. 2. Alcohol-rich phase (ref 1)

Tentative values for the mutual solubilities of 2,4-dimethyl-3-pentanol (1) and water (2)

T/K	Water-rich phase		Alcohol-rich phase	
	g(1)/100g sln	$10^3 x_1$	g(2)/100g sln	x_2
293	0.78	1.2	3.2	0.176
298	0.70	1.1	3.3	0.181
303	0.67	1.0	3.4	0.187

References

1. Ginnings, P.M.; Hauser, M. *J. Am. Chem. Soc.* 1938, *60*, 2581.
2. Ratouis, M.; Dodé, M. *Bull. Soc. Chim. Fr.* 1965, 3318.

COMPONENTS:

(1) 2,4-Dimethyl-3-pentanol; *(diisopropylcarbinol)*; $C_7H_{16}O$; [600-36-2]

(2) Water; H_2O; [7732-18-5]

ORIGINAL MEASUREMENTS:

Ginnings, P.M.; Hauser, M.

J. Am. Chem. Soc. 1938, *60*, 2581-2.

VARIABLES:

Temperature: 20-30°C

PREPARED BY:

A. Maczynski and Z. Maczynska

EXPERIMENTAL VALUES:

Mutual solubility of 2,4-dimethyl-3-pentanol(1) and water(2)

t/°C	g(1)/100g sln		x_1 (compiler)	
	(2)-rich phase	(1)-rich phase	(2)-rich phase	(1)-rich phase
20	0.78	96.79	0.00122	0.8237
25	0.70	96.68	0.00109	0.8186
30	0.67	96.56	0.00104	0.8131

Relative density, d_4

t/°C	Water-rich phase	Alcohol-rich phase
20	0.9974	0.8351
25	0.9965	0.8315
30	0.9955	0.8271

AUXILIARY INFORMATION

METHOD/APPARATUS/PROCEDURE:

The volumetric method was used as described in ref 1.

Both components were introduced in known amounts into a two-bulb graduated and calibrated flask and shaken mechanically in a water-bath at constant temperature. After sufficient time the liquids were allowed to separate and the total volume was measured. Upon centrifugation, the phase separation line was read, and phase volumes were calculated. From the total weights of the components, the total volume, individual phase volumes, and component concentrations in either phase were evaluated.

SOURCE AND PURITY OF MATERIALS:

(1) prepared by Grignard synthesis; distilled from calcium oxide; b.p. range 138.4-138.9°C, d_4^{25} 0.8254; purity not specified.

(2) not specified.

ESTIMATED ERROR:

Solubility: better than 0.1 wt % (type of error not specified)

REFERENCES:

1. Ginnings, P.M.; Baum, R.J. *J. Am. Chem. Soc.* 1937, *59*, 1111.

COMPONENTS:

(1) 2,4-Dimethyl-3-pentanol *(diisopropylcarbinol)*; $C_7H_{16}O$; [600-36-2]

(2) Water; H_2O; [7732-18-5]

ORIGINAL MEASUREMENTS:

Ratouis, M.; Dodé, M.
Bull. Soc. Chim. Fr. 1965, 3318-22.

VARIABLES:

One temperature: 30°C
Ringer solution also studied

PREPARED BY:

S.C. Valvani; S.H. Yalkowsky; A.F.M. Barton

EXPERIMENTAL VALUES:

The proportion of 2,4-dimethyl-3-pentanol (1) in the water-rich phase at equilibrium at 30°C was reported to be 0.61 g(1)/100g sln.

The corresponding mole fraction solubility, calculated by the compilers, is x_1 = 0.00095.

The proportion of (1) in the water-rich phase of a mixture with Ringer solution at 30°C was reported to be 0.53 g(1)/100g sln.

AUXILIARY INFORMATION

METHOD/APPARATUS/PROCEDURE:

In a round bottomed flask, 50 mL of water and a sufficient quantity of alcohol were introduced until two separate layers were formed. The flask assembly was equilibrated by agitation for at least 3h in a constant temperature bath. Equilibrium solubility was attained by first supersaturation at a slightly lower temperature (solubilities of alcohols in water decrease with increasing temperature) and then equilibrating at the desired temperature. The aqueous layer was separated after an overnight storage in a bath. The alcohol content was determined by reacting the aqueous solution with potassium dichromate and titrating the excess dichromate with ferrous sulfate solution in the presence of phosphoric acid and diphenylamine barium sulfonate as an indicator.

SOURCE AND PURITY OF MATERIALS:

(1) laboratory preparation; redistilled with 10:1 reflux ratio
b.p. 139.9-140°C/770.4 mm Hg
n_D^{25} = 1.42265

(2) twice distilled from silica apparatus or ion-exchanged with Sagei A20

ESTIMATED ERROR:

Solubility: relative error of 2 determinations less than 1%

Temperature: ± 0.05°C

REFERENCES:

COMPONENTS:

(1) 4,4-Dimethyl-1-pentanol; $C_7H_{16}O$; [3121-79-7]

(2) Water; H_2O; [7732-18-5]

ORIGINAL MEASUREMENTS:

Ratouis, M.; Dodé, M.

Bull. Soc. Chim. Fr. 1965, 3318-22.

VARIABLES:

Temperature: 25°C and 30°C

Ringer solution also studied

PREPARED BY:

S.C. Valvani; S.H. Yalkowsky; A.F.M. Barton

EXPERIMENTAL VALUES:

Proportion of 4,4-dimethyl-1-pentanol (1) in water-rich phase

$t/°C$	g(1)/100g sln	$10^4\ x_1$ (compilers)
25	0.335	5.20
30	0.325	5.05

The proportion of (1) in the water-rich phase of a mixture with Ringer solution at equilibrium at 30°C was reported to be 0.31 g(1)/100g sln.

AUXILIARY INFORMATION

METHOD/APPARATUS/PROCEDURE:

In a round bottomed flask, 50 mL of water and a sufficient quantity of alcohol were introduced until two separate layers were formed. The flask assembly was equilibrated by agitation for at least 3h in a constant temperature bath. Equilibrium solubility was attained by first supersaturation at a slightly lower temperature (solubilities of alcohols in water decrease with increasing temperature) and then equilibrating at the desired temperature. Aqueous layer was separated after an overnight storage in a bath. The alcohol content was determined by reacting the aqueous solution with potassium dichromate and titrating the excess dichromate with ferrous sulfate solution in the presence of phosphoric acid and diphenylamine barium sulfonate as an indicator.

SOURCE AND PURITY OF MATERIALS:

(1) laboratory preparation; redistilled with 10:1 reflux ratio

b.p. 162.9-163.3°C/755.5 mm Hg

n_D^{25} = 1.41853

(2) twice distilled from silica apparatus or ion-exchanged with Sagei A20

ESTIMATED ERROR:

Solubility: relative error of 2 determinations less than 1%

Temperature: ± 0.05°C

REFERENCES:

COMPONENTS:	EVALUATOR:
(1) 3-Ethyl-3-pentanol; $C_7H_{16}O$; [597-49-9] (2) Water; H_2O; [7732-18-5]	A. Maczynski, Institute of Physical Chemistry of the Polish Academy of Sciences, Warsaw, Poland. November 1982

CRITICAL EVALUATION:

Solubilities in the system comprising 3-ethyl-3-pentanol (1) and water (2) have been reported in two publications (Figure 1). Ginnings and Hauser (ref 1) carried out measurements of the mutual solubilities in the two phases at 293, 298, 303 and 313 K by the volumetric method. The value of the solubility of water in the alcohol-rich phase at 313 K is inconsistent with those obtained at the lower temperature. Ratouis and Dodé (ref 2) determined the solubility of (1) in the water-rich phase at one temperature (303 K) by an analytical method, their value of 1.45 g(1)/100g sln being in excellent agreement with that of 1.50 ± 0.1 g(1)/ 100g sln of ref 1. However, as the comparison involves a single point and as the remaining points are derived from only one source, the data are regarded as tentative.

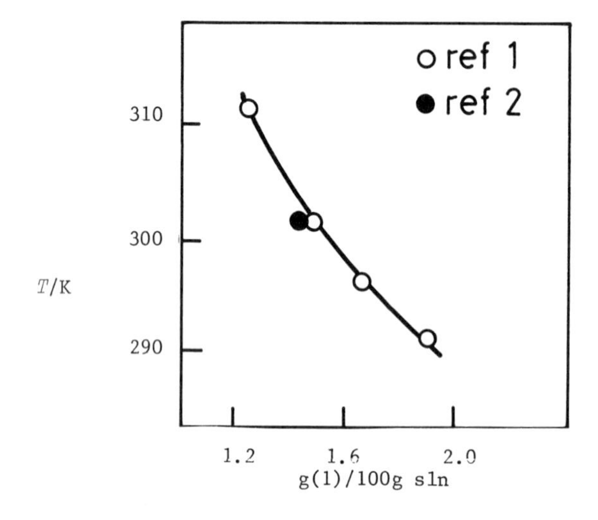

Fig. 1. Solubility of (1) in (2)

Tentative values for the mutual solubilities of 3-ethyl-3-pentanol (1) and water (2)

T/K	Water-rich phase		Alcohol-rich phase	
	g(1)/100g sln	10^3x_1	g(2)/100g sln	x_2
293	1.9	3.0	5.75	0.282
298	1.7	2.6	5.76	0.283
303	1.5	2.4	5.79	0.283
313	1.3	2.0	-	-

References

1. Ginnings, P.M.; Hauser, M. *J. Am. Chem. Soc.* 1938, *60*, 2581.

2. Ratouis, M.; Dodé, M. *Bull. Soc. Chim. Fr.* 1965, 3318.

COMPONENTS:

(1) 3-Ethyl-3-pentanol; $C_7H_{16}O$; [597-49-9]

(2) Water; H_2O; [7732-18-5]

ORIGINAL MEASUREMENTS:

Ginnings, P.M.; Hauser, M.

J. Am. Chem. Soc. 1938, *60*, 2581-2.

VARIABLES:

Temperature: 20-40°C

PREPARED BY:

A. Maczynski and Z. Maczynska

EXPERIMENTAL VALUES:

Mutual solubility of 3-ethyl-3-pentanol(1) and water(2)

$t/^\circ C$	g(1)/100g sln		x_1 (compiler)	
	(2)-rich phase	(1)-rich phase	(2)-rich phase	(1)-rich phase
20	1.91	94.25	0.00301	0.7176
25	1.68	94.24	0.00264	0.7172
30	1.50	94.21	0.00235	0.7161
40	1.26	94.31	0.00197	0.7198

Relative density, d_4

$t/^\circ C$	Water-rich phase	Alcohol-rich phase
20	0.9964	0.8541
25	0.9957	0.8502
30	0.9945	0.8457
40	0.9921	0.8366

AUXILIARY INFORMATION

METHOD/APPARATUS/PROCEDURE:

The volumetric method was used as described in ref 1.

Both components were introduced in known amounts into a two-bulb graduated and calibrated flask and shaken mechanically in a water-bath at constant temperature. After sufficient time the liquids were allowed to separate and the total volume was measured. Upon centrifugation, the phase separation line was read, and phase volumes were calculated. From the total weights of the components, the total volume, individual phase volumes, and component concentrations in either phase were evaluated.

SOURCE AND PURITY OF MATERIALS:

(1) prepared by Grignard synthesis; distilled from calcium oxide; b.p. range 143.1-143.2°C, d_4^{25} 0.8402; purity not specified.

(2) not specified.

ESTIMATED ERROR:

Solubility: better than 0.1 wt % (type of error not specified)

REFERENCES:

1. Ginnings, P.M.; Baum, R.J. *J. Am. Chem. Soc.* 1937, *59*, 1111.

COMPONENTS:

(1) 3-Ethyl-3-pentanol; $C_7H_{16}O$; [597-49-9]

(2) Water; H_2O; [7732-18-5]

ORIGINAL MEASUREMENTS:

Ratouis, M.; Dodé, M.

Bull. Soc. Chim. Fr. 1965, 3318-22.

VARIABLES:

One temperature: 30°C

Ringer solution also studied

PREPARED BY:

S.C. Valvani; S.H. Yalkowsky; A.F.M. Barton

EXPERIMENTAL VALUES:

The proportion of 3-ethyl-3-pentanol (1) in the water-rich phase at equilibrium at 30°C was reported to be 1.45 g(1)/100g sln.

The corresponding mole fraction solubility, calculated by the compilers, is x_1 = 0.00228.

The proportion of (1) in the water-rich phase of a mixture with Ringer solution at 30°C was reported to be 1.32 g(1)/100g sln.

AUXILIARY INFORMATION

METHOD/APPARATUS/PROCEDURE:

In a round bottomed flask, 50 mL of water and a sufficient quantity of alcohol were introduced until two separate layers were formed. The flask assembly was equilibrated by agitation for at least 3h in a constant temperature bath. Equilibrium solubility was attained by first supersaturation at a slightly lower temperature (solubility of alcohols in water decreases with increasing temperature) and then equilibrating at the desired temperature. The aqueous layer was separated after an overnight storage in a bath. The alcohol content was determined by reacting the aqueous solution with potassium dichromate and titrating the excess dichromate with ferrous sulfate solution in the presence of phosphoric acid and diphenylamine barium sulfonate as an indicator.

SOURCE AND PURITY OF MATERIALS:

(1) laboratory preparation; redistilled with 10:1 reflux ratio

b.p. 140-140.2°C/761 mm Hg

n_D^{25} = 1.42716

(2) twice distilled from silica apparatus or ion-exchanged with Sagei A20

ESTIMATED ERROR:

Solubility: relative error of 2 determinations less than 1%

Temperature: ± 0.05°C

REFERENCES:

COMPONENTS:	EVALUATOR:
(1) 2-Methyl-2-hexanol; $C_7H_{16}O$; [625-23-0] (2) Water; H_2O; [7732-18-5]	Z. Maczynska, Institute of Physical Chemistry of the Polish Academy of Sciences, Warsaw, Poland. November 1982

CRITICAL EVALUATION:

Solubilities in the system comprising 2-methyl-2-hexanol (1) and water (2) have been reported in two publications. Ginnings and Hauser (ref 1) carried out measurements of the mutual solubilities in the two phases at 293, 298 and 303 K by the volumetric method (Figures 1 and 2). Ratouis and Dodé (ref 2) determined the solubility of (1) in the water-rich phase at one temperature (303 K) by an analytical method, and their value of 0.82 g(1)/100g sln is in good agreement with the value 0.87 ± 0.1 g(1)/100g sln of ref 1. However, as the comparison involves a single point and as the other five points are derived from only one source, the data are regarded as being tentative.

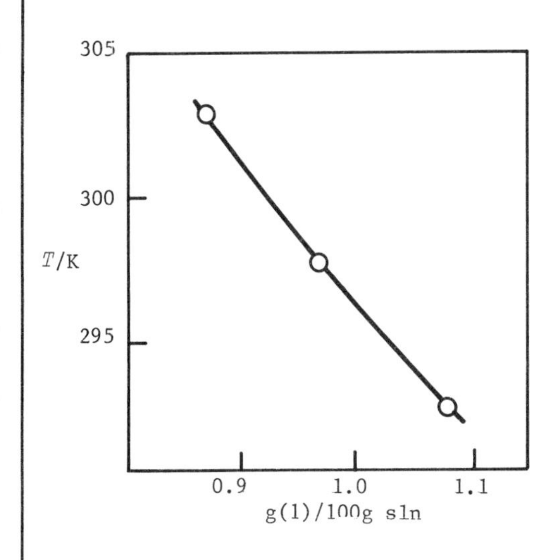

Fig. 1. Water-rich phase (ref 1)

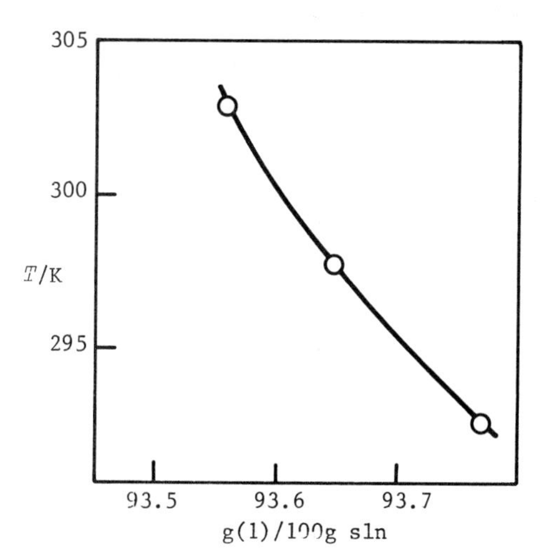

Fig. 2. Alcohol-rich phase (ref 1)

Tentative values for the mutual solubilities of 2-methyl-2-hexanol (1) and water (2)

T/K	Water-rich phase		Alcohol-rich phase	
	g(1)/100g sln	$10^3 x_1$	g(2)/100g sln	x_2
293	1.1	1.7	6.2	0.300
298	1.0	1.5	6.3	0.303
303	0.9	1.3	6.4	0.307

References

1. Ginnings, P.M.; Hauser, M. *J. Am. Chem. Soc.* 1938, *60*, 2581.

2. Ratouis, M.; Dodé, M. *Bull. Soc. Chim. Fr.* 1965, 3318.

COMPONENTS:

(1) 2-Methyl-2-hexanol; $C_7H_{16}O$; [625-23-0]

(2) Water; H_2O; [7732-18-5]

ORIGINAL MEASUREMENTS:

Ginnings, P.M.; Hauser, M.

J. Am. Chem. Soc. 1938, *60*, 2581-2.

VARIABLES:

Temperature: 20-30°C

PREPARED BY:

A. Maczynski and Z. Maczynska

EXPERIMENTAL VALUES:

Mutual solubility of 2-methyl-2-hexanol(1) and water(2)

$t/^{\circ}C$	g(1)/100g sln		x_1 (compiler)	
	(2)-rich phase	(1)-rich phase	(2)-rich phase	(1)-rich phase
20	1.08	93.77	0.00169	0.7000
25	0.97	93.65	0.00152	0.6957
30	0.87	93.56	0.00134	0.6925

Relative density, d_4

$t/^{\circ}C$	Water-rich phase	Alcohol-rich phase
20	0.9967	0.8268
25	0.9958	0.8233
30	0.9946	0.8199

AUXILIARY INFORMATION

METHOD/APPARATUS/PROCEDURE:

The volumetric method was used as described in ref 1.

Both components were introduced in known amounts into a two-bulb graduated and calibrated flask and shaken mechanically in a water-bath at constant temperature. After sufficient time the liquids were allowed to separate and the total volume was measured. Upon centrifugation, the phase separation line was read, and phase volumes were calculated. From the total weights of the components, the total volume, individual phase volumes, and component concentrations in either phase were evaluated.

SOURCE AND PURITY OF MATERIALS:

(1) prepared by Grignard synthesis; distilled from calcium oxide; b.p. range 143.0-143.2°C, d_4^{25} 0.8093; purity not specified.

(2) not specified.

ESTIMATED ERROR:

Solubility: better than 0.1 wt % (type of error not specified)

REFERENCES:

1. Ginnings, P.M.; Baum, R.J. *J. Am. Chem. Soc.* 1937, *59*, 1111.

COMPONENTS:

(1) 2-Methyl-2-hexanol; $C_7H_{16}O$; [625-23-0]

(1) Water; H_2O; [7732-18-5]

ORIGINAL MEASUREMENTS:

Ratouis, M.; Dodé, M.

Bull. Soc. Chim. Fr. 1965, 3318-22.

VARIABLES:

One temperature: 30°C

Ringer solution also studied

PREPARED BY:

S.C. Valvani; S.H. Yalkowsky; A.F.M. Barton

EXPERIMENTAL VALUES:

The proportion of 2-methyl-2-hexanol (1) in the water-rich phase at equilibrium at 30°C was reported to be 0.82 g(1)/100g sln.

The corresponding mole fraction solubility, calculated by the compilers, is x_1 = 0.00128.

The proportion of (1) in the water-rich phase of a mixture with Ringer solution at 30°C was reported to be 0.78 g(1)/100g sln.

AUXILIARY INFORMATION

METHOD/APPARATUS/PROCEDURE:

In a round bottomed flask, 50 mL of water and a sufficient quantity of alcohol were introduced until two separate layers were formed. The flask assembly was equilibrated by agitation for at least 3h in a constant temperature bath. Equilibrium solubility as attained by first supersaturation at a slightly lower temperature (solubility of alcohols in water decreases with increasing temperature) and then equilibrating at the desired temperature. The aqueous layer was separated after an overnight storage in a bath. The alcohol content was determined by reacting the aqueous solution with potassium dichromate and titrating the excess dichromate with ferrous sulfate solution in the presence of phosphoric acid and diphenylamine barium sulfonate as an indicator.

SOURCE AND PURITY OF MATERIALS:

(1) laboratory preparation; redistilled with 10:1 reflux ratio

b.p. 143.6-143.7°C/765 mm Hg

n_D^{25} = 1.41614

(2) twice distilled from silica apparatus or ion-exchanged with Sagei A20

ESTIMATED ERROR:

Solubility: relative error of 2 determinations less than 1%

Temperature: ± 0.05°C

REFERENCES:

COMPONENTS:

(1) 2-Methyl-3-hexanol; $C_7H_{16}O$; [617-29-8]

(2) Water; H_2O; [7732-18-5]

ORIGINAL MEASUREMENTS:

Ratouis, M.; Dodé, M.

Bull. Soc. Chim. Fr. 1965, 3318-22.

VARIABLES:

Temperature: 25°C and 30°C

Ringer solution also studied

PREPARED BY:

S.C. Valvani; S.H. Yalkowsky; A.F.M. Barton

EXPERIMENTAL VALUES:

Proportion of 2-methyl-3-hexanol (1) in water-rich phase

t/°C	g(1)/100g sln	$10^4 x_1$ (compilers)
25	0.57	8.9
30	0.555	8.64

The proportion of (1) in the water-rich phase of a mixture with Ringer solution at equilibrium at 30°C was reported to be 0.50 g(1)/100g sln.

AUXILIARY INFORMATION

METHOD/APPARATUS/PROCEDURE:

In a round bottomed flask, 50 mL of water and a sufficient quantity of alcohol were introduced until two separate layers were formed. The flask assembly was equilibrated by agitation for at least 3h in a constant temperature bath. Equilibrium solubility was attained by first supersaturation at a slightly lower temperature (solubility of alcohols in water decreases with increasing temperature) and then equilibrating at the desired temperature. The aqueous layer was separated after an overnight storage in a bath. The alcohol content was determined by reacting the aqueous solution with potassium dichromate and titrating the excess dichromate with ferrous sulfate solution in the presence of phosphoric acid and diphenylamine barium sulfonate as an indicator.

SOURCE AND PURITY OF MATERIALS:

(1) laboratory preparation; redistilled with 10:1 reflux ratio

b.p. 146-147°C/755 mm Hg

n_D^{25} = 1.41942

(2) twice distilled from silica apparatus or ion-exchanged with Sagei A20

ESTIMATED ERROR:

Solubility: relative error of 2 determinations less than 1%.

Temperature: ± 0.05°C

REFERENCES:

COMPONENTS:

(1) 3-Methyl-3-hexanol; $C_7H_{16}O$; [597-96-6]

(2) Water; H_2O; [7732-18-5]

EVALUATOR:

Z. Maczynska, Institute of Physical Chemistry of the Polish Academy of Sciences, Warsaw, Poland.

November 1982

CRITICAL EVALUATION:

Solubilities in the system comprising 3-methyl-3-hexanol (1) and water (2) have been reported in two publications (Figures 1 and 2). Ginnings and Hauser (ref 1) carried out measurements of the mutual solubilities in the two phases at 293, 298 and 303 K by the volumetric method. Ratouis and Dodé (ref 2) determined the solubility of (1) in the water-rich phase at one temperature (303 K) by an analytical method, their value of 1.05 g(1)/100g sln being in excellent agreement with that of 1.1 ± 0.1 g(1)/100g sln in ref 1. However, as the comparison involves a single point and as the other five points are derived from only one source, the data are regarded as being tentative.

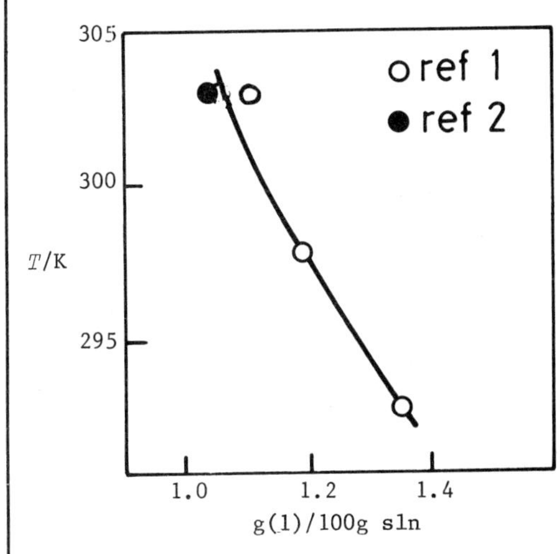

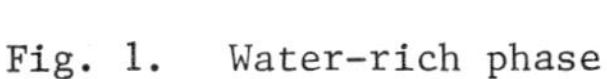
Fig. 1. Water-rich phase

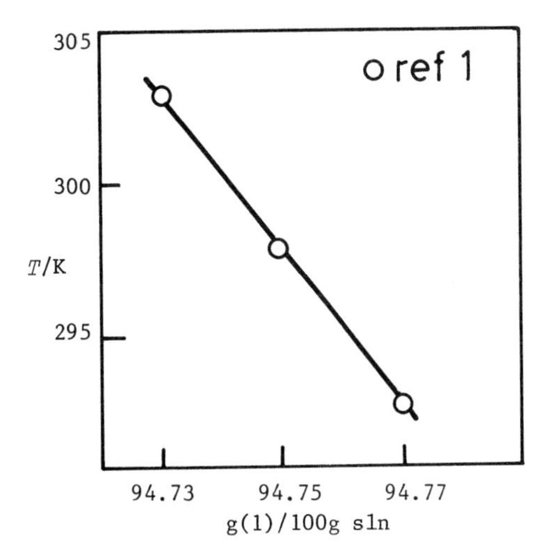

Fig. 2. Alcohol-rich phase

Tentative values for the mutual solubilities of 3-methyl-3-hexanol (1) and water (2)

T/K	Water-rich phase		Alcohol-rich phase	
	g(1)/100g sln	10^3x_1	g(2)/100g sln	x_2
293	1.4	2.1	5.2	0.26
298	1.2	1.9	5.3	0.26
303	1.1	1.7	5.3	0.26

References

1. Ginnings, P.M.; Hauser, M. *J. Am. Chem. Soc.* 1938, *60*, 2581.

2. Ratouis, M.; Dodé, M. *Bull. Soc. Chim. Fr.* 1965, 3318.

COMPONENTS:

(1) 3-Methyl-3-hexanol; $C_7H_{16}O$; [597-96-6]

(2) Water; H_2O; [7732-18-5]

ORIGINAL MEASUREMENTS:

Ginnings, P.M.; Hauser, M.

J. Am. Chem. Soc. 1938, *60*, 2581-2.

VARIABLES:

Temperature: 20-30°C

PREPARED BY:

A. Maczynski and Z. Maczynska

EXPERIMENTAL VALUES:

Mutual solubility of 3-methyl-3-hexanol(1) and water(2)

$t/^{\circ}C$	g(1)/100g sln		x_1 (compiler)	
	(2)-rich phase	(1)-rich phase	(2)-rich phase	(1)-rich phase
20	1.35	94.77	0.00212	0.7374
25	1.19	94.75	0.00186	0.7366
30	1.07	94.73	0.00167	0.7359

Relative density, d_4

$t/^{\circ}C$	Water-rich phase	Alcohol-rich phase
20	0.9965	0.8348
25	0.9958	0.8312
30	0.9946	0.8272

AUXILIARY INFORMATION

METHOD/APPARATUS/PROCEDURE:

The volumetric method was used as described in ref 1.

Both components were introduced in known amounts into a two-bulb graduated and calibrated flask and shaken mechanically in a water-bath at constant temperature. After sufficient time the liquids were allowed to separate and the total volume was measured. Upon centrifugation, the phase separation line was read, and phase volumes were calculated. From the total weights of the components, the total volume, individual phase volumes, and component concentrations in either phase were evaluated.

SOURCE AND PURITY OF MATERIALS:

(1) prepared by Grignard synthesis; distilled from calcium oxide; b.p. range 142.7-142.9°C, d_4^{25} 0.8202; purity not specified.

(2) not specified.

ESTIMATED ERROR:

Solubility: better than 0.1 wt % (type of error not specified)

REFERENCES:

1. Ginnings, P.M.; Baum, R.J. *J. Am. Chem. Soc.* 1937, *59*, 1111.

COMPONENTS:

(1) 3-Methyl-3-hexanol; $C_7H_{16}O$; [597-96-6]

(2) Water; H_2O; [7732-18-5]

ORIGINAL MEASUREMENTS:

Ratouis, M.: Dodé, M.

Bull. Soc. Chim. Fr. 1965, 3318-22.

VARIABLES:

One temperature: 30°C

Ringer solution also studied

PREPARED BY:

S.C. Valvani; S.H. Yalkowsky; A.F.M. Barton

EXPERIMENTAL VALUES:

The proportion of 3-methyl-3-hexanol (1) in the water-rich phase at equilibrium at 30°C was reported to be 1.05 g(1)/100g sln.

The corresponding mole fraction solubility calculated by the compilers is x_1 = 0.00164.

The proportion of (1) in the water-rich phase of a mixture with Ringer solution at 30°C was reported to be 0.97 g(1)/100g sln.

AUXILIARY INFORMATION

METHOD/APPARATUS/PROCEDURE:

In a round bottomed flask, 50 mL of water and a sufficient quantity of alcohol were introduced until two separate layers were formed. The flask assembly was equilibrated by agitation for at least 3h in a constant temperature bath. Equilibrium solubility was attained by first supersaturation at a slightly lower temperature (solubility of alcohols in water decreases with increasing temperature) and then equilibrating at the desired temperature. The aqueous layer was separated after an overnight storage in a bath. The alcohol content was determined by reacting the aqueous solution with potassium dichromate and titrating the excess dichromate with ferrous sulfate solution in the presence of phosphoric acid and diphenylamine barium sulfonate as an indicator.

SOURCE AND PURITY OF MATERIALS:

(1) laboratory preparation; redistilled with 10:1 reflux ratio

b.p. 75.5°C/60 mm Hg

n_D^{25} = 1.42076

(2) twice distilled from silica apparatus or ion-exchanged with Sagei A20

ESTIMATED ERROR:

Solubility: relative error of 2 determinations less than 1%

Temperature: ± 0.05°C

REFERENCES:

COMPONENTS:

(1) 5-Methyl-2-hexanol; $C_7H_{16}O$; [627-59-8]

(2) Water; H_2O; [7732-18-5]

ORIGINAL MEASUREMENTS:

Ratouis, M.; Dodé, M.

Bull. Soc. Chim. Fr. 1965, 3318-22.

VARIABLES:

Temperature: 25°C and 30°C

Ringer solution also studied

PREPARED BY:

S.C. Valvani; S.H. Yalkowsky; A.F.M. Barton

EXPERIMENTAL VALUES:

Proportion of 5-methyl-2-hexanol (1) in water-rich phase

t/°C	g(1)/100g sln	$10^4 x_1$ (compilers)
25	0.49	7.6
30	0.47	7.3

The proportion of (1) in the water-rich phase of a mixture with Ringer solution at equilibrium at 30°C was reported to be 0.37 g(1)/100g sln.

AUXILIARY INFORMATION

METHOD/APPARATUS/PROCEDURE:

In a round bottomed flask, 50 mL of water and a sufficient quantity of alcohol were introduced until two separate layers were formed. The flask assembly was equilibrated by agitation for at least 3h in a constant temperature bath. Equilibrium solubility was attained by first supersaturation at a slightly lower temperature (solubility of alcohols in water decreases with increasing temperature) and then equilibrating at the desired temperature. The aqueous layer was separated after an overnight storage in a bath. The alcohol content was determined by reacting the aqueous solution with potassium dichromate and titrating the excess dichromate with ferrous sulfate solution in the presence of phosphoric acid and diphenylamine barium sulfonate as an indicator.

SOURCE AND PURITY OF MATERIALS:

(1) laboratory preparation; redistilled with 10:1 reflux ratio

b.p. 152.6-153.5°C/760.5 mm Hg

n_D^{25} = 1.41718

(2) twice distilled from silica apparatus or ion-exchanged with Sagei A20

ESTIMATED ERROR:

Solubility: relative error of 2 determinations less than 1%

Temperature: ± 0.05°C

REFERENCES:

COMPONENTS:	EVALUATOR:
(1) 1-Heptanol (*n-heptyl alcohol*); $C_7H_{16}O$; [111-70-6] (2) Water; H_2O; [7732-18-5]	G.T. Hefter and A.F.M. Barton, Murdoch University, Perth, Western Australia. June, 1983

CRITICAL EVALUATION:

Solubilities in the 1-heptanol (1) - water (2) system have been reported in the following publications.

Reference	T/K	Solubility	Method
Fuhner (ref 1)	343-403	(1) in (2)	synthetic
Butler *et al.* (ref 2)	298	(1) in (2)	interferometric
Addison (ref 3)	293	(1) in (2)	surface tension
Booth and Everson (ref 4)	298	(1) in (2)	titration
Harkins and Oppenheimer (ref 5)	-	(1) in (2)	turbidimetric
Donahue and Bartell (ref 6)	298	mutual	analytical
Erichsen (ref 7)	273-526	mutual	synthetic
Erichsen (ref 8)	273-323	(1) in (2)	synthetic
Kinoshita *et al.* (ref 9)	298	(1) in (2)	surface tension
Rao *et al.* (ref 10)	303	mutual	turbidity
Ratouis and Dodé (ref 11)	303	(1) in (2)	analytical
Hanssens (ref 12)	298	(1) in (2)	refractometric
Krasnov and Gartseva (ref 13)	285,313	(2) in (1)	analytical
Ionin and Shanina (ref 14)	298	(2) in (1)	titration
Vochten and Petre (ref 15)	288-333	(1) in (2)	surface tension
Hill and White (ref 16)	279-308	(1) in (2)	interferometric
Korenman *et al.* (ref 17)	298	mutual	analytical
Nishino and Nakamura (ref 18)	280-350	mutual	synthetic
Tokunaga *et al.* (ref 19)	288-313	(2) in (1)	analytical

The original data are compiled in the data sheets immediately following this Critical Evaluation.

The 1-heptanol-water system is interesting in that unlike most of the higher 1-alkanols the data for the water-rich phase is more extensive and self-consistent than the data for the alcohol-rich phase.

In preparing this Critical Evaluation use has been made of the fact that the solubility of 1-heptanol in water is sufficiently low to enable weight/volume data (ref 6, 12, 15, 17) to be converted to weight/weight values by assuming the water-rich phase has the same density as pure water. The same is not true for the alcohol-rich phase and thus such data (ref 6, 12, 17) have been excluded from further consideration. The data of Booth and Everson (ref 4) given in volume/volume units and the graphical data of Nishino and Nakamura (ref 18) were also excluded.

In the water-rich phase the converted data of Korenman *et al.* (ref 17) and the data of Rao *et al.* (ref 10) disagree markedly with all other values and have therefore been rejected. The data of Harkins and Oppenheimer (ref 5) are also rejected as the temperature was not specified.

All other data are included in the Tables below.

(continued next page)

COMPONENTS:	EVALUATOR:
(1) 1-Heptanol (*n-heptyl alcohol*); $C_7H_{16}O$; [111-70-6]	G.T. Hefter and A.F.M. Barton, Murdoch University, Perth, Western Australia.
(2) Water; H_2O; [7732-18-5]	June, 1983

CRITICAL EVALUATION (continued)

Values obtained by the Evaluator by graphical interpolation or extrapolation from the data sheets are indicated by an asterisk(*). "Best" values have been obtained by simple averaging. The uncertainty limits (σ_n) attached to the "best" values do not have statistical significance and should be regarded only as a convenient representation of the spread of reported values and not as error limits. The letter (*R*) designates "Recommended" data. Data are "Recommended" if two or more apparently reliable studies are in reasonable agreement ($\leq \pm$ 5% relative).

For convenience the two phases will be further discussed separately.

<u>The solubility of 1-heptanol (1) in water (2)</u>

The various data available are generally in excellent agreement and enable solubilities to be *Recommended* over an unusually wide range of temperature (see particularly ref 1 and ref 7). Above 403 K, the data of Erichsen (ref 7) are the only values available and hence should be considered as *Tentative*.

<u>Recommended (*R*) and tentative solubilities of 1-heptanol (1) in water (2)</u>

T/K	Solubility, g(1)/100g sln — Reported values	"Best" values ($\pm\sigma_n$)	
273	0.34 (ref 7), 0.34 (ref 8)	0.34	
283	0.26 (ref 7), 0.30 (ref 8), 0.203* (ref 16)	0.25 ± 0.04	
293	0.172 (ref 3), 0.20 (ref 7), 0.20 (ref 8), 0.15* (ref 15), 0.175* (ref 16)	0.17 ± 0.02	(*R*)
298	0.181 (ref 2), 0.180 (ref 6), 0.18* (ref 7), 0.17 (ref 9), 0.17 (ref 12), 0.168* (ref 16)	0.174 ± 0.005	(*R*)
303	0.16 (ref 7), 0.16 (ref 8), 0.17 (ref 11), 0.15* (ref 15), 0.162* (ref 16)	0.160 ± 0.007	(*R*)
313	0.13 (ref 7), 0.13 (ref 8), 0.13 (ref 15)	0.13	(*R*)
323	0.11 (ref 7), 0.12 (ref 8), 0.14 (ref 15)	0.12 ± 0.01	(*R*)
333	0.11 (ref 7), 0.15 (ref 15)	0.13 ± 0.02	
343	0.125 (ref 1), 0.15 (ref 7)	0.14 ± 0.01	(*R*)
353	0.17 (ref 1), 0.19 (ref 7)	0.18 ± 0.01	(*R*)
363	0.225 (ref 1), 0.23 (ref 7)	0.23	(*R*)
373	0.285 (ref 1), 0.29 (ref 7)	0.29	(*R*)
383	0.335 (ref 1), 0.35 (ref 7)	0.34 ± 0.01	(*R*)
393	0.43 (ref 1), 0.43 (ref 7)	0.43	(*R*)
403	0.515 (ref 1), 0.53 (ref 7)	0.52 ± 0.01	(*R*)
413	0.65 (ref 7)	0.65	
423	0.80 (ref 7)	0.80	
433	0.98 (ref 7)	1.0	
443	1.23 (ref 7)	1.2	
453	1.60 (ref 7)	1.6	

(continued next page)

COMPONENTS:	EVALUATOR:
(1) 1-Heptanol (*n-heptyl alcohol*); $C_7H_{16}O$; [111-70-6] (2) Water; H_2O; [7732-18-5]	G.T. Hefter and A.F.M. Barton, Murdoch University, Perth, Western Australia. June, 1983

CRITICAL EVALUATION (continued)

T/K	Solubility, g(1)/100g sln Reported values	"Best" values ($\pm\sigma_n$)
463	2.08 (ref 7)	2.1
473	2.64 (ref 7)	2.6
483	3.48 (ref 7)	3.5
493	4.68 (ref 7)	4.7
503	6.70 (ref 7)	6.7
513	10.10 (ref 7)	10.1

The solubility of water (2) in 1-heptanol (1)

Data for the alcohol-rich phase, unlike the water-rich phase considered above, is rather limited. Even those data which are available are not in good agreement and with the exception of the value at 35°C are regarded as *Tentative* only. This phase warrants thorough re-investigation.

Recommended (*R*) and tentative solubilities of water (2) in 1-heptanol (1)

T/K	Solubility, g(2)/100g sln Reported values	"Best" values ($\pm\sigma_n$)	
273	3.90 (ref 7)		
283	4.30 (ref 7)		
293	4.75 (ref 7), 5.82 (ref 19)	5.3 ± 0.5	
298	5.3 (ref 6), 5.0* (ref 7), 5.8 (ref 14), 5.85 (ref 19)	5.4 ± 0.4	
303	5.20 (ref 7), 5.5 (ref 10), 5.90 (ref 19)	5.5 ± 0.3	
313	5.75 (ref 7), 5.90 (ref 13), 6.02 (ref 19)	5.9 ± 0.1	(*R*)
323	6.30 (ref 7)	6.3	
333	6.95 (ref 7)	7.0	
343	7.60 (ref 7)	7.6	
353	8.35 (ref 7)	8.4	
363	9.10 (ref 7)	9.1	
373	9.85 (ref 7)	9.9	
383	10.75 (ref 7)	10.8	
393	11.65 (ref 7)	11.7	
403	12.65 (ref 7)	12.7	
413	13.75 (ref 7)	13.8	
423	14.90 (ref 7)	14.9	
433	16.20 (ref 7)	16.2	
443	17.65 (ref 7)	17.7	
453	19.25 (ref 7)	19.3	
463	21.10 (ref 7)	21.1	

(continued next page)

COMPONENTS:	EVALUATOR:
(1) 1-Heptanol (*n-heptyl alcohol*); $C_7H_{16}O$; [111-70-6] (2) Water; H_2O; [7732-18-5]	G.T. Hefter and A.F.M. Barton, Murdoch University, Perth, Western Australia. June, 1983.

CRITICAL EVALUATION (continued)

T/K	Solubility, g(1)/100g sln	
	Reported values	"Best" values ($\pm\sigma_n$)
473	23.20 (ref 7)	23.2
483	25.80 (ref 7)	25.8
493	29.15 (ref 7)	29.2
503	33.75 (ref 7)	33.8
513	41.50 (ref 7)	41.5

The upper critical solution temperature

The UCST appears to have been reported only by Erichsen (ref 7) who gave a value of 248.5° C (521.7 K).

Representative data for the mutual solubilities of 1-heptanol and water are plotted in Figure 1.

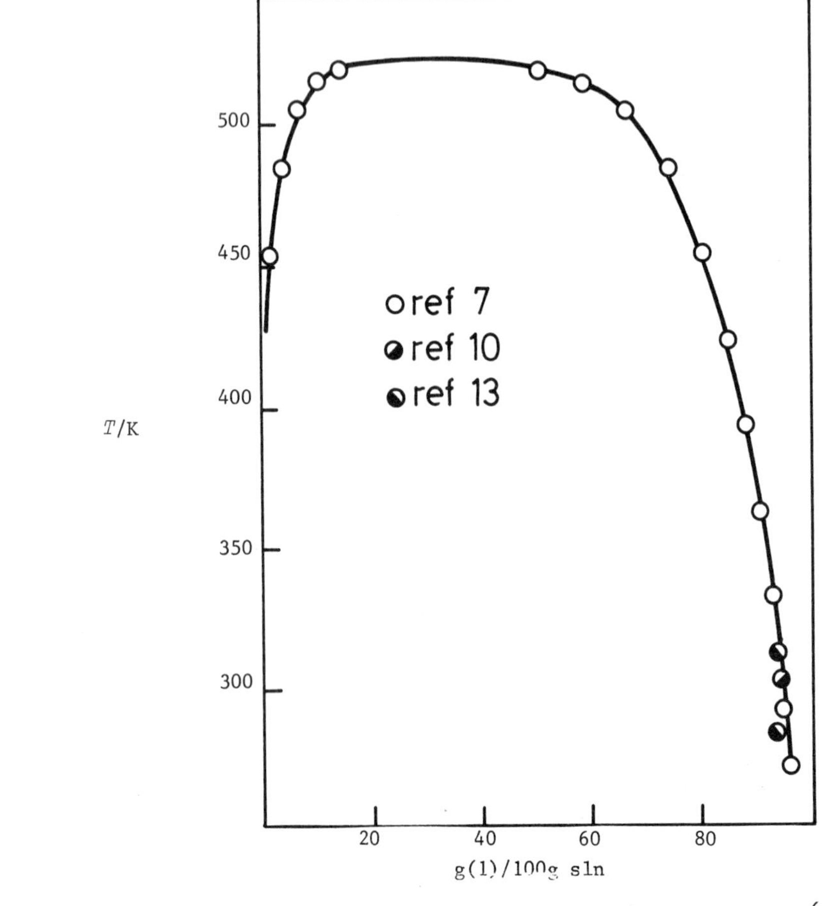

Fig. 1. Mutual solubility of (1) and (2)

(continued next page)

COMPONENTS:

(1) 1-Heptanol (*n-heptyl alcohol*); $C_7H_{16}O$; [111-70-6]

(2) Water; H_2O; [7732-18-5]

EVALUATOR:

G.T. Hefter and A.F.M. Barton, Murdoch University, Perth, Western Australia. June, 1983.

CRITICAL EVALUATION (continued)

References

1. Führer, H. *Ber. Dtsch. Chem. Ges.* 1924, *57*, 510.
2. Butler, J.A.V.; Thomson, D.W.; Maclennan, W.H. *J. Chem Soc.* 1933, 674.
3. Addison, C.C. *J. Chem. Soc.* 1945, 98.
4. Booth, H.S.; Everson, H.E. *Ind. Eng. Chem.* 1948, *40*, 1491.
5. Harkins, W.D.; Oppenheimer, H. *J. Am. Chem. Soc.* 1949, *71*, 808.
6. Donahue, D.J.; Bartell, F.E. *J. Phys. Chem.* 1952, *56*, 480.
7. Erichsen, L. von. *Brennst. Chem.* 1952, *33*, 166.
8. Erichsen, L. von. *Naturwissenschaften* 1952, *39*, 41.
9. Kinoshita, K.; Ishikawa, H.; Shinoda, K. *Bull. Chem. Soc. Jpn.* 1958, *31*, 1081.
10. Rao, K.S.; Rao, M.V.R.; Rao, C.V. *J. Sci. Ind. Res.* 1961, *20B*, 283.
11. Ratouis, M.; Dode, M. *Bull. Soc. Chim. Fr.* 1965, 3318.
12. Hanssens, I. *Associatie van normale alcoholen en hun affiniteit voor water en organische solventen*, Doctoraatsproefschrift, Leuven, 1969; Huyskens, P.; Mullens, J.; Gomez, A.; Tack, J. *Bull. Soc. Chim. Belg.* 1975, *84*, 253.
13. Krasnov, K.S.1 Gartseva, L.A. *Izv. Vysshykh Uchebn. Zavednii Khim. Tekhnol.* 1970, *13*, 952.
14. Ionin, M.V.; Shanina, P.I. *Zh. Neorg. Khim.* 1972, *17*, 1444; *Russ. J. Inorg. Chem.* 1972, *17*, 747.
15. Vochten, R.; Petre, G. *J. Colloid Interface Sci.* 1973, *42*, 320.
16. Hill, D.J.T.; White, L.R. *Aust. J. Chem.* 1974, *27*, 1905.
17. Korenman, I.M.; Gorokhov, A.A.; Polozenko, G.N. *Zhur. Fiz. Khim.* 1974, *48*, 1010; *Russ. J. Phys. Chem.* 1974, *48*, 1065. *Zhur. Fiz. Khim.* 1975, *49*, 1490; *Russ. J. Phys. Chem.* 1975, *49*, 545.
18. Nishino, N.; Nakamura, M. *Bull. Chem. Soc. Jpn.* 1978, *51*, 1617; 1981, *54*, 545.
19. Tokunaga, S.; Manabe, M.; Koda, M. *Niihama Kogyo Koto Semmon Gakko Kiyo, Rikogaku Hen (Memoirs Niihama Technical College, Sci. and Eng.)* 1980, *16*, 96.

COMPONENTS:

(1) 1-Heptanol; $C_7H_{16}O$; [111-70-6]

(2) Water; H_2O; [7732-18-5]

ORIGINAL MEASUREMENTS:

Fühner, H.

Ber. Dtsch. Chem. Ges. 1924, *57*, 510-5.

VARIABLES:

Temperature: 70-130°C

PREPARED BY:

A. Maczynski, Z. Maczynska and A. Szafranski

EXPERIMENTAL VALUES:

Solubility of 1-heptanol (1) in water (2)

$t/^{\circ}C$	g(1)/100g sln	x_1 (compiler)
70	0.125	0.000194
80	0.17	0.000264
90	0.225	0.000349
100	0.285	0.000443
110	0.335	0.000521
120	0.43	0.000669
130	0.515	0.000802

AUXILIARY INFORMATION

METHOD/APPARATUS/PROCEDURE:

Rothmund's synthetic method was used (ref 1).

Small amounts of (1) and (2) were sealed in a glass tube and heated with shaking in an oil bath to complete dissolution. The solution was cooled until a milky turbidity appeared and this temperature was adopted as the equilibrium temperature.

SOURCE AND PURITY OF MATERIALS:

(1) source not specified; specially purified, but no details provided.

ESTIMATED ERROR:

Not specified.

REFERENCES:

1. Rothmund, V. *Z. physik. Chem.* 1898, *26*, 433.

COMPONENTS:

(1) 1-Heptanol; $C_7H_{16}O$; [111-70-6]

(2) Water; H_2O [7732-18-5]

ORIGINAL MEASUREMENTS:

Butler, J.A.V.; Thomson, D.W.; Maclennan, W.H.

J. Chem. Soc. 1933, 674-86.

VARIABLES:

One temperature: 25°C

PREPARED BY:

S.H. Yalkowsky; S.C. Valvani; A.F.M. Barton

EXPERIMENTAL VALUES:

The proportion of 1-heptanol (1) in the water-rich phase at equilibrium at 25°C was reported to be 0.181 g(1)/100g sln, the mean of seven determinations (0.179, 0.184, 0.180, 0.182, 0.180, 0.178, 0.182 g(1)/100g sln).

The corresponding mole fraction solubility was reported as x_1 = 0.000281.

AUXILIARY INFORMATION

METHOD/APPARATUS/PROCEDURE:

An analytical method was used, with a U-tube apparatus having two internal stoppers. Suitable quantities of (1) and (2) were placed in one of the connected vessels and shaken in a thermostat for some hours. The liquid was allowed to separate into two layers, the heavier aqueous layer being separated by raising the stopper and allowing part of the liquid to run into the connected vessel. A weighed portion of the separated sln was diluted with about an equal quantity of (2) and the resulting sln compared with calibration slns in an interferometer. To avoid the possibility of reading the position of the wrong fringe, 2 cells (1 cm and 5 cm) were used. The method was unsuitable for analysis of alcohol-rich slns, as no stoppered interferometer cell was available.

SOURCE AND PURITY OF MATERIALS:

(1) BDH;
repeatedly fractionated in vacuum with a Hempel column, the middle fractions being refluxed with Ca and refractionated;
b.p. 87.5 - 87.6°C/9 mm Hg, 175.6/760 mm Hg
d_4^{25} 0.81960, n_D^{20} 1.42337

(2) not stated.

ESTIMATED ERROR:

Solubility: the result is the mean of seven determinations agreeing within 0.003 g(1)/100g sln.

Temperature: not stated (but in related experiments it was ± 0.03°C)

REFERENCES:

COMPONENTS:

(1) 1-Heptanol; $C_7H_{16}O$; [111-70-6]

(2) Water; H_2O; [7732-18-5]

ORIGINAL MEASUREMENTS:

Addison, C.C.

J. Chem. Soc. 1945, 98-106.

VARIABLES:

One temperature: 20°C

PREPARED BY:

S.H. Yalkowsky; S.C. Valvani; A.F.M. Barton

EXPERIMENTAL VALUES:

The proportion of 1-heptanol (1) in the water-rich phase at equilibrium at 20°C was reported to be 0.172 g(1)/100g sln.

The corresponding mole fraction solubility, calculated by the compilers, is x_1 = 0.000267.

AUXILIARY INFORMATION

METHOD/APPARATUS/PROCEDURE:

The surface tension method was used. Sufficient excess of (1) was added to 100 mL of (2) in a stoppered flask to form a separate lens on the surface. The mixture was swirled gently, too vigorous an agitation being avoided as this gave a semi-permanent emulsion and incorrect readings. After settling, a small sample of the clear aqueous sln was withdrawn into a drop weight pipet and the surface tension determined. The swirling was continued until a constant value was obtained. The surface tension-concentration curve was known, and only a slight extrapolation (logarithmic scale) was necessary to find the concentration corresponding to the equilibrium value.

SOURCE AND PURITY OF MATERIALS:

(1) impure alcohols were purified by fractional distillation, the middle fraction from a distillation being redistilled;

b.p. 176.5°C d_4^{20} 0.8220

n_D^{20} 1.4241

(2) not stated

ESTIMATED ERROR:

Solubility: ± 0.5%

REFERENCES:

COMPONENTS:

(1) 1-Heptanol; $C_7H_{16}O$; [111-70-6]

(2) Water; H_2O; [7732-18-5]

ORIGINAL MEASUREMENTS:

Booth, H.S.; Everson, H.E.
Ind. Eng. Chem. 1948, *40*, 1491-3.

VARIABLES:

One temperature: 25°C

Sodium xylene sulfonate

PREPARED BY:

S.H. Yalkowsky; S.C. Valvani; A.F.M. Barton

EXPERIMENTAL VALUES:

It was reported that the solubility of 1-heptanol (1) in water (2) was 0.40 mL(1)/100mL (2) at 25.0°C.

The solubility in 40% sodium xylene sulfonate solution as solvent was also studied. A gel began to form at 20 mL(1)/100mL sodium xylene sulfonate solution, and the amount of gel increased on successive additions. At 80 mL(1)/100mL sln the gel began to redissolve and at 120 mL(1)/100mL it completely disappeared.

AUXILIARY INFORMATION

METHOD/APPARATUS/PROCEDURE:

A known volume of (2) or aqueous solvent (usually 50mL) in a tightly stoppered Babcock tube was thermostatted. Successive measured quantities of (1) were added and equilibrated until a slight excess of (1) remained. The solution was centrifuged, returned to the thermostat bath for 10 mins, and the volume of excess (1) measured directly. This was a modification of the method described in ref 1.

SOURCE AND PURITY OF MATERIALS:

(1) "CP or highest grade commercial"

(2) distilled.

ESTIMATED ERROR:

Solubility: within 0.1 mL(1)/100mL (2)

REFERENCES:

1. Hanslick, R.S. *Dissertation*, Columbia University, 1935.

COMPONENTS:	ORIGINAL MEASUREMENTS:
(1) 1-Heptanol; $C_7H_{16}O$; [111-70-6] (2) Water; H_2O; [7732-18-5]	Harkins, W.D.; Oppenheimer, H. *J. Am. Chem. Soc.* 1949, *71*, 808-11.
VARIABLES:	PREPARED BY:
One temperature: assumed to be ambient	S.H. Yalkowsky, S.C. Valvani; A.F.M. Barton

EXPERIMENTAL VALUES:

The solubility of 1-heptanol (1) in water was reported to be 0.0103 mol(1)/kg (2).

Corresponding values calculated by the compilers are mass percentage 0.120 g(1)/100g sln and mole fraction $10^4 x_1$ = 1.86.

The temperature was not stated, but is assumed to be ambient.

AUXILIARY INFORMATION

METHOD/APPARATUS/PROCEDURE:	SOURCE AND PURITY OF MATERIALS:
A turbidimetric method was used, with the aid of a photometer (turbidity increased rapidly in the presence of the emulsified second phase). The study was concerned with the effect of long-chain electrolytes (soaps) on solubility. The components were weighed into a glass ampoule, and shaken vigorously for at least 48h. Equilibrium was approached from both undersaturation and supersaturation.	(1) Eastman Kodak Co; or Conneticut Hard Rubber Co; purified by fractional distillation. (2) not stated
	ESTIMATED ERROR:
	not stated
	REFERENCES:

COMPONENTS:

(1) 1-Heptanol; $C_7H_{16}O$; [111-70-6]

(2) Water; H_2O; [7732-18-5]

ORIGINAL MEASUREMENTS:

Donahue, D.J.; Bartell, F.E.

J. Phys. Chem. 1952, *56*, 480-4.

VARIABLES:

One temperature: 25°C

PREPARED BY:

A.F.M. Barton

EXPERIMENTAL VALUES:

	Density	Mutual solubilities	
	g mL^{-1}	x_1	g(1)/100g sln (compiler)
Alcohol-rich phase	0.8268	0.733	94.7
Water-rich phase	0.9967	0.000280[a]	0.180

[a] From ref 1.

AUXILIARY INFORMATION

METHOD/APPARATUS/PROCEDURE:

Mixtures were placed in glass stoppered flasks and were shaken intermittently for at least 3 days in a water bath. The organic phase was analyzed for water content by the Karl Fischer method and the aqueous phase was analyzed interferometrically. The solubility measurements formed part of a study of water-organic liquid interfacial tensions.

SOURCE AND PURITY OF MATERIALS:

(1) "best reagent grade"; fractional distillation

(2) purified

ESTIMATED ERROR:

Temperature: ± 0.1°C

REFERENCES:

(1) Butler, J.A.V.; Thomson, D.W.; Maclennan, W.H. *J. Chem. Soc.* 1933, 674.

COMPONENTS:

(1) 1-Heptanol; $C_7H_{16}O$; [111-70-6]

(2) Water; H_2O; [7732-18-5]

ORIGINAL MEASUREMENTS:

Erichsen, L. von

Brennst. Chem. 1952, *33*, 166-72.

VARIABLES:

Temperature: 0-245°C

PREPARED BY:

S.H. Yalkowsky and Z. Maczynska

EXPERIMENTAL VALUES:

Mutual solubility of 1-heptanol and water

$t/^{\circ}C$	(2)-rich phase		(1)-rich phase	
	g(1)/100 sln	x_1	g(1)/100g sln	x_1
0	0.34	0.00053	96.10	0.7926
10	0.26	0.00046	95.70	0.7753
20	0.20	0.00031	95.25	0.7567
30	0.16	0.00025	94.80	0.7387
40	0.13	0.00020	94.25	0.7177
50	0.11	0.00018	93.70	0.6977
60	0.11	0.00018	93.05	0.6750
70	0.15	0.00023	92.40	0.6535
80	0.19	0.00030	91.65	0.6299
90	0.23	0.00036	90.90	0.6076
100	0.29	0.00047	90.15	0.5868
110	0.35	0.00055	89.25	0.5629
120	0.43	0.00067	88.35	0.5405
130	0.53	0.00087	87.35	0.5171
140	0.65	0.0010	86.25	0.4931
150	0.80	0.0012	85.10	0.4697
160	0.98	0.0015	83.80	0.4450
170	1.23	0.0019	82.35	0.4198
180	1.60	0.0025	80.75	0.3941
190	2.08	0.0032	78.90	0.3671
200	2.64	0.0042	76.80	0.3392
210	3.48	0.0056	74.20	0.3084
220	4.68	0.0076	70.85	0.2731
230	6.70	0.0110	66.25	0.2334
240	10.10	0.0171	58.50	0.1794
245	13.96	0.0245	50.65	0.1373

The UCST is 248.5°C

AUXILIARY INFORMATION

METHOD/APPARATUS/PROCEDURE:

The synthetic method was used.

The measurements were carried out in 2 ml glass ampules. These were placed in an aluminium block equipped with two glass windows. Cloud points were measured with a thermocouple wound up around the ampule. Each measurement was repeated twice.

SOURCE AND PURITY OF MATERIALS:

(1) Merck, or Ciba, or industrial product; distilled and chemically free from isomers; b.p. 175.9-176-0°C (758 mm Hg) n_D^{20} 1.4249.

(2) not specified.

ESTIMATED ERROR:

Not specified.

REFERENCES:

COMPONENTS:

(1) 1-Heptanol; $C_7H_{16}O$; [111-70-6]

(2) Water; H_2O; [7732-18-5]

ORIGINAL MEASUREMENTS:

Erichsen, L. von

Naturwissenschaften 1952, *39*, 41-2.

VARIABLES:

Temperature: 0-50°C

PREPARED BY:

A. Maczynski and Z. Maczynska

EXPERIMENTAL VALUES:

Solubility of 1-heptanol (1) in water (2)

$t/^{\circ}C$	x_1	g(1)/100g sln (compiler)
0	0.00053	0.34
10	0.00046	0.30
20	0.00031	0.20
30	0.00025	0.16
40	0.00020	0.13
50	0.00018	0.12

AUXILIARY INFORMATION

METHOD/APPARATUS/PROCEDURE:

The synthetic method was used.

No details reported in the paper.

SOURCE AND PURITY OF MATERIALS:

(1) not specified.

(2) not specified.

ESTIMATED ERROR:

Not specified.

REFERENCES:

COMPONENTS:

(1) 1-Heptanol; $C_7H_{16}O$; [111-70-6]

(2) Water; H_2O; [7732-18-5]

ORIGINAL MEASUREMENTS:

Kinoshita, K.; Ishikawa, H.; Shinoda, K.
Bull. Chem. Soc. Jpn. 1958, *31*, 1081-4.

VARIABLES:

One temperature: 25°C

PREPARED BY:

S.H. Yalkowsky; S.C. Valvani; A.F.M. Barton

EXPERIMENTAL VALUES:

The equilibrium concentration of 1-heptanol (1) in the water-rich phase at 25°C was reported to be 0.0146 mol(1)/L, the mass percentage solubility was reported as 0.17 g(1)/100g sln, and the corresponding mole fraction solubility, calculated by the compiler, is x_1 = 0.00026.

AUXILIARY INFORMATION

METHOD/APPARATUS/PROCEDURE:

The surface tension in aqueous solutions of alcohols monotonously decreases up to their saturation cencentration and remains constant in the heterogeneous region (ref 1-4). Surface tension was measured by the drop weight method, using a tip 6 mm in diameter, the measurements being carried out in a water thermostat. From the (surface tension) - (logarithm of concentration) curves the saturation points were determined as the intersections of the curves with the horizontal straight lines passing through the lowest experimental points.

SOURCE AND PURITY OF MATERIALS:

(1) Kahlbaum 'pure grade'

(2) not stated

ESTIMATED ERROR:

Temperature: ± 0.05°C

Solubility: within 4%

REFERENCES:

(1) Motylewski, S. *Z. Anorg. Chem.* 1904, *38*, 410.

(2) Taubamann, A. *Z. physik. Chem.* 1932, *A161*, 141.

(3) Zimmerman, H.K., Jr. *Chem. Rev.* 1952, *51*, 25.

(4) Shinoda, K.; Yamanaka, T.; Kinoshita, K. *J. Phys. Chem.* 1959, *63*, 648.

COMPONENTS:

(1) 1-Heptanol; $C_7H_{16}O$; [111-70-6]

(2) Water; H_2O; [7732-18-5]

ORIGINAL MEASUREMENTS:

Rao, K.S.; Rao, M.V.R.; Rao, C.V.

J. Sci. Ind. Res. 1961, *20B*, 283-6.

VARIABLES:

One temperature: 30°C

PREPARED BY:

A. Maczynski

EXPERIMENTAL VALUES:

The solubility of 1-heptanol in water at 30°C was reported to be 0.1 g(1)/100 sln.

The corresponding mole fraction, x_1, calculated by the compiler is 0.0002.

The solubility of water in 1-heptanol at 30°C was reported to be 5.5 g(2)/100g sln.

The corresponding mole fraction, x_2, calculated by the compiler is 0.27.

AUXILIARY INFORMATION

METHOD/APPARATUS/PROCEDURE:

The method of appearance and disappearance of turbidity described in ref 1 was used.

No details were reported in the paper.

SOURCE AND PURITY OF MATERIALS:

(1) British Drug House; used as received; n^{30} 1.4193, d^{30} 0.8210 g/mL

(2) distilled; free from carbon dioxide.

ESTIMATED ERROR:

Not specified.

REFERENCES:

1. Othmer, D.F.; White, R.E.; Trueger, E. *Ind. Eng. Chem.* 1941, *33*, 1240.

COMPONENTS:

(1) 1-Heptanol; $C_7H_{16}O$; [111-70-6]

(2) Water; H_2O; [7732-18-5]

ORIGINAL MEASUREMENTS:

Ratouis, M.; Dodé, M.

Bull. Soc. Chim. Fr. 1965, 3318-22

VARIABLES:

One temperature: 30°C

Ringer solution also studied

PREPARED BY:

S.C. Valvani; S.H. Yalkowsky; A.F.M. Barton

EXPERIMENTAL VALUES:

The proportion of 1-heptanol (1) in the water-rich phase at equilibrium at 30°C was reported to be 0.17 g(1)/100g sln.

The corresponding mole fraction solubility, calculated by the compilers, is x_1 = 0.00026.

The proportion of (1) in the water-rich phase of a mixture with Ringer solution at 30°C was reported to be 0.155 g(1)/100g sln.

AUXILIARY INFORMATION

METHOD/APPARATUS/PROCEDURE:

In a round bottomed flask, 50 mL of water and a sufficient quantity of alcohol were introduced until two separate layers were formed. The flask assembly was equilibrated by agitation for at least 3h in a constant temperature bath. Equilibrium solubility was attained by first supersaturation at a slightly lower temperature (solubility of alcohols in water decrease with increasing temperature) and then equilibrating at the desired temperature. The aqueous layer was separated after an overnight storage in a bath. The alcohol content was determined by reacting the aqueous solution with potassium dichromate and titrating the excess dichromate with ferrous sulfate solution in the presence of phosphoric acid and diphenylamine barium sulfonate as an indicator.

SOURCE AND PURITY OF MATERIALS:

(1) redistilled with 10:1 reflux ratio

b.p. 176.6°C/760 mm Hg

n_D^{25} = 1.42219

(2) twice distilled from silica apparatus or ion-exchanged with Sagei A20

ESTIMATED ERROR:

Solubility: relative error of 2 determinations less than 1%

Temperature: ± 0.05°C

REFERENCES:

COMPONENTS:

(1) 1-Heptanol; $C_7H_{16}O$; [111-70-6]

(2) Water; H_2O; [7732-18-5]

ORIGINAL MEASUREMENTS:

Hanssens, I.
Associatie van normale alcoholen en hun affiniteit voor water en organische solventen
Doctoraatsproefschrift, Leuven, 1969.
Huyskens, P.; Mullens, J.; Gomez, A.; Tack, J;
Bull. Soc. Chim. Belg. 1975, *84*, 253-62.

VARIABLES:

One temperature: 298 K

PREPARED BY:

M.C. Haulait-Pirson; A.F.M. Barton

EXPERIMENTAL VALUES:

The concentration of 1-heptanol (1) in the water-rich phase was reported as 0.01498 mol(1)/L sln, and the concentration of water (2) in the alcohol-rich phase was reported as 6.760 mol(2)/L sln.

The corresponding solubilities on a mass/volume basis, calculated by the compilers, are 1.7 g(1)/L sln, and 121.8 g(2)/L sln, respectively.

(The temperature was unspecified in the Thesis, but reported as 298 K in the 1975 published paper).

AUXILIARY INFORMATION

METHOD/APPARATUS/PROCEDURE:

(1) and (2) were equilibrated using a cell described in ref 1. The Rayleigh M75 interference refractometer with the cell M160 for liquids was used for the determination of the concentrations. Cell thicknesses were 1, 3 and 10 cm depending on the concentration range. Standard solutions covering the whole range of concentrations investigated were used for the calibration.

SOURCE AND PURITY OF MATERIALS:

(1) Merck p.a.

(2) distilled

ESTIMATED ERROR:

Solubility ± 0.00036 - 0.05 mol(1)/L sln, depending on the concentration

REFERENCES:

1. Meeussen, E.; Huyskens, P. *J. Chim. Phys.* 1966, *63*, 845.

COMPONENTS:

(1) 1-Heptanol; $C_7H_{16}O$; [111-70-6]

(2) Water; H_2O; [7732-18-5]

ORIGINAL MEASUREMENTS:

Krasnov, K.S.; Gartseva, L.A.

Izv. Vysshykh Uchebn. Zavedenii Khim. Khim. Tekhnol. 1970, *13(7)*, 952-6.

VARIABLES:

Temperature: 12 and 40°C

PREPARED BY:

A. Maczynski and Z. Maczynska

EXPERIMENTAL VALUES:

Solubility of water in 1-heptanol

$t/^{\circ}C$	g(2)/100g sln	x_2 (compiler)
12	5.75	0.282
40	5.90	0.288

AUXILIARY INFORMATION

METHOD/APPARATUS/PROCEDURE:

The analytical method was used.

The saturated mixture of (1) and (2) was placed in a thermostat and phases were allowed to separate. Then (2) was determined in the organic layer by Karl Fischer analysis.

SOURCE AND PURITY OF MATERIALS:

(1) CP reagent;
source not specified;
distilled;
no isomers by GLC;
d_4^{25} 0.8180.

(2) not specified.

ESTIMATED ERROR:

Temperature: ± 0.05°C
Solubility: ± 0.05 wt %
(type of error not specified)

REFERENCES:

COMPONENTS:

(1) 1-Heptanol; $C_7H_{16}O$; [111-70-6]

(2) Water; H_2O; [7732-18-5]

ORIGINAL MEASUREMENTS:

Ionin, M.V.; Shanina, P.I.

Zh. Neorg. Khim. 1972, *17*, 1444-9;

**Russ. J. Inorg. Chem.* 1972, *17*, 747-50.

VARIABLES:

One temperature: 25°C

PREPARED BY:

A.F.M. Barton

EXPERIMENTAL VALUES:

The proportion of 1-heptanol (1) in the alcohol-rich phase at 25°C was reported to be 94.2 g(1)/100g sln.

The corresponding mole fraction value, calculated by the compiler, is x_1 = 0.716.

AUXILIARY INFORMATION

METHOD/APPARATUS/PROCEDURE:

The procedures and reagents described in ref 2 were used. The binodal curve of the hydrogen-chloride-water-1-heptanol ternary system was determined by isothermal titration. The value reported above is the zero hydrogen chloride point on this binodal curve.

SOURCE AND PURITY OF MATERIALS:

(1) purified as recommended in ref 1.

b.p. 175.5 - 176°C,

d_4^{25} 0.8190 n_D^{25} 1.4203

ESTIMATED ERROR:

The binodal curve was determined with an accuracy of 0.5%

REFERENCES:

(1) Weissberger, A.; Proskauer, E.S.; Riddick, J.A.; Toops, E.E., Jr. *Organic Solvents : Physical Properties and Methods Methods of Purification*, Russian edition, Inostr. Lt., Moscow, 1958.

(2) Ionin, M.V.; Shanina, P.I. *Zh. Obshch. Khim.* 1967, *37*, 749; *J. Gen. Chem. USSR.* 1967, *37*, 708.

COMPONENTS:

(1) 1-Heptanol; $C_7H_{16}O$; [111-70-6]

(2) Water; H_2O; [7732-18-5]

ORIGINAL MEASUREMENTS:

Vochten, R.; Petre, G.

J. Colloid Interface Sci. 1973, *42*, 320-7.

VARIABLES:

Temperature: 15-60°C

PREPARED BY:

S.H. Yalkowsky; S.C. Valvani; A.F.M. Barton

EXPERIMENTAL VALUES:

Solubility of 1-heptanol (1) in water

$t/°C$	mol(1)/L sln	g(1)/100g sln (compilers)[a]	10^4x_1 (compilers)[a]
15	0.017	0.20	3.1
20	0.013	0.15	2.3
30	0.013	0.15	2.3
40	0.011	0.13	2.0
50	0.012	0.14	2.2
60	0.013	0.15	2.3

[a] Assuming a solution density equal to that of water.

AUXILIARY INFORMATION

METHOD/APPARATUS/PROCEDURE:

The solubility was obtained from the surface tension of saturated solutions, measured by the static method of Wilhelmy (platinum plate). The apparatus consisted of an electrobalance (R.G. Cahn) connected with a high impedance null detector (Fluke type 845 AR). An all-Pyrex vessel was used.

SOURCE AND PURITY OF MATERIALS:

(1) purified by distillation and preparative gas chromatography; b.p. 175.8°C/760 mm Hg

(2) triply distilled from permanganate solution

ESTIMATED ERROR:

Temperature: ± 0.1°C
Solubility (probably standard deviation): ± 0.001 mol(1)/L sln .

REFERENCES:

COMPONENTS:

(1) 1-Heptanol; $C_7H_{16}O$; [111-70-6]

(2) Water; H_2O; [7732-18-5]

ORIGINAL MEASUREMENTS:

Hill, D.J.T.; White. L.R.

Aust. J. Chem. 1974, *27*, 1905-16.

VARIABLES:

Temperature: 279 - 308 K

PREPARED BY:

A. Maczynski

EXPERIMENTAL VALUES:

Solubility of 1-heptanol in water

T/K	x_1	g(1)/100g sln (compiler)
279.15	0.0003423	0.2204
283.35	0.0003134	0.2020
283.69	0.0003103	0.1998
288.23	0.0002900	0.1868
291.09	0.0002785	0.1794
293.18	0.0002717	0.1750
295.05	0.0002662	0.1715
297.11	0.0002615	0.1685
298.14	0.0002602	0.1676
298.22	0.0002585	0.1665
299.19*	0.0002565	0.1652
301.17*	0.0002544	0.1639
303.29	0.0002519	0.1623
303.31	0.0002522	0.1625
306.05	0.0002499	0.1610
308.05	0.0002491	0.1605

* Based on one determination only

AUXILIARY INFORMATION

METHOD/APPARATUS/PROCEDURE:

The interferometric method was used.

The saturated solutions of (1) were prepared in an equilibrium flask and solution concentrations were determined from their refractive index using a Zeiss interferometer and appropriate calibration curves. Duplicate determinations were always within the error limits calculated from the curve fit to the calibration measurements (0.5% at the 95% confidence level).

Numerous technical details were reported in the paper.

SOURCE AND PURITY OF MATERIALS:

(1) Fluka puriss grade; dried over molecular sieves and purified by vacuum distillation; b.p. 257.2K (20 mm Hg), n_D^{25} 1.4213.

(2) freshly double-distilled.

ESTIMATED ERROR:

Temperature: ± 0.02 °C
Solubility: < 0.5% (accuracy at 95% confidence level)

REFERENCES:

COMPONENTS:	ORIGINAL MEASUREMENTS:
(1) 1-Heptanol; $C_7H_{16}O$; [111-70-6] (2) Water; H_2O; [7732-18-5]	Korenman, I.M.; Gorokhov, A.A.; Polozenko, G.N. *Zhur. Fiz. Khim.* 1974, *48*, 1010-2; *Russ. J. Phys. Chem.* 1974, *48*, 1065-7; *Zhur. Fiz. Khim.* 1975, *49*, 1490-3; *Russ. J. Phys. Chem.* 1975, *49*, 877-8.
VARIABLES:	PREPARED BY:
One temperature: 25°C	A.F.M. Barton

EXPERIMENTAL VALUES:

The equilibrium concentration of 1-heptanol (1) in the water-rich phase at 25.0°C was reported to be 0.058 mol(1)/L sln, and the concentration of water (2) in the alcohol-rich phase was reported to be 3.59 mol(2)/L sln.

The corresponding solubilities on a mass/volume basis, calculated by the compiler, are 6.7 g(1)/L sln, and 64.7 g(2)/L sln, respectively.

AUXILIARY INFORMATION

METHOD/APPARATUS/PROCEDURE:

The two liquids were shaken in a closed vessel at 25.0 ± 0.1°C until equilibrium was established. The soly of (1) in the aqueous phase was determined in a Tsvet-1 chromatograph with a flame-ionisation detector. The sorbent was a polythethylene glycol adipate deposited on Polychrom-1 (10% of the mass of the carrier). The 1m column had an internal diameter 4mm, its temperature was 140°C, and the flow of the carrier gas (nitrogen) was 50 mL min^{-1}. The soly of water in the alcohol was determined on a UKh-2 universal chromatograph under isothermal conditions (150°C) with a heat-conductivity detector. The 1m by 6 mm column was filled with Polysorb. The carrier gas was helium (50 mL min^{-1}). The study formed part of an investigation of salting-out by alkali halides of higher alcohol-water systems.

SOURCE AND PURITY OF MATERIALS:

Not stated

ESTIMATED ERROR:

Temperature: ± 0.1°C

Solubility: not stated; the results reported are the arithmetic means from four sets of experiments.

REFERENCES:

COMPONENTS:

(1) 1-Heptanol; $C_7H_{16}O$; [111-70-6]

(2) Water; H_2O; [7732-18-5]

ORIGINAL MEASUREMENTS:

Tokunaga, S.; Manabe, M.; Koda, M.
Niihama Kogyo Koto Semmon Gakko Kiyo, Rikogaku Hen (Memoirs Niihama Technical College, Sci. and Eng.) 1980, *16*, 96-101.

VARIABLES:

Temperature: 15-40°C

PREPARED BY:

A.F.M. Barton

EXPERIMENTAL VALUES:

Solubility of water (2) in the alcohol-rich phase

t/°C	g(2)/100g sln	x_2	mol(1)/mol(2)
15	5.73	0.282	2.54
20	5.82	0.285	2.52
25	5.85	0.286	2.49
30	5.90	0.288	2.46
35	6.01	0.292	2.44
40	6.02	0.293	2.41

AUXILIARY INFORMATION

METHOD/APPARATUS/PROCEDURE:

The mixtures of 1-heptanol (~5 mL) and water (~10 mL) were stirred magnetically in a stoppered vessel and allowed to stand for 10-12 h in a water thermostat. The alcohol-rich phase was analyzed for water by Karl Fischer titration.

SOURCE AND PURITY OF MATERIALS:

(1) distilled;
no impurities detectable by gas chromatography.

(2) deionized;
distilled prior to use.

ESTIMATED ERROR:

Temperature: ± 0.1°C

Solubility: each result is the mean of three determinations.

REFERENCES:

COMPONENTS:

(1) 2-Heptanol; *(n-pentylmethylcarbinol)*
$C_7H_{16}O$; [543-49-7]

(2) Water; H_2O; [7732-18-5]

ORIGINAL MEASUREMENTS:

Ratouis, M.; Dodé, M.

Bull. Soc. Chim. Fr. 1965, 3318-22.

VARIABLES:

One temperature: 30°C

PREPARED BY:

S.C. Valvani; S.H. Yalkowsky; A.F.M. Barton

EXPERIMENTAL VALUES:

The proportion of 2-heptanol (1) in the water-rich phase at equilibrium at 30°C was reported to be 0.33 g(1)/100g sln.

The corresponding mole fraction solubility, calculated by the compiler, is x_1 = 0.00051.

AUXILIARY INFORMATION

METHOD/APPARATUS/PROCEDURE:

In a round bottomed flask, 50 mL of water and a sufficient quantity of alcohol were introduced until two separate layers were formed. The flask assembly was equilibrated by agitation for at least 3h in a constant temperature bath. Equilibrium solubility was attained by first supersaturation at a slightly lower temperature (solubility of alcohols in water decreases with increasing temperature) and then equilibrating at the desired temperature. The aqueous layer was separated after an overnight storage in a bath. The alcohol content was determined by reacting the aqueous solution with potassium dichromate and titrating the excess dichromate with ferrous sulfate solution in the presence of phosphoric acid and diphenylamine barium sulfonate as an indicator.

SOURCE AND PURITY OF MATERIALS:

(1) laboratory preparation;
redistilled with 10:1 reflux ratio
b.p. 68-69°C/22 mm Hg
n_D^{25} = 1.41888

(2) twice distilled from silica apparatus or ion-exchanged with Sagei A20

ESTIMATED ERROR:

Solubility: relative error of 2 determinations less than 1%

Temperature: ± 0.05°C

REFERENCES:

COMPONENTS:

(1) 2-Heptanol *(n-pentylmethylcarbinol)*; $C_7H_{16}O$; [543-49-7]

(2) Water; H_2O; [7732-18-5]

ORIGINAL MEASUREMENTS:

Nishino, N.; Nakamura, M.
Bull. Chem. Soc. Jpn. 1978, *51*, 1617-20; 1981, *54*, 545-8.

VARIABLES:

Temperature: 275-360 K

PREPARED BY:

G.T. Hefter

EXPERIMENTAL VALUES:

The mutual solubility of (1) and (2) in mole fractions are reported over the temperature range in graphical form. Graphical data are also presented for the heat of evaporation of (1).

AUXILIARY INFORMATION

METHOD/APPARATUS/PROCEDURE:

The turbidimetric method was used. Twenty to thirty glass ampoules containing aqueous solutions of *ca.* 5 cm^3 of various concentrations near the solubility at room temperature were immersed in a water theromostat. The distinction between clear and turbid ampoules was made after equilibrium was established (*ca.* 2h). The smooth curve drawn to separate the clear and turbid regions was regarded as the solubility curve.

SOURCE AND PURITY OF MATERIALS:

(1) G.R. grade (various commercial sources given); dried over calcium oxide; kept in ampoules over magnesium powder

(2) Deionized, refluxed for 15 h with potassium permanganate then distilled.

ESTIMATED ERROR:

Not stated

REFERENCES:

COMPONENTS:	EVALUATOR:
(1) 3-Heptanol; $C_7H_{16}O$; [589-82-2]	A. Maczynski, Institute of Physical Chemistry of the Polish Academy of Sciences, Warsaw, Poland.
(2) Water; H_2O [7732-18-5]	November 1982

CRITICAL EVALUATION:

Solubilities in the system comprising 3-heptanol (1) and water (2) have been reported in two publications: Crittenden and Hixon (ref 1) determined the mutual solubility at 298 K, presumably by the titration method; and Ratouis and Dodé determined the proportion of 3-heptanol in the aqueous phase at 298 K and 303 K by an analytical method.

The value 0.4 g(1)/100g sln at 298 K in ref 1 is in good agreement with the value 0.43 g(1)/100g sln given in ref 2.

Recommended and tentative values for the solubility of 3-heptanol (1) in the water-rich phase

T/K	g(1)/100g sln		$10^4 x_1$
298	0.43	(recommended)	6.7
303	0.41	(tentative)	6.4

Since there is only one value available for the proportion of water in the alcohol-rich phase, the value of ref 1 is regarded as tentative.

The tentative value for the solubility of water (2) in 3-heptanol (1) at 298 K is 3 g(2)/100g sln or x_2 = 0.17.

References

1. Crittenden, E.D., Jr.; Hixon, A.N. *Ind. Eng. Chem.* 1954, *46*, 265.

2. Ratouis, M., Dodé, M. *Bull. Soc. Chim. Fr.* 1965, 3318.

COMPONENTS:

(1) 3-Heptanol; $C_7H_{16}O$; [589-82-2]

(2) Water; H_2O [7732-18-5]

ORIGINAL MEASUREMENTS:

Crittenden, E.D., Jr.; Hixon, A.N.

Ind. Eng. Chem. 1954, *46*, 265-8.

VARIABLES:

One temperature: 25°C

PREPARED BY:

A. Maczynski

EXPERIMENTAL VALUES:

The solubility of 3-heptanol in water at 25°C was reported to be 0.4 g(1)/100 sln.

The corresponding mole fraction, x_1, calculated by the compiler is 0.0006.

The solubility of water in 3-heptanol at 25°C was reported to be 3.0 g(2)/100g sln.

The corresponding mole fraction, x_2, calculated by the compiler is 0.17.

AUXILIARY INFORMATION

METHOD/APPARATUS/PROCEDURE:

Presumably the titration method described for ternary systems containing HCl was used. In this method the solubility was determined by bringing 100-mL samples of (1) or (2) to a temperature of 25.0 ± 0.1°C and the second component was then added from a calibrated buret, with vigorous stirring, until the solution became permanently cloudy.

SOURCE AND PURITY OF MATERIALS:

(1) source not specified; purified; purity not specified.

(2) not specified.

ESTIMATED ERROR:

Temperature: ± 0.10°C.

REFERENCES:

COMPONENTS:

(1) 3-Heptanol; $C_7H_{16}O$; [589-82-2]

(2) Water; H_2O; [7732-18-5]

ORIGINAL MEASUREMENTS:

Ratouis, M.; Dodé, M.

Bull. Soc. Chim. Fr. 1965, 3318-22.

VARIABLES:

Temperature: 25°C and 30°C

PREPARED BY:

S.C. Valvani; S.H. Yalkowsky; A.F.M. Barton

EXPERIMENTAL VALUES:

Proportion of 3-heptanol (1) in water-rich phase.

$t/^{\circ}C$	g(1)/100g sln	$10^4 x_1$ (compiler)
25	0.43	6.7
30	0.41	6.4

The proportion of (1) in the water-rich phase of a mixture with Ringer solution at equilibrium at 30°C was reported to be 0.38 g(1)/100g sln.

AUXILIARY INFORMATION

METHOD/APPARATUS/PROCEDURE:

In a round bottomed flask, 50 mL of water and a sufficient quantity of alcohol were introduced until two separate layers were formed. The flask assembly was equilibrated by agitation for at least 3h in a constant temperature bath. Equilibrium solubility was attained by first supersaturation at a slightly lower temperature (solubility of alcohols in water decrease with increased temperature) and then equilibrating at the desired temperature. The aqueous layer was separated after an overnight storage in a bath. The alcohol content was determined by reacting the aqueous solution with potassium dichromate and titrating the excess dichromate with ferrous sulfate solution in the presence of phosphoric acid and diphenylamine barium sulfonate as an indicator.

SOURCE AND PURITY OF MATERIALS:

(1) redistilled with 10:1 reflux ratio
b.p. 158.5-158.6°C/768.8 mm Hg
$n_D^{25} = 1.41978$

(2) twice distilled from silica apparatus or ion-exchanged with Sagei A20

ESTIMATED ERROR:

Solubility: relative error of 2 determinations less than 1%

Temperature: ± 0.05°C

REFERENCES:

COMPONENTS:

(1) 4-Heptanol; $C_7H_{16}O$; [589-55-9]

(2) Water; H_2O; [7732-18-5]

ORIGINAL MEASUREMENTS:

Ratouis, M.; Dodé, M.

Bull. Soc. Chim. Fr. 1965, 3318-22.

VARIABLES:

Temperature: 25°C and 30°C

Ringer solution also studied

PREPARED BY:

S.C. Valvani; S.H. Yalkowsky; A.F.M. Barton

EXPERIMENTAL VALUES:

Proportion of 4-heptanol (1) in water-rich phase.

t/°C	g(1)/100g sln	$10^4 x_1$ (compiler)
25	0.47	7.3
30	0.45	7.0

The proportion of (1) in the water-rich phase of a mixture with Ringer solution at equilibrium at 30°C was reported to be 0.37 g(1)/100g sln.

AUXILIARY INFORMATION

METHOD/APPARATUS/PROCEDURE:

In a round bottomed flask, 50 mL of water and a sufficient quantity of alcohol were introduced until two separate layers were formed. The flask assembly was equilibrated by agitation for at least 3h in a constant temperature bath. Equilibrium solubility was attained by first supersaturation at a slightly lower temperature (solubility of alcohols in water decreases with increasing temperature) and then equilibrating at the desired temperature. The aqueous layer was separated after an overnight storage in a bath. The alcohol content was determined by reacting the aqueous solution with potassium dichromate and titrating the excess dichromate with ferrous sulfate solution in the presence of phosphoric acid and diphenylamine barium sulfonate as an indicator.

SOURCE AND PURITY OF MATERIALS:

(1) redistilled with 10:1 reflux ratio
b.p. 154.8-154.9°C/760.7 mm Hg
n_D^{25} = 1.41780

(2) twice distilled from silica apparatus or ion-exchanged with Sagei A20

ESTIMATED ERROR:

Solubility: relative error of 2 determinations less than 1%

Temperature: ± 0.05°C

REFERENCES:

COMPONENTS:

(1) 2,2,3-Trimethyl-3-pentanol; $C_8H_{18}O$; [7294-05-5]

(2) Water; H_2O; [7732-18-5]

ORIGINAL MEASUREMENTS:

Ginnings, P.M.; Coltrane, B.

J. Am. Chem. Soc. 1939, *61*, 525.

VARIABLES:

Temperature: 20-30 °C

PREPARED BY:

A. Maczynski and Z. Maczynska

EXPERIMENTAL VALUES:

Mutual solubility of 2,2,3-trimethyl-3-pentanol(1) and water(2)

$t/^{\circ}C$	g(1)/100g sln		x_1 (compiler)	
	(2)-rich phase	(1)-rich phase	(2)-rich phase	(1)-rich phase
20	0.75	98.02	0.00104	0.8726
25	0.69	97.99	0.00096	0.8708
30	0.64	97.98	0.00089	0.8703

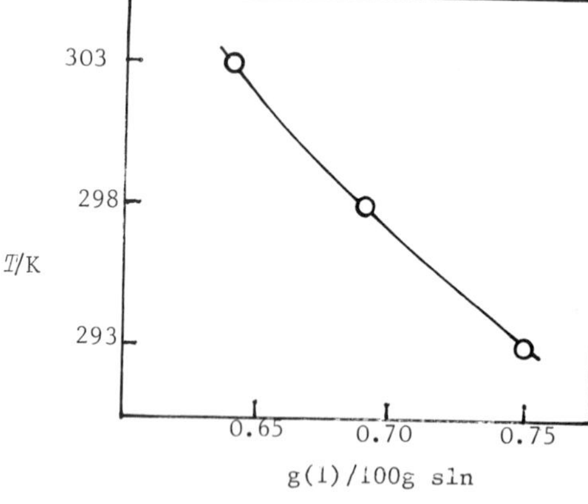

Fig. 1. Water-rich phase

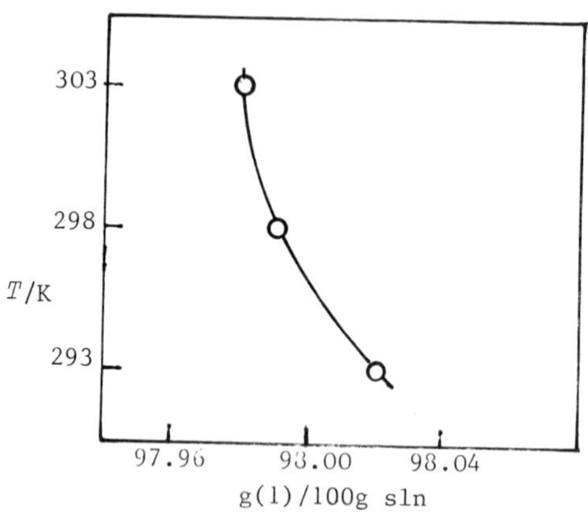

Fig. 2. Alcohol-rich phase

AUXILIARY INFORMATION

METHOD/APPARATUS/PROCEDURE:

The volumetric method was used as described in ref 1.

Both components were introduced in known amounts into a two-bulb graduated and calibrated flask and shaken mechanically in a water-bath at constant temperature. After sufficient time the liquids were allowed to separate and the total volume was measured. Upon centrifugation, the phase separation line was read, and phase volumes were calculated. From the total weights of the components, the total volume, individual phase volumes, and component concentrations in either phase were evaluated.

Relative densities of both phases were also determined.

SOURCE AND PURITY OF MATERIALS:

(1) prepared by Grignard synthesis; purified; b.p. range 153-154 °C, d_4^{25} 0.8420; purity not specified.

(2) not specified.

ESTIMATED ERROR:

Not specified.

REFERENCES:

1. Ginnings, P.M.; Baum, R.J. *J. Am. Chem. Soc.* 1937, *59*, 1111.

COMPONENTS:	EVALUATOR:
(1) 2-Ethyl-1-hexanol; $C_8H_{18}O$; [104-76-7] (2) Water; H_2O; [7732-18-5]	A.F.M. Barton, Murdoch University, Perth, Western Australia November 1982.

CRITICAL EVALUATION:

Only one determination of each of the phases at equilibrium at 298 K has been reported (ref. 1 and 2) so the values can be considered tentative only.

The tentative value for the solubility of 2-ethyl-1-hexanol (1) in the water-rich phase at 298 K is 0.01 g(1)/100 g sln or $10^5 x_1 = 2$.

The tentative value for the solubility of water (2) in the alcohol-rich phase at 298 K is 0.02 g(2)/100 g sln or $x_2 = 0.001$.

References:

1. McBain, J.W.; Richards, P.H. *Ind. Eng. Chem.* 1946, *38*, 642.

2. Hlavaty, K.; Linek, J. *Collect. Czech. Chem. Commun.* 1973, *38*, 374.

COMPONENTS:

(1) 2-Ethyl-1-hexanol; $C_8H_{18}O$; [104-76-7]

(2) Water; H_2O; [7732-18-5]

ORIGINAL MEASUREMENTS:

McBain, J.W.; Richards, P.H.
Ind. Eng. Chem. 1946, *38*, 642-6.

VARIABLES:

One temperature : 25°C

PREPARED BY:

M.C. Haulait-Pirson; A.F.M. Barton

EXPERIMENTAL VALUES:

The solubility of 2-ethyl-1-hexanol in water at 25°C was reported to be 0.013 g(1)/100 g(2). The corresponding mole fraction solubility calculated by the compilers is $10^5 x_1 = 1.7$.

AUXILIARY INFORMATION

METHOD/APPARATUS/PROCEDURE:

10 mL of (2) was measured into glass bottles with plastic caps and known amounts of (1) were added from a microburet reading to 0.001 mL. They were placed on a gentle shaker in an air thermostat overnight. The turbidity was measured in a Barnes turbidimeter (ref. 1), the turbidity rising sharply in the presence of emulsified droplets. The measurements formed part of a study of solubilization by detergents.

SOURCE AND PURITY OF MATERIALS:

(1) "purest obtainable".

(2) unspecified

ESTIMATED ERROR:

unspecified

REFERENCES:

1. McBain, J.W.; Stamberg, O.E. *Rept. to Master Brewers' Assoc. of Am.*, 1942.

COMPONENTS:

(1) 2-Ethyl-1-hexanol; $C_8H_{18}O$; [104-76-7]

(2) Water; H_2O; [7732-18-5]

ORIGINAL MEASUREMENTS:

Hlavaty, K.; Linek, J.

Collect. Czech. Chem. Commun. 1973, *38*, 374-8.

VARIABLES:

One temperature: 24.6°C

PREPARED BY:

A. Maczynski

EXPERIMENTAL VALUES:

The solubility of water in 2-ethyl-1-hexanol at 24.6°C was reported to be 0.020 g(2)/100g sln.

The corresponding mole fraction, x_2, calculated by the compiler, is 0.0014.

AUXILIARY INFORMATION

METHOD/APPARATUS/PROCEDURE:

The titration method was used. The measurements were performed in a titration vessel according to Mertl, ref 1, which can be thermostatted during the titration. Both the vessel and the buret were thermostatted to 24.6 ± 0.1°C. The mixture was agitated vigorously by a magnetic stirrer during the titration.

SOURCE AND PURITY OF MATERIALS:

(1) not specified

(2) not specified

ESTIMATED ERROR:

Solubility: better than ± 0.3 wt %

Temperature: ± 0.1°C

REFERENCES:

1 Mertl, I. Thesis, Institute of Chemical Process Fundamentals, Czechoslovak Academy of Sciences, Prague 1969.

COMPONENTS:	ORIGINAL MEASUREMENTS:
(1) 2-Methyl-2-heptanol; $C_8H_{18}O$; [625-25-2] (2) Water; H_2O; [7732-18-5]	Ratouis, M.; Dodé, M.; *Bull. Soc. Chim. Fr.* 1965, 3318-22.

VARIABLES:	PREPARED BY:
One temperature: 30°C	S.C. Valvani; S.H. Yalkowsky; A.F.M. Barton

EXPERIMENTAL VALUES:

The proportion of 2-methyl-2-heptanol (1) in the water-rich phase at equilibrium at 30°C was reported to be 0.25 g(1)/100 g sln.

The corresponding mole fraction solubility, calculated by the compilers, is $10^4 x_1 = 3.5$.

AUXILIARY INFORMATION

METHOD/APPARATUS/PROCEDURE:

In a round bottomed flask, 50 mL of water and a sufficient quantity of alcohol were introduced until two separate layers were formed. The flask assembly was equilibrated by agitation for at least 3 h in a constant temperature bath. Equilibrium solubility was attained by first supersaturation at a slightly lower temperature (solubility of alcohols in water decreases with increasing temperature) and then equilibrating at the desired temperature. The aqueous layer was separated after an overnight storage in a bath. The alcohol content was determined by reacting the aqueous solution with potassium dichromate and titrating the excess dichromate with ferrous sulfate solution in the presence of phosphoric acid and diphenylamine barium sulfonate as an indicator.

SOURCE AND PURITY OF MATERIALS:

(1) laboratory preparation; redistilled with 10:1 reflux ratio, b.p. 63.1-63.2°C/15mm Hg n_D^{25} = 1.42110

(2) twice distilled from silica apparatus or ion-exchanged with Sagei A20.

ESTIMATED ERROR:

Solubility: relative error of 2 determinations less than 1%.

Temperature: ± 0.05°C

REFERENCES:

COMPONENTS:

(1) 3-Methyl-3-heptanol; $C_8H_{18}O$; [598-06-1]

(2) Water; H_2O; [7732-18-5]

ORIGINAL MEASUREMENTS:

Ratouis, M.; Dode, M.;

Bull. Soc. Chim. Fr. 1965, 3318-22

VARIABLES:

One temperature: 30°C

Ringer solution also studied

PREPARED BY:

S.C. Valvani; S.H. Yalkowsky; A.F.M. Barton

EXPERIMENTAL VALUES:

The proportion of 3-methyl-3-heptanol (1) in the water-rich phase at equilibrium at 30°C was reported to be 0.325 g(1)/100g sln.

The corresponding mole fraction solubility, calculated by the compilers, is 4.5×10^{-4}.

The proportion of (1) in the water-rich phase of a mixture with Ringer solution at equilibrium at 30°C was reported to be 0.29 g(1)/100g sln.

AUXILIARY INFORMATION

METHOD/APPARATUS/PROCEDURE:

In a round bottomed glask, 50 mL of water and a sufficient quantity of alcohol were introduced until two separate layers were formed. The flask assembly was equilibrated by agitation for at least 3 h in a constant temperature bath. Equilibrium solubility was attained by first supersaturation at a slightly lower temperature (solubility of alcohols in water decreases with increasing temperature) and then equilibrating at the desired temperature. The aqueous layer was separated after an overnight storage in a bath. The alcohol content was determined by reacting the aqueous solution with potassium dichromate and titrating the excess dichromate with ferrous sulfate solution in the presence of phosphoric acid and diphenylamine barium sulfonate as an indicator.

SOURCE AND PURITY OF MATERIALS:

(1) laboratory preparation; redistilled with 10:1 reflux ratio;

b.p. 68.8-69.2°C/24mm Hg

n_D^{25} = 1.42670

(2) twice distilled from silica apparatus or ion-exchanged with Sagei A20.

ESTIMATED ERROR:

Solubility: relative error of 2 determinations less than 1%.

Temperature: ± 0.05°C

REFERENCES:

COMPONENTS:

(1) 1-Octanol (*1-octyl alcohol, capryl alcohol*); $C_8H_{18}O$; [111-87-5]

(2) Water; H_2O; [7732-18-5]

EVALUATOR:

G.T. Hefter, Murdoch University, Perth, Western Australia.
June 1983.

CRITICAL EVALUATION:

Solubilities in the system comprising 1-octanol (1) and water (2) have been reported in the following publications:

Reference	T/K	Solubility	Method
Butler *et al.* (ref 1)	298	(1) in (2)	interferometric
Sobotka and Glick (ref 2)	303	(1) in (2)	not stated
Addison (ref 3)	293	(1) in (2)	surface tension
McBain and Richards (ref 4)	298	(1) in (2)	titration
Erichsen (ref 5)	273-523	(2) in (1)	synthetic
Crittenden and Hixon (ref 6)	298	mutual	titration
Kinoshita *et al.* (ref 7)	298	(1) in (2)	surface tension
Shinoda *et al.* (ref 8)	298	(1) in (2)	surface tension
Ababi and Popa (ref 9)	298	(2) in (1)	turbidimetric
Rao *et al.* (ref 10)	303	mutual	turbidimetric
Vochten and Petre (ref 11)	288	(1) in (2)	surface tension
Lavrova and Lesteva (ref 12)	313,333	mutual	titration
Zhuravleva *et al.* (ref 13)	280-388	mutual	polythermic
Tokunaga *et al.* (ref 14)	288-313	(2) in (1)	analytical

The original data are compiled in the data sheets immediately following this Critical Evaluation.

Considering its widespread use in solvent extraction there is surprisingly little information available on the mutual solubilities of 1-octanol and water especially for the water-rich phase. Insufficient data have been reported to define the upper critical solution temperature. It will be clear from the following Tables and discussion that this system warrants a thorough reinvestigation.

In preparing this Critical Evaluation use has been made of the fact that the solubility of 1-octanol in water is sufficiently low to enable weight/volume (ref 11) and volume/volume (ref 2) solubilities to be converted to weight/weight values by assuming the water-rich phase has the same density as pure water. Values obtained by the Evaluator by graphical interpolation or extrapolation from the data sheets are indicated by an asterisk (*). "Best" values have been obtained by simple averaging. The uncertainty limits (σ_n) attached to the "best" values do not have statistical significance and should be regarded only as a convenient representation of the spread of reported values and not as error limits. The letter (R) designates "Recommended" data Data are "Recommended" if two or more apparently reliable studies are in reasonable agreement (≤ ±5% relative).

For convenience the two phases will be further discussed separately.

(continued next page)

COMPONENTS:	EVALUATOR:
(1) 1-Octanol (*1-octyl alcohol, capryl alcohol*); $C_8H_{18}O$; [111-87-5] (2) Water; H_2O; [7732-18-5]	G.T. Hefter, Murdoch University, Perth, Western Australia. June 1983.

CRITICAL EVALUATION (continued)

Water-rich phase

All the available data are summarized in the Table below. With the exception of the value at 298 K most of the data are mutually inconsistent and should therefore be regarded as *very tentative*.

Tentative and Recommended (*R*) values for the solubility of 1-octanol (1) in water (2)

T/K	Solubility, g(1)/100g sln Reported values	"Best" values ($\pm\sigma_n$)
288	0.053 (ref 11)	
293	0.042 (ref 3)	
298	0.0586 (ref 1), 0.059 (ref 4), 0.05 (ref 6), 0.049 (ref 7)	0.054 ± 0.005(*R*)
303	0.076 (ref 2), 0.1 (ref 10)	0.09 ± 0.01
313	0.06 (ref 12)	
333	0.06 (ref 12)	
368	1.0 (ref 13)	
388	0.5 (ref 13)	

It should be noted that the recommended value at 298 K is slightly higher than the value 0.046 g(1)/100g sln predicted by the equation given in the Editor's Preface.

Alcohol-rich phase

Sufficient data are available for the solubility of water in 1-octanol for a realistic Evaluation to be made.

The data of Crittenden and Hixon (ref 6) and Zhuravleva *et al.* (ref 13) are consistently lower than all other values and have therefore been rejected. All other data are included in the Table below. With the exception of the values otherwise indicated all data are regarded as *Tentative* and, as may be seen from the Table, are largely the values of Erichsen (ref 5).

Tentative and Recommended (*R*) values for the solubility of water (2) in 1-octanol (1)

T/K	Solubility, g(2)/100g sln Reported values	"Best" values ($\pm\sigma_n$)
273	3.40 (ref 5)	3.4
283	3.85 (ref 5), 4.62* (ref 14)	4.2 ± 0.4
293	4.30 (ref 5), 4.75 (ref 14)	4.5 ± 0.2 (*R*)
298	4.45* (ref 5), 4.7 (ref 9), 4.78 (ref 14)	4.6 ± 0.1 (*R*)
303	4.75 (ref 5), 4.6 (ref 10), 4.81 (ref 14)	4.7 ± 0.1 (*R*)
313	5.35 (ref 5), 5.10 (ref 12), 4.93 (ref 14)	5.1 ± 0.2 (*R*)

(continued next page)

AWW-M

COMPONENTS:	EVALUATOR:
(1) 1-Octanol (*1-octyl alcohol, capryl alcohol*); $C_8H_{18}O$; [111-87-5] (2) Water; H_2O: [7732-18-5]	G.T. Hefter, Murdoch University, Perth, Western Australia. June 1983.

CRITICAL EVALUATION (continued)

T/K	Solubility, g(2)/100g sln Reported values	"Best" values ($\pm\sigma_n$)
323	5.90 (ref 5)	5.9
333	6.45 (ref 5), 6.0 (ref 12)	6.2 ± 0.2 (*R*)
343	7.10 (ref 5)	7.1
353	7.70 (ref 5)	7.7
363	8.35 (ref 5)	8.4
373	9.10 (ref 5)	9.1
393	10.65 (ref 5)	10.7
413	12.45 (ref 5)	12.5
433	14.60 (ref 5)	14.6
453	17.10 (ref 5)	17.1
473	20.10 (ref 5)	20.1
493	23.90 (ref 5)	23.9
513	29.95 (ref 5)	30.0

The "best" values of the mutual solubility of 1-octanol and water are plotted in Figure 1 (see following page).

References

1. Butler, J.A.V.; Thomson, D.W.; Maclennan, W.H. *J. Chem. Soc.* 1933, 674.
2. Sobotka, H.; Glick, D. *J. Biol. Chem.* 1934, *105*, 199.
3. Addison, C.C. *J. Chem. Soc.* 1945, 98.
4. McBain, J.W.; Richards, P.H. *Ind. Eng. Chem.* 1946, *38*, 642.
5. Erichsen, L.von *Brennst. Chem.* 1952, *33*, 166.
6. Crittenden, E.D., Jr.; Hixon, A.N. *Ind. Eng. Chem.* 1954, *46*, 265.
7. Kinoshita, K.; Ishikawa H.; Shinoda, K. *Bull. Chem. Soc. Jpn.* 1958, *31*, 1081.
8. Shinoda, K.; Yamanaka, T.; Kinoshita, K. *J. Phys. Chem.* 1959, *63*, 648.
9. Ababi, V.; Popa, A. *An. Stiint. Univ. "Al. I. Cuza" Iasi.* 1960, *6*, 929.
10. Rao, K.S.; Rao, M.V.R.; Rao, C.V. *J. Sci. Ind. Res.* 1961, *20B*, 283.
11. Vochten, R.; Petre, G. *J. Colloid Interface Sci.* 1973, *42*, 320.
12. Lavrova, O.A.; Lesteva, T.M. *Zh. Fiz. Khim.* 1976, *50*, 1617; *Dep. Doc. VINITI* 3818-75.
13. Zhuravleva, I.K.; Zhuravlev, E.F.; Lomakina, N.G. *Zh. Fiz. Khim.* 1977, *51*, 1700.
14. Tokunaga, S.; Manabe, M.; Koda, M. *Niihama Kogyo Koto Semmon Gakko Kiyo, Rikagaku Hen (Memoirs Niihama Technical College, Sci. and Eng.)* 1980, *16*, 96.

(continued next page)

COMPONENTS:	EVALUATOR:
(1) 1-Octanol (*1-octyl alcohol, capryl alcohol*); $C_8H_{18}O$; [111-87-5]	G.T. Hefter, Murdoch University, Perth, Western Australia.
(2) Water; H_2O; [7732-18-5]	June 1983

CRITICAL EVALUATION: (continued)

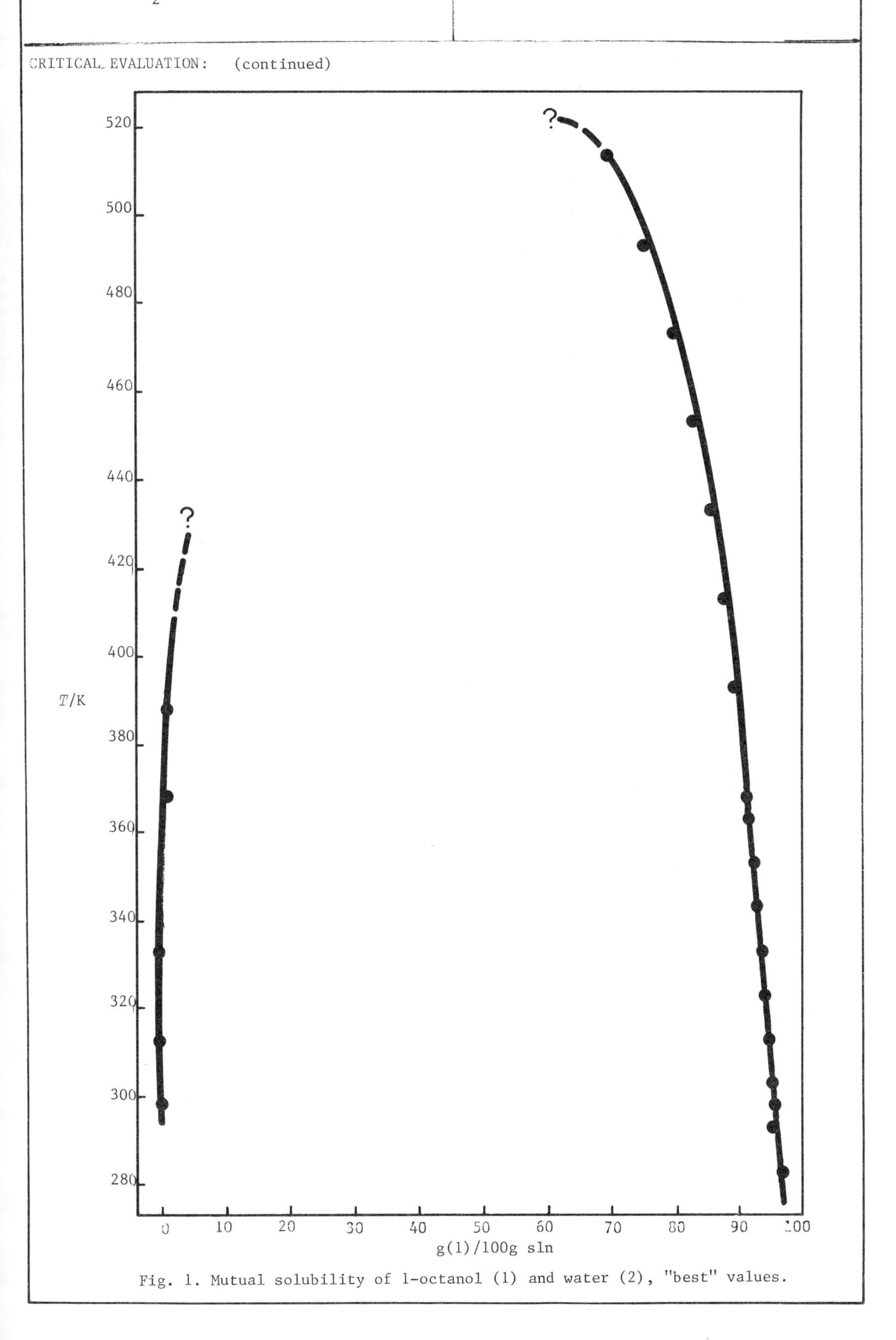

Fig. 1. Mutual solubility of 1-octanol (1) and water (2), "best" values.

COMPONENTS:

(1) 1-Octanol (*capryl alcohol*); $C_8H_{18}O$; [111-87-5]

(2) Water; H_2O; [7732-18-5]

ORIGINAL MEASUREMENTS:

Butler, J.A.V.; Thomson, D.W.; Maclennan, W.H.

J. Chem. Soc. 1933, 674-86.

VARIABLES:

One temperature: 25°C

PREPARED BY:

S.H. Yalkowsky; S.C. Valvani; A.F.M. Barton

EXPERIMENTAL VALUES:

The proportion of 1-octanol (1) in the water-rich phase at equilibrium at 25°C was reported to be 0.0586 g(1)/100g sln, the mean of six determinations (0.0571, 0.0606, 0.0590, 0.0587, 0.0582, 0.0580 g(1)/100g sln).

The corresponding mole fraction solubility was reported as $10^5 x_1 = 8.11$.

AUXILIARY INFORMATION

METHOD/APPARATUS/PROCEDURE:

An analytical method was used, with a U-tube apparatus having two internal stoppers. Suitable quantities of (1) and (2) were placed in one of the connected vessels and shaken in a thermostat for some hours. The liquid was allowed to separate into two layers, the heavier aqueous layer being separated by raising the stopper and allowing part of the liquid to run into the connected vessel. A weighed portion of the separated sln was diluted with about an equal quantity of (2) and the resulting sln compared with calibration slns in an interferometer. To avoid the possibility of reading the position of the wrong fringe, 1 cells (1 cm and 5 cm) were used. The method was unsuitable for analysis of alcohol-rich slns, as no stoppered interferometer cell was available.

SOURCE AND PURITY OF MATERIALS:

(1) B.D.H;
repeated fractionated in vacuum with a Hempel column, the middle fractionation being refluxed with Ca and refractionated;

b.p. 94.80-94.85°C/8 mm Hg, 194.5°C/760 mm Hg

d_4^{25} 0.82238 n_D^{20} 1.42937

(2) not stated

ESTIMATED ERROR:

Solubility: the result is the mean of six determinations agreeing within 0.002 g(1)/100g sln.

Temperature: not stated (but in related experiment is was ± 0.03°C).

REFERENCES

COMPONENTS:

(1) 1-Octanol *(capryl alcohol)*; $C_8H_{18}O$; [111-87-5]

(2) Water; H_2O; [7732-18-5]

ORIGINAL MEASUREMENTS:

Sobotka, H.; Glick, D.

J. Biol. Chem. 1934, *105*, 199-219.

VARIABLES:

One temperature: 30°C

Effect of enzyme solutions also studied

PREPARED BY:

S.H. Yalkowsky; S.C. Valvani; A.F.M. Barton

EXPERIMENTAL VALUES:

The solubility of 1-octanol (1) in water at 30°C was reported to be 0.093 mL(1)/100mL (2).

The corresponding mass percentage and mole fraction solubilities, calculated by the compiler using a density of 0.82 g mL^{-1}, are 0.076 g(1)/100g sln and 10^4x_1 = 1.05.

The solubility of (1) in pancreas globulin solution (total solids 0.29 mg/100mL) was 0.111 mL(1)/100mL sln.

The solubility of (1) in liver albumin solution (total solids 0.50 mg/100mL) was 0.104 mL(1)/100mL sln.

AUXILIARY INFORMATION

METHOD/APPARATUS/PROCEDURE:

The method was described in ref 1. A few mL of (1) was stained with a minute quantity of Sudan IV, a water-insoluble lipoid-soluble dye. It was then added dropwise from a microburet with a capillary tip to 100, 250 or 500 mL of (2) of constant temperature in a narrow-mouthed glass-stoppered stock bottle, which was shaken after each addition. While the added (1) was dissolved, the Sudan dye was wholly or partly dissolved, imparting a pink tinge to the aqueous sln. When the water was saturated with (1), the second phase consisted of transparent droplets. The end point could be improved by adding the Sudan IV (1-5 mg) to the water, and the alcohol added dropwise. When saturation was reached, one additional drop converted the floating, jagged indicator particles into dark transparent droplets.

SOURCE AND PURITY OF MATERIALS:

(1) Not stated

(2) Distilled

ESTIMATED ERROR:

Solubility: ± 0.001 mL(1)/100mL (2)

REFERENCES:

1. Sobotka, H.; Kahn, J.
J. Am. Chem. Soc. 1931, *53*, 2935.

COMPONENTS:	ORIGINAL MEASUREMENTS:
(1) 1-Octanol *(capryl alcohol)*; $C_8H_{18}O$; [111-87-5] (2) Water; H_2O; [7732-18-5]	Addison, C.C. *J. Chem. Soc.* 1945, 98-106.
VARIABLES:	PREPARED BY:
One temperature: 20°C	S.H. Yalkowsky; S.C. Valvani; A.F.M. Barton

EXPERIMENTAL VALUES:

The proportion of 1-octanol (1) in the water-rich phase at equilibrium at 20°C was reported to be 0.042 g(1)/100g sln.

The corresponding mole fraction solubility, calculated by the compilers is $10^5 x_1 = 5.8$.

AUXILIARY INFORMATION

METHOD/APPARATUS/PROCEDURE:

The surface tension method was used. Sufficient excess of (1) was added to 100 mL of (2) in a stoppered flask to form a separate lens on the surface. The mixture was swirled gently, too vigorous an agitation being avoided as this gave a semi-permanent emulsion and incorrect readings. After settling, a small sample of the clear aqueous sln was withdrawn into a drop weight pipet and the surface tension determined. The swirling was continued until a constant value was obtained. The surface tension-concentration curve was known, and only a slight extrapolation (logarithmic scale) was necessary to find the concentration corresponding to the equilibrium value.

SOURCE AND PURITY OF MATERIALS:

(1) impure alcohols were purified by fractional distillation, the middle fraction from a distillation being redistilled;

b.p. 194.5°C d_4^{20} 0.8244

n_D^{20} 1.4291

(2) not stated

ESTIMATED ERROR:

Solubility: ± 0.5%

REFERENCES:

COMPONENTS:

(1) 1-Octanol (*capryl alcohol*); $C_8H_{18}O$; [111-87-5]

(2) Water; H_2O; [7732-18-5]

ORIGINAL MEASUREMENTS:

McBain, J.W.; Richards, P.H.

Ind. Eng. Chem. 1946, *38*, 642-6.

VARIABLES:

One temperature: 25°C

PREPARED BY:

S.H. Yalkowsky; S.C. Valvani; A.F.M. Barton

EXPERIMENTAL VALUES:

The proportion of 1-octanol (1) in the water-rich phase at equilibrium at 25°C was reported to be 0.059 g(1)/100g (2).

The corresponding mole fraction solubility, calculated by the compilers, is $10^5 x_1 = 8.1$.

AUXILIARY INFORMATION

METHOD/APPARATUS/PROCEDURE:

10 mL of (2) were measured into glass bottles with plastic caps and known amounts of (1) were added from a microburet reading to 0.001 mL. They were placed in a gentle shaker in an air thermostat overnight. The turbidity was measured in a Barnes Turbidimeter (ref 1), the turbidity rising sharply in the presence of emulsified droplets. The measurements formed part of a study of solubilization by detergents.

SOURCE AND PURITY OF MATERIALS:

(1) "purest obtainable"

(2) not specified

ESTIMATED ERROR:

Not specified

REFERENCES:

1. McBain, J.W.; Stamberg, O.E. *Rept. to Master Brewers' Assoc. of Am.*, 1942.

COMPONENTS:

(1) 1-Octanol (*capryl alcohol*); $C_8H_{18}O$; [111-87-5]

(2) Water; H_2O; [7732-18-5]

ORIGINAL MEASUREMENTS:

Erichsen, L. von

Brennst. Chem. 1952, *33*, 166-72.

VARIABLES:

Temperature: 0-260 °C

PREPARED BY:

S.H. Yalkowsky and Z. Maczynska

EXPERIMENTAL VALUES:

Proportion of 1-octanol (1) in alcohol-rich phase

$t/^{\circ}C$	g(1)/100g sln	x_1
0	96.60	0.7972
10	96.15	0.7756
20	95.70	0.7549
30	95.25	0.7351
40	94.65	0.7100
50	94.10	0.6882
60	93.55	0.6674
70	92.90	0.6442
80	92.30	0.6239
90	91.65	0.6030
100	90.90	0.5802
110	90.15	0.5644
120	89.35	0.5372
130	88.50	0.5157
140	87.55	0.4932
150	86.50	0.4700
160	85.40	0.4474
170	84.20	0.4244
180	82.90	0.4015
190	81.45	0.3780
200	79.90	0.3550
210	78.15	0.3311
220	76.10	0.3059
230	73.45	0.2769
240	70.05	0.2445
250	65.50	0.2081

AUXILIARY INFORMATION

METHOD/APPARATUS/PROCEDURE:

The synthetic method was used.

The measurements were carried out in 2 mL glass ampules. These were placed in an aluminium block equipped with two glass windows. Cloud points were measured with a thermocouple wound up around the ampule. Each measurement was repeated twice.

SOURCE AND PURITY OF MATERIALS:

(1) Merck, or Ciba, or industrial product; distilled and chemically free from isomers; b.p. 195.7-195.8 °C (759 mm Hg), n_D^{20} 1.4292.

(2) not specified.

ESTIMATED ERROR:

Not specified.

REFERENCES:

COMPONENTS:

(1) 1-Octanol *(capryl alcohol)*; $C_8H_{18}O$; [111-87-5]

(2) Water; H_2O; [7732-18-5]

ORIGINAL MEASUREMENTS:

Crittenden, E.D., Jr.; Hixon, A.N.

Ind. Eng. Chem. 1954, *46*, 265-8.

VARIABLES:

One temperature: 25°C

PREPARED BY:

A. Maczynski

EXPERIMENTAL VALUES:

The solubility of 1-octanol in water at 25°C was reported to be 0.5 g(1)/100g sln.

The corresponding mole fraction, x_1, calculated by the compiler is 0.0007.

The solubility of water in 1-octanol at 25°C was reported to be 1.7 g(2)/100g sln.

The corresponding mole fraction, x_2, calculated by the compiler is 0.111.

AUXILIARY INFORMATION

METHOD/APPARATUS/PROCEDURE:

Presumably the titration method described for ternary systems containing HCl was used. In this method the solubility was determined by bringing 100 mL samples of (1) or (2) to a temperature 25.0 ± 0.1°C and the second component was then added from a calibrated buret with vigorous stirring, until the solution became permanently cloudy.

SOURCE AND PURITY OF MATERIALS:

(1) source not specified; purified; purity not specified.

(2) not specified.

ESTIMATED ERROR:

Temperature: ± 0.10°C.

REFERENCES:

COMPONENTS:	ORIGINAL MEASUREMENTS:
(1) 1-Octanol *(capryl alcohol)*; $C_8H_{18}O$; [111-87-5] (2) Water; H_2O; [7732-18-5]	Kinoshita, K.; Ishikawa, H.; Shinoda, K; *Bull. Chem. Soc. Jpn.* 1958, *31*, 1081-4.
VARIABLES:	**PREPARED BY:**
One temperature: 25°C	S.H. Yalkowsky; S.C. Valvani; A.F.M. Barton

EXPERIMENTAL VALUES:

The equilibrium of 1-octanol (1) in the water-rich phase at 25.0°C was reported to be 0.0038 mol(1)/L^{-1} and the mass percentage solubility was reported as 0.049 g(1)/100g sln.

The corresponding mole fraction solubility, calculated by the compilers, is $10^5 x_1 = 6.7$.

AUXILIARY INFORMATION

METHOD/APPARATUS/PROCEDURE:

The surface tension in aqueous solutions of alcohols monotonously decreases up to their saturation concentration and remains constant in the heterogeneous region (ref 1-4). Surface tension was measured by the drop weight method, using a tip 6 mm in diameter, the measurements being carried out in a water thermostat. From the (surface tension)-(logarithm of concentration) curves the saturation points were determined as the intersections of the curves with the horizontal straight lines passing through the lowest experimental points.

SOURCE AND PURITY OF MATERIALS:

(1) purified by vacuum distillation through 50-100 cm column;
b.p. 96°C/16mm Hg

(2) not stated

ESTIMATED ERROR:

Temperature: ± 0.05°C

Solubility: within 4%

REFERENCES:

1. Motylewski, S. *Z. Anorg. Chem.* 1904, *38*, 410.
2. Taubamann, A. *Z. physick. Chem.* 1932, *A161*, 141.
3. Zimmerman, H.K. Jr. *Chem. Rev.* 1952, *51*, 25.
4. Shinoda, K.; Yamanaka, T.; Kinoshita, K. *J. Phys. Chem.* 1959, *63*, 648.

COMPONENTS:	ORIGINAL MEASUREMENTS:
(1) 1-Octanol (*capryl alcohol*); $C_8H_{18}O$; [111-87-5] (2) Water; H_2O; [7732-18-5]	Shinoda, K.; Yamanaka, T.; Kinoshita, K. *J. Phys. Chem.* 1959, *63*, 648-50.
VARIABLES:	**PREPARED BY:**
One temperature: 25°C	A.F.M. Barton

EXPERIMENTAL VALUES:

The equilibrium concentration of 1-octanol (1) in the water-rich phase at 25°C was reported to be 0.0038 mol(1)/L sln.

AUXILIARY INFORMATION

METHOD/APPARATUS/PROCEDURE:

Surface tension was measured by the drop weight method, with a tip 0.249 cm in diameter, in an air thermostat. Solubility was determined by the turbidity change and/or the break in the (surface tension)-(logarithm of concentration) plot.

SOURCE AND PURITY OF MATERIALS:

(1) Kao Soap Co.
fractional distillation in 100 cm column;
b.p. 96°C/16 mm Hg

(2) not stated

ESTIMATED ERROR:

Temperature: ± 0.2°C

REFERENCES:

COMPONENTS:

(1) 1-Octanol *(capryl alcohol)*; $C_8H_{18}O$; [111-87-5]

(2) Water; H_2O; [7732-18-5]

ORIGINAL MEASUREMENTS:

Ababi, V.; Popa, A.

An. Stiint. Univ. "Al. I. Cuza" Iasi. 1960, *6*, 929-42.

VARIABLES:

One temperature: 25°C

PREPARED BY:

A. Maczynski

EXPERIMENTAL VALUES:

The solubility of water in 1-octanol at 25°C was reported to be 4.7 g(2)/100g sln.

The corresponding mole fraction, x_2, calculated by the compiler is 0.26.

AUXILIARY INFORMATION

METHOD/APPARATUS/PROCEDURE:

The turbidimetric method was used.

Ternary solubilities were described in the paper but nothing reported on binary solubilities.

SOURCE AND PURITY OF MATERIALS:

(1) Merck analytical reagent; used as received,

(2) not specified.

ESTIMATED ERROR:

Not specified.

REFERENCES:

COMPONENTS:

(1) 1-Octanol *(capryl alcohol)*; $C_8H_{18}O$; [111-87-5]

(2) Water; H_2O; [7732-18-5]

ORIGINAL MEASUREMENTS:

Rao, K.S.; Rao, M.V.R.; Rao, C.V.

J. Sci. Ind. Res. 1961, *20B*, 283-6.

VARIABLES:

One temperature: 30°C

PREPARED BY:

A. Maczynski

EXPERIMENTAL VALUES:

The solubility of 1-octanol in water at 30°C was reported to be 0.1 g(1)/100g sln.

The corresponding mole fraction, x_1, calculated by the compiler is 0.0001.

The solubility of water in 1-octanol at 30°C was reported to be 4.6 g(2)/100g sln.

The corresponding mole fraction, x_2, calculated by the compiler is 0.268.

AUXILIARY INFORMATION

METHOD/APPARATUS/PROCEDURE:

The method of appearance and disappearance of turbidity described in ref 1 was used.

No details were reported in the paper.

SOURCE AND PURITY OF MATERIALS:

(1) Naarden and Co.; used as recived; n^{30} 1.4245, d^{30} 0.8215 g/mL.

(2) distilled; free from carbon dioxide.

ESTIMATED ERROR:

Not specified.

REFERENCES:

1. Othmer, D.F.; White, R.E.; Trueger, E. *Ind. Eng. Chem.* 1941, *33*, 1240.

COMPONENTS:

(1) 1-Octanol (*capryl alcohol*); $C_8H_{18}O$; [111-87-5]

(2) Water; H_2O; [7732-18-5]

ORIGINAL MEASUREMENTS:

Vochten, R.; Petre, G.

J. Colloid Interface Sci. 1973, *42*, 320-7.

VARIABLES:

One temperature: 15°C

PREPARED BY:

S.H. Yalkowsky; S.C. Valvani; A.F.M. Barton

EXPERIMENTAL VALUES:

The equilibrium concentration of 1-octanol (1) in the water-rich phase at 15°C was reported to be 0.0041 mol(1)/L.

The corresponding mass percentage solubility, calculated by the compilers with the assumption of a solution density equal to that of water, is 0.053 g(1)/100g sln, and the mole fraction solubility is $10^5 x_1$ = 7.3.

AUXILIARY INFORMATION

METHOD/APPARATUS/PROCEDURE:

The solubility was obtained from the surface tension of saturated solutions, measured by the static method of Wilhelmy (platinum plate). The apparatus consisted of an electrobalance (R.G. Cahn) connected with a high impedance null detector (Fluke type 845 AR). An all-Pyrex vessel was used.

SOURCE AND PURITY OF MATERIALS:

(1) purified by distillation and preparative gas chromatography;

b.p. 195.0°C/760 mm Hg

(2) triply distilled from permanganate solution.

ESTIMATED ERROR:

Temperature: ± 0.1°C

Solubility: (probably standard deviation) ± 0.0001 mol(1) L^{-1} sln.

REFERENCES:

COMPONENTS:

(1) 1-Octanol *(capryl alcohol)*; $C_8H_{18}O$; [111-87-5]

(2) Water; H_2O; [7732-18-5]

ORIGINAL MEASUREMENTS:

Lavrova, O.A.; Lesteva, T.M.

Zh. Fiz. Khim., 1976, *50*, 1617; *Dep. Doc. VINITI*, 3813-75.

VARIABLES:

Temperature: 40 and 60°C

PREPARED BY:

A. Maczynski

EXPERIMENTAL VALUES:

Mutual solubility of 1-octanol (1) and water (2)

$t/°C$	g(1)/100g sln		x_1 (compiler)	
	(2)-rich phase	(1)-rich phase	(2)-rich phase	(1)-rich phase
40	0.06	94.90	0.00008	0.720
60	0.06	94.0	0.00008	0.685

AUXILIARY INFORMATION

METHOD/APPARATUS/PROCEDURE:

The titration method was used.

No details were reported in the paper.

SOURCE AND PURITY OF MATERIALS:

(1) source not specified;
distilled with heptane;
purity 99.97 wt %, 0.03 wt % of water,
n_D^{20} 1.4291, d_4^{20} 0.8148
b.p. 195.0°C.

(2) not specified.

ESTIMATED ERROR:

Not specified.

REFERENCES:

COMPONENTS:

(1) 1-Octanol *(capryl alcohol)*; $C_8H_{18}O$; [111-87-5]

(2) Water; H_2O; [7732-18-5]

ORIGINAL MEASUREMENTS:

Zhuravleva, I.K.; Zhuravlev, E.F.; Lomakina, N.G.

**Zh. Fiz. Khim.* 1977, *51*, 1700-7;
Russ. J. Phys. Chem. 1977, *51*, 994-8.

VARIABLES:

Temperature: 7-115 °C

PREPARED BY:

A. Maczynski

EXPERIMENTAL VALUES:

Solubility of water (2) in 1-octanol (1)

$t/^{\circ}C$	g(2)/100g sln	x_2 (compiler)
7.0	2.5	0.15
23.0	3.5	0.21
35.0	4.0	0.23
52.0	5.0	0.28
74.0	6.0	0.32

Solubility of 1-octanol (1) in water (2)

$t/^{\circ}C$	g(1)/100g sln	x_2 (compiler)
95.0	1.0	0.001
115.0	0.5	0.001

AUXILIARY INFORMATION

METHOD/APPARATUS/PROCEDURE:

The "polythermic" method (ref 1) was used.

No details were reported in the paper.

The results formed part of a report on the ternary system including nitromethane.

SOURCE AND PURITY OF MATERIALS:

(1) source not specified; purity not specified, n_D^{20} 1.4395, d^{20} 0.8240

(2) n_D^{20} 1.333.

ESTIMATED ERROR:

Not specified.

REFERENCES:

1. Alekseev, W.F. *Zh. russk. khim. o-va,* 1876, *8*, 249.

COMPONENTS:

(1) 1-Octanol (*capryl alcohol*); $C_8H_{18}O$; [111-87-5]

(2) Water; H_2O; [7732-18-5]

ORIGINAL MEASUREMENTS:

Tokunaga, S.; Manabe, M.; Koda, M. *Niihama Kogyo Koto Semmon Gakko Kiyo, Rikogaku Hen (Memoirs Niihama Technical College, Sci. and Eng.)* 1980, *16*, 96-101.

VARIABLES:

Temperature: 15-40 °C

PREPARED BY:

A.F.M. Barton

EXPERIMENTAL VALUES:

Proportion of water (2) in the alcohol-rich phase

$t/^{\circ}C$	g(2)/100g sln	x_2	mol(1)/mol(2)
15	4.68	0.262	2.82
20	4.75	0.265	2.78
25	4.78	0.266	2.75
30	4.81	0.267	2.72
35	4.88	0.271	2.70
40	4.93	0.273	2.67

AUXILIARY INFORMATION

METHOD/APPARATUS/PROCEDURE:

The mixtures of 1-octanol (~5 mL) and water (~10 mL) were stirred magnetically in a stoppered vessel and allowed to stand for 10-12 h in a water thermostat. The alcohol-rich phase was analyzed for water by Karl Fischer titration.

SOURCE AND PURITY OF MATERIALS:

(1) distilled; no impurities detectable by gas chromotography

(2) deionized; distilled prior to use

ESTIMATED ERROR:

Temperature: ± 0.1 °C

Solubility: each result is the mean of three determinations

REFERENCES:

COMPONENTS:	EVALUATOR:
(1) 2-Octanol; $C_8H_{18}O$; [123-96-6] (2) Water; H_2O; [7732-18-5]	A. Maczynski, Institute of Physical Chemistry of the Polish Academy of Sciences, Warsaw, Poland; A.F.M. Barton, Murdoch University, Perth, Western Australia. November 1982

CRITICAL EVALUATION:

The proportion of 2-octanol (1) in the water-rich phase has been reported in three publications: Mitchell (ref 1) at 288 K and 298 K by an interferometric method; Addison (ref 2) at 293 K from surface tension measurements; and Crittenden and Hixon (ref 3) at 298 K, presumably by a titration method. The values in refs 1 and 2 are in reasonable agreement, but about one-third of those reported in ref 3. Since the data of Crittenden and Hixon (ref 3) is the most recent, because their results for other systems appear reliable, and because the earlier works identified the alcohol only as "sec-octyl alcohol", the measurement at 298 K of ref 3 is selected as the tentative value.

<u>The tentative value for the solubility of 2-octanol in water at 298 K is 0.4 g(1)/100g sln, $x_1 = 6 \times 10^{-4}$.</u>

The proportion of water (2) in the alcohol-rich phase has been measured only by Crittenden and Hixon (ref 3).

<u>The tentative value for the solubility of water in 2-octanol at 298 K is 3.7 g(2)/100g sln or $x_2 = 0.22$.</u>

<u>References</u>

1. Mitchell, S. *J. Chem. Soc.* <u>1926</u>, 1333.
2. Addison, C.C. *J. Chem. Soc.* <u>1945</u>, 98.
3. Crittenden, E.D., Jr.; Hixon, A.N. *Ind. Eng. Chem.* <u>1954</u>, *46*, 265.

COMPONENTS:

(1) 2-Octanol; $C_8H_{18}O$; [123-96-6]
(identified in this publication only as sec-octyl alcohol)

(2) Water; H_2O; [7732-18-5]

ORIGINAL MEASUREMENTS:

Mitchell S.;
J. Chem. Soc. 1926, 1333-6.

VARIABLES:

Temperature: 15°C and 25°C

PREPARED BY:

S.H. Yalkowsky; S.C. Valvni; A.F.M. Barton

EXPERIMENTAL VALUES:

Solubility of *sec*-octyl alcohol (1) in water-rich phase

t/°C	g(1)/L sln	10^3 mol (1) (compilers)	g(1)/100 g sln (compilers)[a]
15	1.508	11.6	0.15
25	1.280	9.8	0.13

[a] Assuming a solution density equal to that of water

AUXILIARY INFORMATION

METHOD/APPARATUS/PROCEDURE:

An interferometric method was used. A plot of % saturation against compensator reading was linear, and the composition of the saturated solution could be obtained by extrapolation.

SOURCE AND PURITY OF MATERIALS:

Not stated

ESTIMATED ERROR:

Not stated

REFERENCES:

COMPONENTS:

(1) 2-Octanol; $C_8H_{18}O$; [123-96-6]
(identified in this publication only as sec-octyl alcohol)

(2) Water; H_2O; [7732-18-5]

ORIGINAL MEASUREMENTS:

Addison, C.C.

J. Chem. Soc. 1945, 98-106.

VARIABLES:

One temperature: 20°C

PREPARED BY:

S.H. Yalkowsky; S.C. Valvani; A.F.M. Barton

EXPERIMENTAL VALUES:

The proportion of *sec*-octyl alcohol (1) in the water-rich phase at equilibrium at 20°C was reported to be 0.106 g(1)/100g sln.

The corresponding mole fraction solubility, calculated by the compilers, is $10^4 x_1 = 1.46$.

AUXILIARY INFORMATION

METHOD/APPARATUS/PROCEDURE:

The surface tension method was used. Sufficient excess of (1) was added to 100 mL of (2) in a stoppered flask to form a separate lens on the surface. The mixture was swirled gently, too vigorous an agitation being avoided as this gave a semi-permanent emulsion and incorrect readings. After settling, a small sample of the clear aqueous sln was withdrawn into a drop weight pipet and the surface tension determined. The swirling was continued until a constant value was obtained. The surface tension-concentration curve was known, and only a slight extrapolation (logarithmic scale) was necessary to find the concentration corresponding to the equilibrium value.

SOURCE AND PURITY OF MATERIALS:

(1) impure alcohols were purified by fractional distillation, the middle fraction from a distillation being redistilled;

b.p. 178.0°C d_4^{20} 0.8200

n_D^{20} 1.4255

(2) not stated

ESTIMATED ERROR:

Solubility: ± 0.5%

REFERENCES:

COMPONENTS:

(1) 2-Octanol; $C_8H_{18}O$; [123-96-6]

(2) Water; H_2O; [7732-18-5]

ORIGINAL MEASUREMENTS:

Crittenden, E.D.,Jr.; Hixon, A.N.;

Ind. Eng. Chem. 1954, *46*, 265-8.

VARIABLES:

One temperature: 25°C

PREPARED BY:

A. Maczynski

EXPERIMENTAL VALUES:

The solubility of 2-octanol in water at 25°C was reported to be 0.4 g(1)/100g sln.

The corresponding mole fraction, x_1, calculated by the compiler is 0.0006.

The solubility of water in 2-octanol at 25°C was reported to be 3.7 g(2)/100 sln.

The corresponding mole fraction, x_2, calculated by the compiler is 0.22.

AUXILIARY INFORMATION

METHOD/APPARATUS/PROCEDURE:

Presumably the titration method described for ternary systems containing HCl was used. In this method the solubility was determined by bringing 100-mL samples of (1) or (2) to a temperature of 25.0 ± 0.1°C and the second component was then added from a calibrated buret, with vigorous stirring, until the solution became permanently cloudy.

SOURCE AND PURITY OF MATERIALS:

(1) source not specified;
purified;
purity not specified.

(2) not specified.

ESTIMATED ERROR:

Temperature: ± 0.10

REFERENCES:

COMPONENTS:

(1) 2,2-Diethyl-1-pentanol; $C_9H_{20}O$; [14202-62-1]

(2) Water; H_2O; [7732-18-5]

ORIGINAL MEASUREMENTS:

Vochten, R.; Petre, G.

J. Colloid Interface Sci. 1973, *42*, 320-7.

VARIABLES:

One temperature: 15°C

PREPARED BY:

S.H. Yalkowsky; S.C. Valvani; A.F.M. Barton

EXPERIMENTAL VALUES:

The equilibrium concentration of 2,2-diethyl-1-pentanol (1) in the water-rich phase at 15°C was reported to be 0.0038 mol(1)/L sln.

The corresponding mass percentage solubility, calculated by the compilers with the assumption of a solution density equal to that of water, is 0.055 g(1)/100g sln, and the mole fraction solubility is $10^5 x_1 = 6.8$.

AUXILIARY INFORMATION

METHOD/APPARATUS/PROCEDURE:

The solubility was obtained from the surface tension of saturated solutions, measured by the static method of Wilhelmy (platinum plate). The apparatus consisted of an electrobalance (R.G. Cahn) connected with a high impedance null detector (Fluke type AR). An all-Pyrex vessel was used.

SOURCE AND PURITY OF MATERIALS:

(1) purified by distillation and preparative gas chromatography;
b.p. 192.0°C/760 mm Hg

(2) triply distilled from permanganate solution

ESTIMATED ERROR:

Temperature: ± 0.1°C

Solubility: (probably standard deviation) ± 0.0001 mol(1)/L sln.

REFERENCES:

COMPONENTS:	EVALUATOR:
(1) 2,6-Dimethyl-4-heptanol; $C_9H_{20}O$; [108-82-7] (2) Water; H_2O; [7732-18-5]	A. Maczynski, Institute of Physical Chemistry of the Polish Academy of Sciences, Warsaw, Poland. November 1982

CRITICAL EVALUATION:

The solubilities of 2,6-dimethyl-4-heptanol (1) and water (2) have been reported in two publications. Crittenden and Hixon (ref 1) determined the mutual solubilities at 298 K, presumably by the titration method, Vochten and Petre (ref 2) determined the proportion of (1) in the water-rich phase at 288 K from surface tension measurements. Data from the two references cannot be compared directly, and the information should be regarded as very tentative.

Tentative values of the mutual solubilities of 2,6-dimethyl-4-heptanol (1) and water (2)

T/K	g(1)/100g sln		x_1	
	water-rich phase	alcohol-rich phase	water-rich phase	alcohol-rich phase
288	0.05	-	0.00006	-
298	0.1	99.0	0.0001	0.93

References

1. Crittenden, E.D., Jr.; Hixon, A.N. *Ind. Eng. Chem.* 1954, *46*, 265.

2. Vochten, R.; Petre, G. *J. Colloid Interface Sci.* 1973, *42*, 320.

COMPONENTS:

(1) 2,6-Dimethyl-4-heptanol; $C_9H_{20}O$; [108-82-7]

(2) Water; H_2O; [7732-18-5]

ORIGINAL MEASUREMENTS:

Crittenden, E.D., Jr., Hixon, A.N.

Ind. Eng. Chem. 1954, *46*, 265-8.

VARIABLES:

One temperature: 25°C

PREPARED BY:

A. Maczynski

EXPERIMENTAL VALUES:

The solubility of 2,6-dimethyl-4-heptanol in water at 25°C was reported to be 0.1 g(1)/100 sln.

The corresponding mole fraction, x_1, calculated by the compiler is 0.0001.

The solubility of water in 2,6-dimethyl-4-heptanol at 25°C was reported to be 1.0 g(2)/100g sln.

The corresponding mole fraction, x_2, calculated by the compiler is 0.07.

AUXILIARY INFORMATION

METHOD/APPARATUS/PROCEDURE:

Presumably the titration method described for ternary systems containing HCl was used. In this method the solubility was determined by bringing 100-ml samples of (1) or (2) to a temperature of 25.0 ± 0.1°C and the second component was then added from a calibrated buret, with vigorous stirring, until the solution became permanently cloudy.

SOURCE AND PURITY OF MATERIALS:

(1) source not specified;
purified;
purity not specified.

(2) not specified.

ESTIMATED ERROR:

Temperature: ± 0.10°

REFERENCES:

COMPONENTS:

(1) 2,6-Dimethyl-4-heptanol; $C_9H_{20}O$; [108-82-7]

(2) Water; H_2O; [7732-18-5]

ORIGINAL MEASUREMENTS:

Vochten, R.; Petre, G.

J. Colloid Interface Sci. 1973, *42*, 320-7.

VARIABLES:

One temperature: 15°C

PREPARED BY:

S.H. Yalkowsky; S.C. Valvani; A.F.M. Barton

EXPERIMENTAL VALUES:

The equilibrium concentration of 2,6-dimethyl-4-heptanol (1) in the water-rich phase at 15°C was reported to be 0.0031 mol(1)/L sln.

The corresponding mass percentage solubility, calculated by the compilers, with the assumption of a solution density equal to that of water, is 0.045 g(1)/100g sln, and the mole fraction solubility is x_1 = 5.6 x 10^{-5}.

AUXILIARY INFORMATION

METHOD/APPARATUS/PROCEDURE:

The solubility was obtained from the surface tension of saturated solutions, measured by the static method of Wilhelmy (platinum plate). The apparatus consisted of an electrobalance (R.G. Cahn) connected with a high impedance null detector (Fluke type 845 AR). An all-Pyrex vessel was used.

SOURCE AND PURITY OF MATERIALS:

(1) purified by distillation and preparative gas chromatography;

b.p. 178.0°C/760 mm Hg

(2) triply distilled from permanganate solution

ESTIMATED ERROR:

Temperature: ± 0.1°C

Solubility: (probably standard deviation) ± 0.0001 mol(1)/L sln.

REFERENCES:

COMPONENTS:

(1) 3,5-Dimethyl-4-heptanol; $C_9H_{20}O$; 19549-79-2

(2) Water; H_2O; 7732-18-5

ORIGINAL MEASUREMENTS:

Vochten, R.; Petre, G.

J. Colloid Interface Sci. 1973, *42*, 320-7.

VARIABLES:

One temperature: 15°C

PREPARED BY:

S.H. Yalkowsky; S.C. Valvani; A.F.M. Barton

EXPERIMENTAL VALUES:

The equilibrium concentration of 3,5-dimethyl-4-heptanol (1) in the water-rich phase at 15°C was reported to be 0.0050 mol(1)/L sln.

The corresponding mass percentage solubility, calculated by the compilers with the assumption of a solution density equal to that of water, is 0.072 g(1)/100g sln, and the mole fraction solubility is $x_1 = 8.9 \times 10^{-5}$.

AUXILIARY INFORMATION

METHOD/APPARATUS/PROCEDURE:

The solubility was obtained from the surface tension of saturated solutions, measured by the static method of Wilhelmy (platinum plate). The apparatus consisted of an electrobalance (R.G. Cahn) connected with a high impedance null detector (Fluke type 845 AR). An all-Pyrex vessel was used.

SOURCE AND PURITY OF MATERIALS:

(1) purified by distillation and preparative gas chromatography;

b.p. 171.0°C/760 mm Hg

(2) triply distilled from permanganate solution

ESTIMATED ERROR:

Temperature: ± 0.1°C

Solubility: (probably standard deviation) ± 0.0001 mol(1)/L sln.

REFERENCES:

COMPONENTS:	ORIGINAL MEASUREMENTS:
(1) 7-Methyl-1-octanol; $C_9H_{20}O$; [2430-22-0] (2) Water; H_2O; [7732-18-5]	Vochten, R.; Petre, G. *J. Colloid Interface Sci.* 1973, *42*, 320-7.
VARIABLES:	PREPARED BY:
One temperature: 15°C	S.H. Yalkowsky; S.C. Valvani; A.F.M. Barton

EXPERIMENTAL VALUES:

The equilibrium concentration of 7-methyl-1-octanol (1) in the water-rich phase at 15°C was reported to be 0.0032 mol(1)/L sln.

The corresponding mass percentage solubility calculated by the compilers with the assumption of a solution density equal to that of water, is 0.046 g(1)/100g sln, and the mole fraction solubility is $x_1 = 5.7 \times 10^{-5}$.

AUXILIARY INFORMATION

METHOD/APPARATUS/PROCEDURE:	SOURCE AND PURITY OF MATERIALS:
The solubility was obtained from the surface tension of saturated solutions, measured by the static method of Wilhemy (platinum plate). The apparatus consisted of an electrobalance (R.G. Cahn) connected with a high impedance null detector (Fluke type 845 AR). An all-Pyrex vessel was used.	(1) purified by distillation and preparative gas chromatography; b.p. 206.0°C/760 mm Hg (2) triply distilled from permanganate solution
	ESTIMATED ERROR:
	Temperature: ± 0.1°C Solubility: (probably standard deviation) ± 0.0001 mol(1)/L sln.
	REFERENCES:

COMPONENTS:	EVALUATOR:
(1) 1-Nonanol (*n-nonyl alcohol*) $C_9H_{20}O$; [143-08-8] (2) Water; H_2O; [7732-18-5]	A. Maczynski, Institute of Physical Chemistry of the Polish Academy of Sciences, Warsaw, Poland; A.F.M. Barton, Murdoch University, Perth, Western Australia November 1982

CRITICAL EVALUATION:

The proportion of 1-nonanol (1) in the water-rich phase was determined by Kinoshita *et al.* (ref 2) at 298 K and by Vochten and Petre (ref 3) at 288 K, both from surface tension measurements. The two values are consistent and the former agrees with that calculated from the equation correlating the solubilities at 298 K of normal aliphatic alcohols (equation 5 in the Editor's Preface), $\log (c/\mathrm{mol(1)/L\ sln} = 2.722 - 0.6988n + 0.006418n^2$, where n is the number of carbon atoms in the alcohol. This equation predicts for 1-nonanol 0.00090 mol(1)/L sln or 0.013 g(1)/100g sln.

<u>The recommended value for the solubility of 1-nonanol in water at 298K is 1.0×10^{-3} mol(1)/L sln or 0.014 g(1)/100g sln.</u>

The proportion of water (2) in the alcohol-rich phase has been reported by Erichsen (ref 1) over a large temperature range. The temperature dependence of these results in the 290-310 K range does not agree with the more recent work of Tokunaga *et al.* (ref 6), although the values agree at 293 K. Sazonov and Chernysheva (ref 4) provided only graphical data of the binary system, and details of the report of Zhuravleva *et al.* (ref 5) were unavailable to the evaluator. Erichsen's values form the basis of the tentative values (Figure 1).

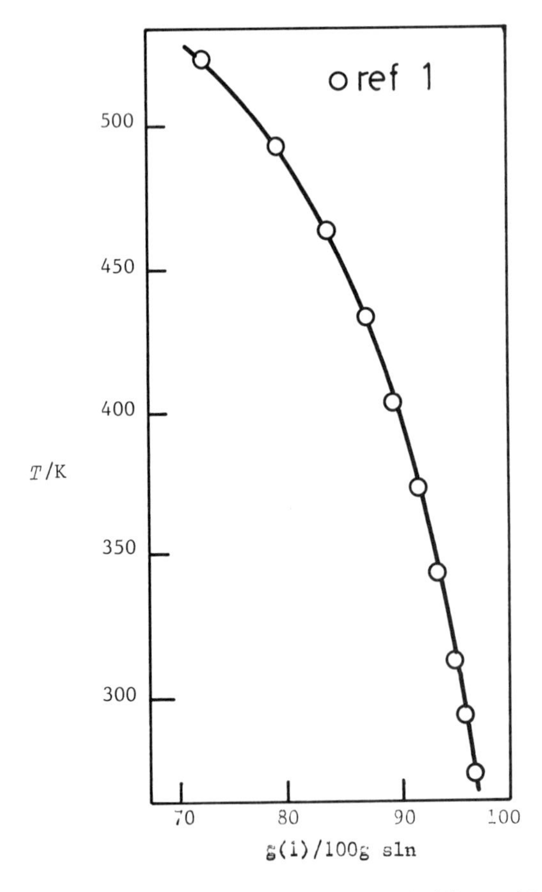

Fig. 1. Solubility of (2) in (1)

(continued next page)

COMPONENTS:	EVALUATOR:
(1) 1-Nonanol (*n-nonyl alcohol*) $C_9H_{20}O$; [143-08-8] (2) Water; H_2O; [7732-18-5]	A. Maczynski, Institute of Physical Chemistry of the Polish Academy of Sciences, Warsaw, Poland; A.F.M. Barton, Murdoch University, Perth, Western Australia. November 1982

CRITICAL EVALUATION: (continued)

Tentative values of the solubility of water (2) in the 1-nonanol-rich phase

T/K	g(2)/100g sln	x_2
273	2.9	0.20
293	3.8	0.24
298	4.0	0.25
303	4.3	0.26
313	4.9	0.29
343	6.5	0.35
373	8.2	0.42
403	10.4	0.48
433	13	0.55
463	17	0.61
493	21	0.68
523	28	0.76

References

1. Erichsen, L. von *Brennst. Chem.* 1952, *33*, 166.
2. Kinoshita, K.; Ishikawa, H.; Shinoda, K. *Bull. Chem. Soc. Jpn.* 1958, *31*, 1081.
3. Vochten, R.; Petre, G. *J. Colloid Interface Sci.* 1973, *42*, 320.
4. Sazonov, V.P.; Chernysheva, M.F. *Zh. Obshch. Khim.* 1976, *46*, 993.
5. Zhuravleva, I.K.; Zhuravlev, E.F.; Khotkovskaya, T.L. *Zhur. Prikl. Khim.* 1976, *49*, 2586; *Dep. Doc. VINITI* 1951-76.
6. Tokunaga, S.; Manabe, M.; Koda, M. *Niihama Kogyo Koto Semmon Gakko Kiyo, Rikogaku Hen (Memoirs Niihama Technical College, Sci and Eng.)* 1980, *16*, 96.

COMPONENTS:

(1) 1-Nonanol; $C_9H_{20}O$; [143-08-8]

(2) Water; H_2O; [7732-18-5]

ORIGINAL MEASUREMENTS:

Erichsen, L. von

Brennst. Chem. 1952, *33*, 166-72.

VARIABLES:

Temperature: 0-250°C

PREPARED BY:

S.H. Yalkowsky and Z. Maczynska

EXPERIMENTAL VALUES:

Solubility of 1-nonanol (1) in alcohol-rich phase

$t/°C$	g(1)/100g sln	x_1
0	97.05	0.8045
10	96.65	0.7815
20	96.20	0.7590
30	95.75	0.7370
40	95.10	0.7150
50	94.75	0.6930
60	94.25	0.6710
70	93.55	0.6495
80	93.10	0.6280
90	92.50	0.6060
100	91.85	0.5850
110	90.95	0.5625
120	90.40	0.5410
130	89.65	0.5195
140	88.80	0.4980
150	87.90	0.4760
160	86.95	0.4540
170	85.90	0.4320
180	84.75	0.4100
190	83.50	0.3875
200	82.15	0.3650
210	80.80	0.3425
220	79.00	0.3195
230	77.10	0.2960
240	74.95	0.2720
250	72.00	0.2430

AUXILIARY INFORMATION

METHOD/APPARATUS/PROCEDURE:

The synthetic method was used.

The measurements were carried out in 2 mL glass ampules. These were placed in an aluminium block equipped with two glass windows. Cloud points were measured with a thermocouple wound up around the ampule. Each measurement was repeated twice.

SOURCE AND PURITY OF MATERIALS:

(1) Merck, or Ciba, or industrial product; distilled and chemical free from isomers;
b.p. 214.7-214.8°C (762 mm Hg)
n_D^{20} 1.4334.

(2) not specified.

ESTIMATED ERROR:

Not specified.

REFERENCES:

COMPONENTS:

(1) 1-Nonanol; $C_9H_{20}O$; [143-08-8]

(2) Water; H_2O; [7732-18-5]

ORIGINAL MEASUREMENTS:

Kinoshita, K.; Ishikawa, H.; Shinoda, K.
Bull. Chem. Soc. Jpn. 1958, *31*, 1081-4.

VARIABLES:

One temperature: 25°C

PREPARED BY:

S.H. Yalkowsky; S.C. Valvani; A.F.M. Barton

EXPERIMENTAL VALUES:

The equilibrium concentration of 1-nonanol (1) in the water-rich phase at 25.0°C was reported to be 0.00097 mol(1)/L sln and the mass percentage solubility was reported as 0.014 g(1)/100g sln.

The corresponding mole fraction solubility, calculated by the compilers, is $x_1 = 1.7 \times 10^{-5}$.

AUXILIARY INFORMATION

METHOD/APPARATUS/PROCEDURE:

The surface tension in aqueous solutions of alcohols monotonously decreases up to their saturation concentration and remains constant in the heterogeneous region (ref 1-4). Surface tension was measured by the drop weight method, using a tip 6 mm in diameter, the measurements being carried out in a water thermostat. From the (surface tension -logarithm of concentration) curves the saturation points were determined as the intersections of the curves with the horizontal straight lines passing through the lowest experimental points.

SOURCE AND PURITY OF MATERIALS:

(1) laboratory preparation from nonanoic acid; purified by vacuum distillation through 50-100 cm column;

b.p. 113-114.5°C/20 mm Hg

(2) not stated

ESTIMATED ERROR:

Temperature: ± 0.05°C

Solubility: within 4%

REFERENCES:

1. Motylewski, S. *Z. Anorg. Chem.* 1904, *38*, 410
2. Taubamann, A. *Z Physik. Chem.* 1932, *A161*, 141.
3. Zimmerman H.K.,Jr. *Chem. Rev.* 1952, *51*, 25.
4. Shinoda, K., Yamanaka, R.; Kinoshita, K. *J. Phys. Chem.* 1959, *63*, 648.

COMPONENTS:	ORIGINAL MEASUREMENTS:
(1) 1-Nonanol; $C_9H_{20}O$; [143-08-8] (2) Water; H_2O; [7732-18-5]	Vochten, R.; Petre, G. *J. Colloid Interface Sci.* 1973, *42*, 320-7.
VARIABLES:	**PREPARED BY:**
One temperature; 15°C	S.H. Yalkowsky; S.C. Valvani; A.F.M. Barton

EXPERIMENTAL VALUES:

The equilibrium concentration of 1-nonanol (1) in the water-rich phase at 15°C was reported to be 0.0010 mol(1)/L sln.

The corresponding mass percentage solubility, calculated by the compilers with the assumption of a solution density equal to that of water, is 0.014 g(1)/100g sln, and the mole fraction solubility is $x_1 = 1.7 \times 10^{-5}$.

AUXILIARY INFORMATION

METHOD/APPARATUS/PROCEDURE:

The solubility was obtained from the surface tension of saturated solutions, measured by the static method of Wilhelmy (platinum plate). The apparatus consisted of an electrobalance (R.G. Cahn) connected with a high impedance null detector (Fluke type 845 AR). An all-Pyrex vessel was used.

SOURCE AND PURITY OF MATERIALS:

(1) purified by distillation and preparative gas chromatography;
b.p. 212.0°C/760 mm Hg

(2) triply distilled from permangante solutions

ESTIMATED ERROR:

Temperature: ± 0.1°C
Solubility (probably standard deviation) ± 0.0001 mol(1)/L sln.

REFERENCES:

COMPONENTS:

(1) 1-Nonanol; $C_9H_{20}O$; [143-08-8]

(2) Water; H_2O [7732-18-5]

ORIGINAL MEASUREMENTS:

Tokunaga, S.; Manabe, M.; Koda, M.

Niihama Kogyo Koto Semmon Gakko Kiyo, Rikogaku Hen (Memoirs Niihama Technical College, Sci. and Eng.) 1980, *16*, 96-101.

VARIABLES:

Temperature: 15-40°C.

PREPARED BY:

A.F.M. Barton

EXPERIMENTAL VALUES:

Proportion of water (2) in the alcohol-rich phase

t/°C	g(2)/100g sln	x_2	mol(1)/mol(2)
15	4.25	0.262	2.81
20	4.27	0.263	2.80
25	4.31	0.265	2.78
30	4.32	0.265	2.77
35	4.36	0.267	2.76
40	4.32	0.265	2.75

AUXILIARY INFORMATION

METHOD/APPARATUS/PROCEDURE:

The mixtures of 1-nonanol (~5 mL) and water (~10 mL) were stirred magnetically in a stoppered vessel and allowed to stand for 10-12 h in a water thermostat. The alcohol-rich phase was analyzed for water by Karl Fischer titration.

SOURCE AND PURITY OF MATERIALS:

(1) distilled;
no impurities detectable by gas chromatography

(2) deionized;
distilled prior to use

ESTIMATED ERROR:

Temperature: ± 0.1°C

Solubility: each result is the mean of three determinations

REFERENCES:

COMPONENTS:

(1) 2-Nonanol; $C_9H_{20}O$; [628-99-9]

(2) Water; H_2O; [7732-18-5]

ORIGINAL MEASUREMENTS:

Vochten, R.; Petre, G.

J. Colloid Interface Sci. 1973, *42*, 320-7.

VARIABLES:

One temperature: 15°C

PREPARED BY:

S.H. Yalkowsky; S.C. Valvani; A.F.M. Barton

EXPERIMENTAL VALUES:

The equilibrium concentration of 2-nonanol (1) in the water-rich phase at 15°C was reported to be 0.0018 mol(1)/L sln.

The corresponding mass percentage solubility, calculated by the compilers with the assumption of a solution density equal to that of water, is 0.026 g(1)/100g sln, and the mole fraction solubility is $x_1 = 3.2 \times 10^{-5}$.

AUXILIARY INFORMATION

METHOD/APPARATUS/PROCEDURE:

The solubility was obtained from the surface tension of saturated solutions, measured by the static method of Wilhelmy (platinum plate). The apparatus consisted of an electrobalance (R.G. Cahn) connected with a high impedance null detector (Fluke type 845 AR). An all-Pyrex vessel was used.

SOURCE AND PURITY OF MATERIALS:

(1) purified by distillation and preparative gas chromatography;

b.p. 193.0°C/760mm Hg

(2) triply distilled from permanganate solution

ESTIMATED ERROR:

Temperature: ± 0.1°C

Solubility (probably standard deviation): ± 0.0001 mol(1)/L sln .

REFERENCES:

COMPONENTS:

(1) 3-Nonanol; $C_9H_{20}O$; [624-51-1]

(2) Water; H_2O; [7732-18-5]

ORIGINAL MEASUREMENTS:

Vochten, R.; Petre, G.

J. Colloid Interface Sci. 1973, *42*, 320-7.

VARIABLES:

One temperature: 15°C

PREPARED BY:

S.H. Yalkowsky; S.C. Valvani; A.F.M. Barton

EXPERIMENTAL VALUES:

The equilibrium concentration of 3-nonanol (1) in the water-rich phase at 15°C was reported to be 0.0022 mol(1)/L sln.

The corresponding mass percentage solubility, calculated by the compilers with the assumption of a solution density equal to that of water, is 0.032 g(1)/100g sln, and the mole fraction solubility is $x_1 = 3.9 \times 10^{-5}$.

AUXILIARY INFORMATION

METHOD/APPARATUS/PROCEDURE:

The solubility was obtained from the surface tension of saturated solutions, measured by the static method of Wilhelmy (platinum plate). The apparatus consisted of an electrobalance (R.G. Cahn) connected with a high impedance null detector (Fluke type 845 AR). An all-Pyrex vessel was used.

SOURCE AND PURITY OF MATERIALS:

(1) purified by distillation and preparative gas chromatography;
b.p. 194.5°C/760 mm Hg

(2) triply distilled from permanganate solution

ESTIMATED ERROR:

Temperature: ± 0.1°C

Solubility (probably standard deviation): ± 0.0001 mol(1)/L sln.

REFERENCES:

COMPONENTS:	ORIGINAL MEASUREMENTS:
(1) 4-Nonanol; $C_9H_{20}O$; [5932-79-6] (2) Water; H_2O; [7732-18-5]	Vochten, R.; Petre, G. *J. Colloid Interface Sci.* 1973, *42*, 320-7.
VARIABLES:	**PREPARED BY:**
One temperature: 15°C	S.H. Yalkowsky; S.C. Valvani; A.F.M. Barton

EXPERIMENTAL VALUES:

The equilibrium concentration of 4-nonanol (1) in the water-rich phase at 15°C was reported to be 0.0026 mol(1)/L sln.

The corresponding mass percentage solubility, calculated by the compilers with the assumption of a solution density equal to that of water, is 0.038 g(1)/100g sln, and the mole fraction solubility is $x_1 = 4.7 \times 10^{-5}$.

AUXILIARY INFORMATION

METHOD/APPARATUS/PROCEDURE:	SOURCE AND PURITY OF MATERIALS:
The solubility was obtained from the surface tension of saturated solutions, measured by the static method of Wilhelmy (platinum plate). The apparatus consisted of an electrobalance (R.G. Cahn) connected with a high impedance null detector (Fluke type 845 AR). An all-Pyrex vessel was used.	(1) purified by distillation and preparative gas chromatography; b.p. 192.0°C/760 mm Hg (2) triply distilled from permanganate solution
	ESTIMATED ERROR: Temperature: ± 0.1°C Solubility (probably standard deviation): ± 0.0001 mol(1)/L sln.
	REFERENCES:

COMPONENTS:

(1) 5-Nonanol; $C_9H_{20}O$; [623-93-8]

(2) Water; H_2O; [7732-18-5]

ORIGINAL MEASUREMENTS:

Vochten, R.; Petre, G.

J. Colloid Interface Sci. 1973, *42*, 320-7.

VARIABLES:

One temperature: 15°C

PREPARED BY:

S.H. Yalkowsky; S.C. Valvani; A.F.M. Barton

EXPERIMENTAL VALUES:

The equilibrium concentration of 5-nonanol (1) in the water-rich phase at 15°C was reported to be 0.0032 mol(1)/L sln.

The corresponding mass percentage solubility, calculated by the compilers with the assumption of a solution density equal to that of water, is 0.046 g(1)/100g sln, and the mole fraction solubility is $x_1 = 5.7 \times 10^{-5}$.

AUXILIARY INFORMATION

METHOD/APPARATUS/PROCEDURE:

The solubility was obtained from the surface tension of saturated solutions, measured by the static method of Wilhelmy (platinum plate). The apparatus consisted of an electrobalance (R.G. Cahn) connected with a high impedance null detector (Fluke type 845 AR). An all-Pyrex vessel was used.

SOURCE AND PURITY OF MATERIALS:

(1) purified by distillation and preparative gas chromatography;

b.p. 193°C/760 mm Hg

(2) triply distilled from permanganate solution

ESTIMATED ERROR:

Temperature: ± 0.1°C

Solubility (probably standard deviation): ± 0.0001 mol(1)/L sln.

REFERENCES:

COMPONENTS:	EVALUATOR:
(1) 1-Decanol; $C_{10}H_{22}O$; [112-30-1] (2) Water; H_2O; [7732-18-5]	A. Maczynski, Institute of Physical Chemistry of the Polish Academy of Sciences, Warsaw, Poland; and A.F.M. Barton, Murdoch University, Perth, Western Australia. November 1982

CRITICAL EVALUATION:

The proportion of 1-decanol (1) in the water rich phase has been reported in six publications:

Reference	T/K	g(1)/100g sln	Method
Stearns *et al.* (ref 1)	298	0.005	turbidimetric
Addison and Hutchinson (ref 2)	293	0.0036	surface tension
Harkins and Oppenheimer (ref 3)	-	0.005	turbidimetric
Kinoshita *et al.* (ref 5)	298	0.0037	surface tension
Vochten and Petre (ref 6)	288	0.0032	surface tension
Zhuravleva *et al.* (ref 7)	375	1.0	polythermic
	394	0.8	

The equation correlating the solubilities at 298K of normal aliphatic alcohols with alcohol chain length (equation 5 in the Editors's Preface) is

$$\log (c/\text{mol(1) L}^{-1}\ \text{sln}) = 2.722 - 0.6988n + 0.006418\, n^2$$

where n is the number of carbon atoms in the alochol. This equation predicts for 1-decanol 2.4×10^{-4} mol(1)/L sln, or 0.0038 g(1)/100g sln.

The data of ref 2,5, and 6 are in mutual agreement, conform with the results for other 1-alkanols, and provide the basis for recommended values. The higher temperature values of ref 7 are considered tentative only.

Recommended and tentative values of the solubility of 1-decanol (1) in water

T/K	g(1)/100g sln		x_1
288	0.0032	(recommended)	3.0×10^{-6}
293	0.0036	(recommended)	3.8×10^{-6}
298	0.0037	(recommended)	4.2×10^{-6}
375	1	(tentative)	1×10^{-3}
393	0.8	(tentative)	9×10^{-4}

The proportion of water (2) in the alcohol-rich phase in equilibrium with the water-rich phase has been reported in three publications:

Reference	T/K	Method
Erichsen (ref 4)	273-533	synthetic
Zhuravleva *et al.* (ref 7)	284-356	polythermic
Tokunaga *et al.* (ref 8)	288-313	analytical

The data of ref 4 span a wide temperature range, they are in good agreement at two of the three temperatures of ref 7 (although they are not very consistent with ref 8) and have been used as the basis for the following tentative values.

(continued next page)

COMPONENTS:	EVALUATOR:
(1) 1-Decanol; $C_{10}H_{22}O$; [112-30-1] (2) Water; H_2O; [7732-18-5]	A. Maczynski, Institute of Physical Chemistry of the Polisy Academy of Sciences, Warsaw, Poland; and A.F.M. Barton, Murdoch University, Perth, Western Australia. November 1982

CRITICAL EVALUATION: (continued)

Tentative values of the solubility of water (2) in 1-decanol-rich phase

T/K	g(2)/100g sln	x_2
273	2.6	0.19
293	3.4	0.24
298	3.6	0.25
313	4.2	0.28
333	5.1	0.33
353	6.1	0.37
373	7.3	0.41
393	8.6	0.45
413	10	0.50
433	12	0.54
453	14	0.58
473	16	0.63
493	19	0.67
513	23	0.72
523	25	0.75
533	28	0.77

References

1. Stearns, R.S.; Oppenheimer, H.; Simon, E.; Harkins, W.D. *J. Chem. Phys.* 1947, *15*, 496.
2. Addison, C.C.; Hutchinson, S.K. *J. Chem. Soc.* 1949, 3387.
3. Harkins, W.D.; Oppenheimer, H. *J. Am. Chem. Soc.* 1949, *71*, 808.
4. Erichsen, L. von *Brennst. Chem.* 1952, *33*, 166.
5. Kinoshita, K.; Ishikawa, H.; Shinoda, K. *Bull. Chem. Soc. Jpn.* 1958, *31*, 1081.
6. Vochten, R.; Petre, G. *J. Colloid Interface Sci.* 1973, *42*, 320.
7. Zhuravleva, I.K.; Zhuravlev, E.F.; Lomakina, N.G. *Zh. Fiz. Khim.* 1977, *51*, 1700.
8. Tokunaga, S.; Manabe, M.; Koda, M. *Niihama Kogyokota Semmon Gakko Kiyo, Rikogaku Hen (Memoirs Niihama Technical College, Sci. and Eng.)* 1980, *16*, 96.

COMPONENTS:	ORIGINAL MEASUREMENTS:
(1) 1-Decanol; $C_{10}H_{22}O$; [112-30-1] (2) Water; H_2O; [7732-18-5]	Stearns, R.S.; Oppenheimer, H.; Simon, E.; Harkins, W.D. *J. Chem. Phys.* 1947, *15*, 496-507.
VARIABLES:	PREPARED BY:
One temperature: 25°C	A. Maczynski

EXPERIMENTAL VALUES:

The solubility of 1-decanol in water at 25 ± 3°C was reported to be 0.005 g(1)/100g soln.

The corresponding mole fraction, x_1, calculated by the compiler 6×10^{-6}.

AUXILIARY INFORMATION

METHOD/APPARATUS/PROCEDURE:	SOURCE AND PURITY OF MATERIALS:
The turbidimetric method was used. (A turbidimetric method in which a photometer was used to determine turbidity in a soap solution was described in the paper but nothing was reported regarding the turbidity in water).	(1) not specified. (2) not specified.
	ESTIMATED ERROR: Temperature ± 3°C
	REFERENCES:

COMPONENTS:	ORIGINAL MEASUREMENTS:
(1) 1-Decanol; $C_{10}H_{22}O$; [112-30-1] (2) Water; H_2O; [7732-18-5]	Addison, C.C.; Hutchinson, S.K. *J. Chem. Soc.* 1949, 3387-95.
VARIABLES:	PREPARED BY:
One temperature: 20°C	A.F.M. Barton

EXPERIMENTAL VALUES:

The proportion of 1-decanol (1) in the water-rich phase at equilibrium at 20°C was reported to be 0.0036 g(1)/100g sln.

The corresponding mole fraction solubility, calculated by the compilers, is $x_1 = 4.0 \times 10^{-6}$.

AUXILIARY INFORMATION

METHOD/APPARATUS/PROCEDURE:	SOURCE AND PURITY OF MATERIALS:
The surface tension of the saturated water-rich phase was determined by vertical plate and expanding drop measurements, and extrapolation of a surface-tension-composition curve provided the solubility value.	(1) Lights Ltd; b.p. 231°C/749 mm Hg, m.p. 6.0°C; no variation in properties of five fractions of vacuum distillation. (2) distilled.
	ESTIMATED ERROR:
	REFERENCES:

COMPONENTS:	ORIGINAL MEASUREMENTS:
(1) 1-Decanol; $C_{10}H_{22}O$; [112-30-1] (2) Water; H_2O; [7732-18-5]	Harkins, W.D.; Oppenheimer. H. *J. Am. Chem. Soc.* 1949, *71*, 808-11.
VARIABLES:	PREPARED BY:
One temperature: assumed to be ambient	S.C. Yalkowsky; S.C.Valvani; A.F.M. Barton

EXPERIMENTAL VALUES:

The solubility of 1-decanol (1) in water was reported to be 0.0003 mol(1)/kg (2).

Corresponding values calculated by the compilers are mass percentage 0.005 g(1)/100g sln and mole fraction $x_1 = 5 \times 10^{-6}$.

The temperature was not stated, but is assumed to be ambient .

AUXILIARY INFORMATION

METHOD/APPARATUS/PROCEDURE:	SOURCE AND PURITY OF MATERIALS:
A turbidimetric method was used, with the aid of a photometer (turbidity increased rapidly in the presence of the emulsified second phase). The study was concerned with the effect of long-chain electrolytes (soaps) on solubility. The components were weighed into a glass ampoule and shaken vigorously for at least 48 h. Equilibrium was approached from both undersaturation and supersation.	(1) Eastman Kodak Co. or Connecticut Hard Rubber Co.; purified by fractional distillation. (2) not stated
	ESTIMATED ERROR:
	Not stated
	REFERENCES:

COMPONENTS:

(1) 1-Decanol; $C_{10}H_{22}$; [112-30-1]

(2) Water; H_2O; [7732-18-5]

ORIGINAL MEASUREMENTS:

Erichsen, L. von

Brennst. Chem. 1952, *33*, 166-72.

VARIABLES:

Temperature: 0-260°C

PREPARED BY:

S. H. Yalkowsky and Z. Maczynska

EXPERIMENTAL VALUES:

Proportion of 1-decanol (1) in alcohol-rich phase

t/°C	g(1)/100g sln	x_1
0	97.35	0.8065
10	96.95	0.7850
20	96.60	0.7640
30	96.20	0.7425
40	95.75	0.7205
50	95.30	0.6985
60	94.85	0.6765
70	94.35	0.6550
80	93.85	0.6335
90	93.25	0.6115
100	92.65	0.5900
110	92.05	0.5680
120	91.35	0.5460
130	90.60	0.5235
140	89.85	0.5020
150	89.05	0.4805
160	88.15	0.4590
170	87.20	0.4370
180	86.15	0.4150
190	85.05	0.3930
200	83.80	0.3710
210	82.45	0.3485
220	80.90	0.3255
230	79.20	0.3025
240	77.20	0.2780
250	74.95	0.2340
260	72.09	0.2270

AUXILIARY INFORMATION

METHOD/APPARATUS/PROCEDURE:

The synthetic method was used.

The measurements were carried out in 2 mL glass ampules. These were placed in an aluminium block equpped with two glass windows. Cloud points were measured with a thermocouple wound up around the ampule. Each measurement was repeated twice.

SOURCE AND PURITY OF MATERIALS:

(1) Merck, or Ciba, or industrial product; distilled and chemically free from isomers;
b.p. 236.0-236.1°C (764 mm Hg)
n_D^{20} 1.4375.

(2) not specified.

ESTIMATED ERROR:

Not specified.

REFERENCES:

COMPONENTS:	ORIGINAL MEASUREMENTS:
(1) 1-Decanol; $C_{10}H_{22}O$; 112-30-1 (2) Water; H_2O; 7732-18-5	Kinoshita, K.; Ishikawa, H.; Shinoda, K. *Bull. Chem. Soc. Jpn.* 1958, *31*, 1081-4.
VARIABLES:	PREPARED BY:
One temperature: 25°C	S.H. Yalkowsky; S.C. Valvani; A.F.M. Barton

EXPERIMENTAL VALUES:

The equilibrium concentration of 1-decanol (1) in the water-rich phase at 25°C was reported to be 0.000234 mol(1) /L sln. and the mass percentage solubility was reported as 0.0037 g(1)/100g sln.

The corresponding mole fraction solubility, calculated by the compilers, is $x_1 = 4.0 \times 10^{-6}$.

AUXILIARY INFORMATION

METHOD/APPARATUS/PROCEDURE:

The surface tension in aqueous solutions of alcohols monotonically decreases up to their saturation concentration and remains constant in the heterogeneous region (ref 1-4). Surface tension was measured by the drop weight method, using a tip 6 mm in diameter, the measurements being carried out in a water thermostat. From the (surface tension)-(logarithm of concentration) curves the saturation points were determined as the intersections of the curves with the horizontal straight lines passing through the lowest experimental points.

SOURCE AND PURITY OF MATERIALS:

(1) purified by vacuum distillation through 50-100 cm column;
b.p. 101°C/5 mm Hg

(2) not stated

ESTIMATED ERROR:

Temperature: ± 0.05°C
Solubility : within 4%

REFERENCES:

1. Motylewski, S. *Z.Anorg.Chem.* 1904, *38*, 410
2. Taubamann, A. *Z.physic.Chem.* 1932, *A161*, 141.
3. Zimmerman, H.K.,Jr. *Chem.Rev.* 1952, *51*, 25.
4. Shinoda, K.; Yamanaka, I., Kinoshita, K. *J.Phys.Chem.* 1959, *63*, 648.

COMPONENTS:

(1) 1-Decanol; $C_{10}H_{22}O$; [112-30-1]

(2) Water; H_2O; [7732-18-5]

ORIGINAL MEASUREMENTS:

Vochten, R.; Petre, G.

J. Colloid Interface Sci. 1973, *42*, 320-7.

VARIABLES:

One temperature: 15°C

PREPARED BY:

S.H. Yalkowsky; S.C. Valvani; A.F.M.Barton.

EXPERIMENTAL VALUES:

The equilibrium concentration of 1-decanol (1) in the water-rich phase at 15°C was reported to be 2.0×10^{-4} mol(1)/L sln.

The corresponding mass percentage solubility, calculated by the compilers with the assumption of a solution density equal to that of water, is 0.0032 g(1)/100g sln, and the mole fraction solubility is $x_1 = 3.0 \times 10^{-6}$.

AUXILIARY INFORMATION

METHOD/APPARATUS/PROCEDURE:

The solubility was obtained from the surface tension of saturated solutions, measured by the static method of Wilhelmy (platinum plate). The apparatus consisted of an electrobalance (R.G. Cahn) connected with a high impedance mill detector (FLUKE type 845 AR). An all-pyrex vessel was used.

SOURCE AND PURITY OF MATERIALS:

(1) purified by distillation and preparative gas chromatography;

b.p. 231.0°C/760 mm Hg

(2) triply distilled from permanganate solution.

ESTIMATED ERROR:

Temperature: ± 0.1°C
Solubility (probably standard deviation):
0.1×10^{-4} mol(1)/L sln.

REFERENCES:

AWW-O

COMPONENTS:	ORIGINAL MEASUREMENTS:
(1) 1-Decanol; $C_{10}H_{22}O$; [112-30-1] (2) Water; H_2O; [7732-18-5]	Zhuravleva, I.K.; Zhuravlev, E.F.; Lomakina, N.G. *Zh. Fiz. Khim.* 1977, *51*, 1700-7; *Russ. J. Phys. Chem.* 1977, *51*, 994-8.
VARIABLES:	PREPARED BY:
Temperature: 10.5 - 120.5°C	A. Maczynski

EXPERIMENTAL VALUES:

Solubility of water (2) in 1-decanol (1)

t/°C	g(2)/100g sln	x_2(compiler)
10.5	3.5	0.24
62.5	5.0	0.32
82.5	6.0	0.36

Solubility of 1-decanol (1) in water (2)

t/°C	g(1)/100g sln	x_1(compiler)
102.5	1.0	0.001
120.5	0.8	0.001

AUXILIARY INFORMATION

METHOD/APPARATUS/PROCEDURE:

The "polythermic" method (ref 1) was used.

No details were reported in the paper.

The results formed part of a report on the ternary system including nitromethane.

SOURCE AND PURITY OF MATERIALS:

(1) source not specified; n_D^{20} 1.4370; d_4^{20} 0.8300; purity not specified.

(2) n_D^{20} 1.333.

ESTIMATED ERROR:

Not specified.

REFERENCES:

1. Alekseev, W.F. *Zh. russk. khim. o-va,* 1876, *8*, 249.

COMPONENTS:

(1) 1-Decanol, $C_{10}H_{22}O$; [112-30-1]

(2) Water; H_2O; [7732-18-5]

ORIGINAL MEASUREMENTS:

Tokunaga, S.; Manabe, M.; Koda, M.
Niihama Kogyo Koto Semmon Gakko Kiyo, Rikogakn Hen (Memoirs Niihama Technical College Sci. and Eng.) 1980, *16*, 96-101.

VARIABLES:

Temperature: 15-40°C

PREPARED BY:

A.F.M. Barton

EXPERIMENTAL VALUES:

Proportion of water (2) in the alcohol-rich phase

$t/^{\circ}C$	g(2)/100g sln	x_2	mol (1)/mol (2)
15	3.79	0.257	2.84
20	3.86	0.261	2.87
25	3.83	0.259	2.88
30	3.79	0.257	2.90
35	3.76	0.255	2.92
40	3.70	0.252	2.94

AUXILIARY INFORMATION

METHOD/APPARATUS/PROCEDURE:

The mixtures of 1-decanol (~5 mL) and water (~10 mL) were stirred magnetically in a stoppered vessel and allowed to stand for 10-12 h in a water thermostat. The alcohol-rich phase was analyzed for water by Karl Fischer titration.

SOURCE AND PURITY OF MATERIALS:

(1) Tokyo Kasei, GR > 98%; used without further purification

(2) deionized; distilled prior to use

ESTIMATED ERROR:

Temperature: ± 0.1°C

Solubility: each result is the mean of three determinations

REFERENCES:

COMPONENTS:

(1) 1-Undecanol; $C_{11}H_{24}O$; [112-42-5]

(2) Water; H_2O; [7732-18-5]

ORIGINAL MEASUREMENTS:

Tokunaga, S.; Manabe, M.; Koda, M.
Niihama Kogyo Koto Semmon Gakko Kiyo, Rikogaku Hen (Memoirs Niihama Technical College, Sci. and Eng.) 1980, *16*, 96-101.

VARIABLES:

Temperature: 15-40°C

PREPARED BY:

A.F.M. Barton

EXPERIMENTAL VALUES:

Proportion of water (2) in the alcohol-rich phase

t/°C	g(2)/100g sln	x_2	mol (1)/mol (2)
15	3.45	0.255	2.92
20	3.40	0.252	2.94
25	3.43	0.254	2.97
30	3.39	0.251	3.00
35	3.37	0.250	3.02
40	3.29	0.245	3.04

AUXILIARY INFORMATION

METHOD/APPARATUS/PROCEDURE:

The mixtures of 1-undecanol (~5 mL) and water (~10 mL) were stirred magnetically in a stoppered vessel and allowed to stand for 10-12 h in a water thermostat. The alcohol-rich phase was analyzed for water by Karl Fischer titration.

SOURCE AND PURITY OF MATERIALS:

(1) distilled

(2) deionized;
distilled prior to use

ESTIMATED ERROR:

Temperature: 0.1°C
Solubility: each result is the mean of three determinations

REFERENCES:

COMPONENTS:	EVALUATOR:
(1) 1-Dodecanol *(n-dodecyl alcohol)*; $C_{12}H_{26}O$; [112-53-8] (2) Water; H_2O; [7732-18-5]	A. Maczynski, Institute of Physical Chemistry of the Polish Academy of Sciences, Warsaw, Poland; and A.F.M. Barton, Murdoch University, Perth, Western Australia. November 1982

CRITICAL EVALUATION:

The proportion of 1-dodecanol (1) in the water-rich phase has been reported in two publications: Krause and Lange (ref 1) carried out measurements at 289K, 307K and 322K by the analytical method; and Robb (ref 2) determined one point only at 298K by a film balance technique. These values are in poor agreement, and equation 5 of the Editor's Preface has been used to provide an estimated figure of 2 x 10^{-5} mol(1)/L sln.

$$\log (c/\text{mol(1)/L sln}) = 2.722 - 0.6988n + 0.006418n^2,$$

where n is the number of carbon atoms in the alcohol. The figure lies between that of refs 1 and 2, and is suggested as a tentative value.

The tentative value for the solubility of 1-dodecanol in water at 298K is 2 x 10^{-5} mol(1)/L sln or 4 x 10^{-4} g(1)/100g sln.

The proportion of water (2) in the alcohol-rich phase in equilibrium with the water rich-phase has also been reported in two publications: Zhuravleva *et al.* (ref 3) between 295K and 315K; and Tokunaga *et al.* (ref 4) between 303K and 313K by an analytical method. These reports disagree, not only with regard to the values but also in the sign of the temperature dependence, so only an estimate of the 298K value can be provided.

The tentative value for the solubility of water in 1-dodecanol at 298K is 3 g(2)/100g sln, x_2 = 0.25, with only a small dependence on temperature.

References:

1. Krause, F.P.; Lange, W. *J. Phys. Chem.* 1965, *69*, 3171.
2. Robb, I.D. *Aust. J. Chem.* 1966, *19*, 2281.
3. Zhuravleva, I.K.; Zhuravlev, E.F.; Salamatin, L.N. *Zh. Obshch. Khim.* 1976, *46*, 1210.
4. Tokunaga, S.; Manabe, M.; Koda, M. *Niihama Kogyo Koto Semmon Gakko Kiyo, Rikogaku Hen (Memoirs Niihama Technical College, Sci. and Eng.)* 1980, *16*, 96.

COMPONENTS:

(1) 1-Dodecanol; $C_{12}H_{26}O$ [112-53-8]

(2) Water; H_2O; [7732-18-5]

ORIGINAL MEASUREMENTS:

Krause, F.P.; Lange, W.;
J. Phys. Chem. 1965, *69*, 3171-3.

VARIABLES:

Temperature: 16-49°C

PREPARED BY:

S.H. Yalkowsky; S.C. Valvani; A.F.M. Barton

EXPERIMENTAL VALUES:

Solubility of 1-dodecanol (1) in water (2)

t/°C	$10^6 c_1$/mol(1) (L sln)$^{-1}$	10^4 g(1)/100g sln^a	$10^7 x_1{}^a$
16	9.1	1.7	1.6
34	15.6	2.9	2.8
49	19.3	3.6	3.5

^{a} Calculated by the compilers with the assumption of a solution density equal to that of water.

AUXILIARY INFORMATION

METHOD/APPARATUS/PROCEDURE:

The isotopically labelled 1-dodecanol (1-5 mg) was deposited on the surface film of the water in a closed vessel by evaporating the hexane solvent, and held there virtually stationary for 1-2 weeks in a thermostat box while equilibrium with the stirred water was achieved. Periodically samples were withdrawn, diluted with ethanol, extracted with hexane and counted. (The solubility of 1-hexadecanol, determined after equilibrating from both undersaturated and supersaturated solutions, was found to be the same within accuracy limits. Two solubility determinations with 1-dodecanol at 16°C, one with 1 mg and the other with 5 mg, gave the same results).

SOURCE AND PURITY OF MATERIALS:

(1) C^{14} labelled at position 1 prepared in laboratory; passed through 0.6 x 200 cm column of diglycolic acid polyethylen glycol on 60/80-mesh chromosorb W at 180°C with helium at 50 cm 3 min^{-1}; diluted with 99.5% pure (1); m.p. 25°C free of comparatively water-soluble impurities; stored in hexane solution.

(2) distilled; sterilized by boiling.

ESTIMATED ERROR:

Temperature: uncertain by 1-2°C

Solubility: within ±10%

REFERENCES:

COMPONENTS:

(1) 1-Dodecanol; $C_{12}H_{26}O$; [112-53-8]

(2) Water; H_2O; [7732-18-5]

ORIGINAL MEASUREMENTS:

Robb, I.D.

Aust. J. Chem. 1966, *19*, 2281-4.

VARIABLES:

One temperature: 25°C

PREPARED BY:

S.H. Yalkowsky; S.C. Valvani; A.F.M. Barton

EXPERIMENTAL VALUES:

The equilibrium concentration of 1-dodecanol (1) in the water-rich phase at 25°C was reported to be 2.3×10^{-5} mol(1)/L sln.

The corresponding mass percentage solubility, calculated by the compilers with the assumption of a solution density equal to that of water, is 4.3×10^{-4} g(1)/100g sln, and the mole fraction solubility is $10^7 x_1 = 4.2$.

AUXILIARY INFORMATION

METHOD/APPARATUS/PROCEDURE:

In this method (described in detail in ref 1) crystals of (1) were stirred with (2) for 24 h at 25°C, and a solution filtered (Gooch sintered glass and 350-1000 Å nitrocellulose under N_2 pressure). The alcohol was extracted into hexane, and estimated by its surface properties on a film balance.

SOURCE AND PURITY OF MATERIALS:

(1) Fluka puriss.
m.p. 23-24°C.

(2) in equilibrium with air; pH 5.7 ± 0.1 specific conductivity 0.9-1.0 μS cm^{-1}.

ESTIMATED ERROR:

Solubility: 0.1×10^{-5} mol(1)/L sln on the basis of 16 determinations.

REFERENCES:

1. Robb, I., *Kolloidz. Z. Polymere* 1966, *209*, 162.

COMPONENTS:

(1) 1-Dodecanol; $C_{12}H_{26}O$; [112-53-8]

(2) Water; H_2O; [7732-18-5]

ORIGINAL MEASUREMENTS:

Zhuravleva, I.K.; Zhuravlev, E.F.; Salamatin, L.N.

Zh. Obshch. Khim. 1976, *46*, 1210-4.

VARIABLES:

Temperature: 21.5 - 41.5°C

PREPARED BY:

A. Maczynski

EXPERIMENTAL VALUES:

Solubility of water in 1-dodecanol

$t/^{\circ}C$	g(2)/100g sln	x_2(compiler)
21.5	0.7	0.07
23.7	1.3	0.12
24.2	1.4	0.13
31.0	1.7	0.15
41.5	2.1	0.18

AUXILIARY INFORMATION

METHOD/APPARATUS/PROCEDURE:

Not specified.

SOURCE AND PURITY OF MATERIALS:

(1) source not specified; b.p. 101.0°C, m.p. 22.6°C, n_D^{24} 1.4420, d_4^{20} 0.8310

(2) not specified.

ESTIMATED ERROR:

Not specified.

REFERENCES:

COMPONENTS:

(1) 1-Dodecanol; $C_{12}H_{26}O$; 112-53-8

(2) Water; H_2O; 7732-18-5

ORIGINAL MEASUREMENTS:

Tokunaga, S.; Manabe, M.; Koda, M.
Niihama Kogyo Koto Semmon Gakko Kiyo, Rikogaku Hen (Memoirs Niihama Technical College, Sci. and Eng.) 1980, *16*, 96-101.

VARIABLES:

Temperature: 30-40°C

PREPARED BY:

A.F.M. Barton

EXPERIMENTAL VALUES:

Proportion of water (2) in the alcohol-rich phase

t/°C	g(2)/100g sln	x_2	mol(1)/mol(2)
30	3.09	0.248	3.03
35	3.06	0.246	3.06
40	3.03	0.244	3.09

AUXILIARY INFORMATION

METHOD/APPARATUS/PROCEDURE:

The mixtures of 1-dodecanol (~5 mL) and water (~10 mL) were stirred magnetically in a stoppered vessel and allowed to stand for 10-12 h in a water thermostat. The alcohol-rich phase was analyzed for water by Karl Fischer titration.

SOURCE AND PURITY OF MATERIALS:

(1) distilled

(2) deionized;
distilled prior to use

ESTIMATED ERROR:

Temperature: ±0.1°C

Solubility: each result is the mean of three determinations.

REFERENCES:

COMPONENTS:	EVALUATOR:
(1) 1-Tetradecanol *(n-tetradecyl alcohol)*; $C_{14}H_{30}O$; [112-72-1] (2) Water; H_2O: [7732-18-5]	A. Maczynski, Institute of Physical Chemistry of Polish Academy of Sciences, Warsaw, Poland; and A.F.M. Barton, Murdoch University, Perth, Western Australia November 1982

CRITICAL EVALUATION:

The proportion of 1-tetradecanol (1) in the water-rich phase has been reported in two publications: Robb (ref 1) determined the solubility of (1) in (2) at 298K by a film balance technique; and Hoffman and Anacker (ref 2) used the analytical method at four temperatures between 277K and 334K. The two reports are in marked disagreement, but the result of Robb is supported by comparison with the solubilities of other normal aliphatic alcohols in water, as discussed in the Editor's Preface. If the equation

$$\log (c/\text{mol L}^{-1}) = -0.57n + 2.14$$

is used to relate the solubility of 1-alkanols in water at 298K to the carbon number n (ref 1), the solubility of 1-tetradecanol is found to be 1.5×10^{-6} mol(1)/L sln or 3×10^{-5} g(1)/100g sln. Equation 5 of the Editor's Preface yields a marginally higher value.

The tentative value for the solubility of 1-tetradecanol in water at 298K is 3×10^{-5} g(1)/100g sln or 2×10^{-6} mol(1)/L sln.

The proportion of water in the alcohol-rich phase in equilibrium with the water-rich phase has not been reported in the literature.

References:

1. Robb, I.D. *Aust. J. Chem.* 1966, *19*, 2281.

2. Hoffman, C.S.; Anacker, E.W. *J. Chromatogr.* 1967, *30*, 390.

COMPONENTS:

(1) 1-Tetradecanol; $C_{14}H_{30}O$; [112-72-1]

(2) Water; H_2O; [7732-18-5]

ORIGINAL MEASUREMENTS:

Robb, I.D.

Aust. J. Chem. 1966, *19*, 2281-4.

VARIABLES:

One temperature: 25°C

PREPARED BY:

S.H. Yalkowsky; S.C. Valvani; A.F.M. Barton

EXPERIMENTAL VALUES:

The equilibrium concentration of 1-tetradecanol (1) in the water-rich phase at 25°C was reported to be 1.46×10^{-6} mol(1)/L sln.

The corresponding mass percentage solubility, calculated by the compilers with the assumption of a solution density equal to that of water, is 3.1×10^{-5} g(1)/100g sln, and the mole fraction solubility is $x_1 = 2.6 \times 10^{-8}$.

AUXILIARY INFORMATION

METHOD/APPARATUS/PROCEDURE:

In this method (described in detail in ref 1) crystals of (1) were stirred with (2) for 24h at 25°C, and the solution was then filtered (Gooch sintered glass and 350-1000 Å nitrocellulose under N_2 pressures). The alcohol was extracted into hexane, and estimated by its surface properties on a film balance.

SOURCE AND PURITY OF MATERIALS:

(1) Fluka puriss.
m.p. 38-39°C

(2) in equilibrium with air; pH 5.7 ± 0.1
specific conductivity 0.9-1.0
μS cm^{-1}.

ESTIMATED ERROR:

Solubility: 0.1×10^{-6} mol(1)/L sln on the basis of 9 determinations.

REFERENCES:

1. Robb, I. *Kolloidz. Z. Polymere* 1966 *209*, 162.

COMPONENTS:

(1) 1-Tetradecanol; $C_{14}H_{30}O$; [112-72-1]

(2) Water; H_2O; [7732-18-5]

ORIGINAL MEASUREMENTS:

Hoffman, C.S.; Anacker, E.W.
J. Chromatogr. 1967, *30*, 390-6.

VARIABLES:

Temperature: 4-61 °C

PREPARED BY:

A. Maczynski; A. Szafranski; A.F.M. Barton

EXPERIMENTAL VALUES:

Solubility of 1-tetradecanol (1) in water (2)

$t/^{\circ}C$	$10^6 g(1)\ (L\ sln)^{-1}$	$10^5 g(1)/100g\ sln^a$	$10^8 x_1{}^a$
4	0.194	0.42	0.35
32	1.23	2.6	2.2
45	2.37	5.1	4.3
61	4.49	9.6	7.5

[a] Calculated by the compilers with the assumption of a solution density equal to that of water.

AUXILIARY INFORMATION

METHOD/APPARATUS/PROCEDURE:

The analytical method was used.

Saturated solutions of (1) in (2) free from colloidal particles were prepared by using the Krause and Lange (ref 1) procedure. To 1000 mL (2) containing 50 ppb $AgNO_3$ (to prevent bacteria) in a 2000-mL flask 1-29 mg (1) was added with minimum agitation, the flask stoppered, the solution stirred gently at a constant temperature and sampled. The 2nd sample was retained, weighed and 10-12 mL aliquots collected, each aliquot shaken 18 h with 5 mL hexane. The hexane was evaporated the (1) residue redissolved in 25-200 mL hexane and analyzed by gas-liquid chromatography.

SOURCE AND PURITY OF MATERIALS:

(1) Matheson, Coleman and Bell; 99.5+%; purified by preparative gas chromatography.

(2) double distilled.

ESTIMATED ERROR:

Not specified.

REFERENCES:

1. Krause, F.P.; Lange, W.
J. Phys. Chem. 1965, *69*, 3171.

COMPONENTS:

(1) 1-Pentadecanol; $C_{15}H_{32}O$; [629-76-5]

(2) Water; H_2O; [7732-18-5]

ORIGINAL MEASUREMENTS:

Robb, I.D.

Aust. J. Chem. 1966, *19*, 2281-4.

VARIABLES:

One temperature: 25°C

PREPARED BY:

S.H. Yalkowsky; S.C. Valvani; A.F.M. Barton

EXPERIMENTAL VALUES:

The equilibrium concentration of 1-pentadecanol (1) in the water-rich phase at 25°C was reported to be 4.5×10^{-7} mol(1)/L sln.

The corresponding mass percentage solubility, calculated by the compilers with the assumption of a solution density equal to that of water, is 1.0×10^{-5} g(1)/100g sln, and the mole fraction solubility is $x_1 = 7.9 \times 10^{-9}$.

(The water-solubilities of the higher normal alcohols may also be estimated from the number of carbon atoms, as discussed in the Editor's Preface). This predicts 3.9×10^{-7} mol(1)/L sln.

AUXILIARY INFORMATION

METHOD/APPARATUS/PROCEDURE:

In this method (described in detail in ref 1) crystals of (1) were stirred with (2) for 24 h at 25°C and the solution filtered (Gooch sintered glass and 350-1000 Å nitrocellulose under N_2 pressure). The alcohol was extracted into hexane, and estimated by its surface properties on a film balance.

SOURCE AND PURITY OF MATERIALS:

(1) Fluka purum; m.p. 42-44°C

(2) in equilibrium with air; pH 5.7 ± 0.1 specific conductivity 0.9-1.0 μS cm^{-1}.

ESTIMATED ERROR:

Solubility: 0.3×10^{-7} mol(1)/L sln. on the basis of 8 determinations.

REFERENCES:

1. Robb, I. *Kolloidz. Z. Polymere* 1966, *209*, 162.

COMPONENTS:	EVALUATOR:
(1) 1-Hexadecanol *(n-hexadecyl alcohol)*; $C_{16}H_{34}O$; [36653-82-4] (2) Water; H_2O; [7732-18-5]	A. Maczynski, Institute of Physical Chemistry of the Polish Academy of Sciences, Warsaw, Poland; and A.F.M. Barton, Murdoch University, Perth, Western Australia November 1982

CRITICAL EVALUATION:

The proportion of 1-hexadecanol (1) in the water-rich phase has been reported in three publications: Krause and Lange (ref 1) carried out measurements of the solubility of (1) in (2) at 307K and 328K by the analytical method; Robb (ref 2) determined one point only at 298K by a film balance technique; and Hoffman and Anacker (ref 3) determined the solubility of (1) in (2) at 316K and 334K by the analytical method.

These data from refs 1-3 are in poor agreement. However, the solubilities of normal aliphatic alcohols in water may be correlated as discussed in the Editor's Preface. The solubility of 1-alkanols in water at 298K is related to the carbon number n, by (ref 2)

$$\log (c/\text{mol L}^{-1}) = -0.57n + 2.14$$

For 1-hexadecanol this gives 1×10^{-7} mol(1)/L sln, or 3×10^{-6} g(1)/100g sln, which is slightly lower than Robb's experimental value.

The tentative range of values for the solubility of 1-hexadecanol in water at 298K is thus $(3 \text{ to } 4) \times 10^{-6}$ g(1)/100g sln or $(1\text{-}2) \times 10^{-7}$ mol(1)/L sln.

The proportion of water in the alcohol-rich phase in equilibrium with the water-rich phase has not been reported.

References:

1. Krause, F.P.; Lange, W. *J. Phys. Chem.* 1965, *69*, 3171.

2. Robb, I.D. *Aust. J. Chem.* 1966, *19*, 2281.

3. Hoffman, C.S.; Anacker, E.W. *J. Chromatogr.* 1967, *30*, 390.

COMPONENTS:

(1) 1-Hexadecanol; $C_{16}H_{34}O$; [36653-82-4]

(2) Water; H_2O; [7732-18-5]

ORIGINAL MEASUREMENTS:

Krause, F.P.; Lange, W.

J. Phys. Chem. 1965, *69*, 3171-3.

VARIABLES:

Temperature: 33°C and 127°C

PREPARED BY:

S.H. Yalkowsky; S.C. Valvani; A.F.M. Barton

EXPERIMENTAL VALUES:

Solubility of 1-hexadecanol (1) in water (2)

$t/^{\circ}C$	$10^8 c_1/mol(1)(L\ sln)^{-1}$	10^7 g(1)/100g sln[a]	$10^{10} x_1$[a]
34	3.3	8.0	6.0
55	12.7	31	23

[a] Calculated by the compilers with the assumption of a solution density equal to that of water.

AUXILIARY INFORMATION

METHOD/APPARATUS/PROCEDURE:

The isotopically labelled 1-hexadecanol (0.2 mg) was deposited on the surface film of the water in a closed vessel by evaporating the hexane solvent, and held there virtually stationary for 1-2 weeks in a thermostat box while equilibrium with the stirred bulk water was achieved. Periodically, samples were withdrawn, diluted with ethanol, extracted with hexane and counted. (The solubility determined after equilibrating from both undersaturated and supersaturated solutions, was found to be the same within accuracy limits. Two solubility determinations with 1-dodecanol at 16°C, one with 1 mg and the other with 5 mg, gave the same results).

SOURCE AND PURITY OF MATERIALS:

(1) Isotopes Specialities Co, Burbank, Cal. passed through 0.5 x 75 cm column of 10% Dow-Corning silica-free silicone stopcock grease on 60/80 mesh Chromosorb W at 162°C with helium at 62 $cm^3\ min^{-1}$;

free of comparatively water-soluble impurities, stored in hexane solution; m.p. 40°C

(2) distilled; sterilized by boiling.

ESTIMATED ERROR:

Temperature: uncertain by 1-2°C

Solubility: within ± 10%

REFERENCES:

COMPONENTS:	ORIGINAL MEASUREMENTS:
(1) 1-Hexadecanol; $C_{16}H_{34}O$; [36653-82-4] (2) Water; H_2O; [7732-18-5]	Robb, I.D. *Aust. J. Chem.* 1966, *19*, 2281-4
VARIABLES:	**PREPARED BY:**
One temperature: 25°C	S.H. Yalkowsky; S.C. Valvani; A.F.M. Barton

EXPERIMENTAL VALUES:

This equilibrium concentration of 1-hexadecanol (1) in the water-rich phase at 25°C was reported to be 1.7×10^{-7} mol(1)/L sln.

The corresponding mass percentage solubility, calculated by the compilers with the assumption of a solution density equal to that of water, is 4.1×10^{-6} g(1)/100g sln, and the mole fraction solubility is $x_1 = 3.1 \times 10^{-9}$.

AUXILIARY INFORMATION

METHOD/APPARATUS/PROCEDURE:

In this method (described in detail in ref 1) crystals of (1) were stirred with (2) for 24 h at 25°C and the solution was then filtered (Gooch sintered glass and 350-100 Å nitrocellulose under N_2 pressure). The alcohol was extracted into hexane, and estimated by its surface properties on a film balance.

SOURCE AND PURITY OF MATERIALS:

(1) Fluka puriss., > 99%; m.p. 49-50°C

(2) in equilibrium with air; pH 5.7 ± 0.1 specific conductivity 0.9-1.0 μS cm^{-1}.

ESTIMATED ERROR:

Solubility: 0.2×10^{-7} mol(1)/L sln on the basis of 14 determinations.

REFERENCES:

1. Robb, I. *Kolloidz. Z. Polymere*, 1966, *209*, 162.

COMPONENTS:

(1) 1-Hexadecanol; $C_{16}H_{34}O$; [36653-82-4]

(2) Water; H_2O; [7732-18-5]

ORIGINAL MEASUREMENTS:

Hoffman, C.S.; Anacker, E.W.

J. Chromatogr. 1967, *30*, 390-6.

VARIABLES:

Temperature: 43 and 61°C

PREPARED BY:

A. Maczynski; A. Szafranski; A.F.M. Barton

EXPERIMENTAL VALUES:

Solubility of 1-hexadecanol (1) in water (2)

$t/°C$	$10^7 g(1)(L\ sln)^{-1}$	$10^6 g(1)/100g\ sln^a$	$10^9 x_1{}^a$
43	1.55	3.8	2.8
61	4.06	9.8	7.3

[a] Calculated by the compilers with the assumption of a solution density equal to that of water.

AUXILIARY INFORMATION

METHOD/APPARATUS/PROCEDURE:

The analytical method was used.

Saturated solutions of (1) in (2) free from colloidal particles were prepared by using the Krause and Lange (ref 1) procedure. To 1000 mL (2) containing 50 ppb $AgNO_3$ (to prevent bacteria) in a 2000mL flask 1-25 mg (1) was added with minimum agitation, the flask stoppered, the solution stirred gently at a constant temperature, sampled, the 2nd sample retained, weighed and 80 mL aliquots collected, each aliquot shaken 18 hr with 20 mL hexane, the hexane evaporated, the (1) residue redissolved in 25-200 mL hexane and analyzed by gas-liquid chromatography.

SOURCE AND PURITY OF MATERIALS:

(1) Applied Science Laboratory; 99.8+ % pure by thin layer and gas-liquid chromatography; used as received.

(2) doubly distilled.

ESTIMATED ERROR:

Not specified

REFERENCES:

1. Krause, F.P.; Lange, W. *J. Phys. Chem.* 1965, *69*, 3171.

COMPONENTS:	ORIGINAL MEASUREMENTS:
(1) 1-Heptadecanol; $C_{17}H_{36}O$; [1454-85-9] (2) Water; H_2O; [7732-18-5]	Robb, I.D. *Aust. J. Chem.* 1966, *19*, 2281-4.
VARIABLES:	PREPARED BY:
One temperature: 25°C	A.F.M. Barton

EXPERIMENTAL VALUES:

The equilibrium concentration of 1-heptadecanol (1) in the water-rich phase at 25°C was reported to be no more than 10^{-7} mol(1)/L sln.

The corresponding limiting mass percentage solubility, calculated by the compilers, is 3×10^{-6} g(1)/100g sln, and the mole fraction solubility is $x_1 \leq 7 \times 10^{-10}$.

(The water-solubility of the higher normal alcohols may also be estimated from the number of carbon atoms, as discussed in the Editor's Preface. This predicts 3×10^{-8} mol(1)/L sln).

AUXILIARY INFORMATION

METHOD/APPARATUS/PROCEDURE:

In this method (described in detail in ref 1) crystals of (1) were stirred with (2) for 24 h at 25°C and the solution was then filtered (Gooch sintered glass and 350-1000 Å nitrocellulose under N_2 pressure). The alcohol was extracted into hexane, and estimated by its surface properties on a film balance. The solubility of (1) was found to be too low to be measured by this method.

SOURCE AND PURITY OF MATERIALS:

(1) Fluka purum., > 97%: m.p. 50-53°C.

(2) in equilibrium with air pH 5.7 ± 0.1 specific conductivity 0.9-1.0 μS cm^{-1}.

ESTIMATED ERROR:

Not specified

REFERENCES:

1. Robb, I., *Kolloidz. Z. Polymere* 1966, *209*, 162.

COMPONENTS:

(1) 1-Octadecanol; $C_{18}H_{38}O$ [112-92-5]

(2) Water; H_2O; [7732-18-5]

ORIGINAL MEASUREMENTS:

Krause, F.P.; Lange, W.

J. Phys. Chem. 1965, *69*, 3171-3.

VARIABLES:

Temperature: 34°C and 65°C

PREPARED BY:

S.H. Yalkowsky; S.C. Valvani; A.F.M. Barton

EXPERIMENTAL VALUES:

Solubility of 1-octadecanol (1) in water (2)

$t/^{\circ}C$	$10^9 c_1$/mol(1)(L sln)$^{-1}$	10^5 g(1)/100g sln[a]	$10^9 x_1$[a]
34	4	1.1	7.3
63	22	6.0	40

[a] Calculated by the compilers with the assumption of a solution density equal to that of water.

(The water solubilities of the higher normal alcohols may also be estimated from the number of carbon atoms, as discussed in the Editor's Preface).

AUXILIARY INFORMATION

METHOD/APPARATUS/PROCEDURE:

The isotopically-labelled 1-octadecanol (0.1 mg) was deposited on the surface film of the water in a closed vessel by evaporating the hexane solvent, and held there virtually stationary for 1-2 weeks in a thermostat box while equilibrium with the stored bulk water was achieved. Periodically samples were withdrawn, diluted with ethanol, extracted with hexane and counted. (The solubility of 1-hexadecanol, determined after equilibrating from both undersaturated and supersaturated solutions, was found to be the same within accuracy limits. Two solubility determinations with 1-dodecanol at 16°C, one with 1 mg and the other with 5 mg, gave the same results. No comparable checks were reported for 1-octadecanol).

SOURCE AND PURITY OF MATERIALS:

(1) Isotope Specialities Co., Burbank, Cal., passed through 0.6 x 75 cm column of 20% Tween 80 on 60/80-mesh HMDS-Chromosorb W at 180°C with a helium flow rate of 67 cm^3 min^{-1}; free of comparatively water-soluble impurities; stored in hexane solution m.p. 59°C.

(2) distilled; sterilized by boiling.

ESTIMATED ERROR:

Temperature: uncertain by 1-2°C

Solubility: within ±10%

REFERENCES:

SYSTEM INDEX

Underlined page numbers refer to evaluation text and those not underlined to compiled tables. All compounds are listed as in Chemical Abstracts.

(cont.)

(cont.)

REGISTRY NUMBER INDEX

Underlined page numbers refer to evaluation text and those not underlined to compiled tables.

(cont.)

AUTHOR INDEX

Underlined page numbers refer to evaluation text and those not underlined to compiled tables.

(cont.)

(cont.)

(cont.)

(cont.)